Methods in Molecular Biology

Series Editor
John M. Walker
School of Life and Medical Sciences
University of Hertfordshire
Hatfield, Hertfordshire, UK

For further volumes:
http://www.springer.com/series/7651

For over 35 years, biological scientists have come to rely on the research protocols and methodologies in the critically acclaimed *Methods in Molecular Biology* series. The series was the first to introduce the step-by-step protocols approach that has become the standard in all biomedical protocol publishing. Each protocol is provided in readily-reproducible step-by-step fashion, opening with an introductory overview, a list of the materials and reagents needed to complete . the experiment, and followed by a detailed procedure that is supported with a helpful notes section offering tips and tricks of the trade as well as troubleshooting advice. These hallmark features were introduced by series editor Dr. John Walker and constitute the key ingredient in, each and every volume of the *Methods in Molecular Biology* series. Tested and trusted, comprehensive and reliable, all protocols from the series are indexed in PubMed.

Apomixis

Methods and Protocols

Edited by

Tatyana Radoeva and Peter J. van Dijk

Keygene N.V., Wageningen, The Netherlands

Humana Press

Editors
Tatyana Radoeva
Keygene N.V.
Wageningen, The Netherlands

Peter J. van Dijk
Keygene N.V.
Wageningen, The Netherlands

ISSN 1064-3745 ISSN 1940-6029 (electronic)
Methods in Molecular Biology
ISBN 978-1-0716-5000-4 ISBN 978-1-0716-5001-1 (eBook)
https://doi.org/10.1007/978-1-0716-5001-1

This Humana imprint is published by the registered company Springer Science+Business Media, LLC, part of Springer Nature.
The registered company address is: 1 New York Plaza, New York, NY 10004, U.S.A.

Mature cypselas of dandelion (*Taraxacum officinale*)

Preface

Breeding systems are like the operating systems of plant species: they govern how genetic information is processed and distributed to the next generation, shaping the diversity and adaptability of populations. There are three major breeding systems in flowering plants: self-fertilization, cross-fertilization, and apomixis. Apomixis is clonal reproduction through seeds. In this process, the key steps of sexual reproduction—chromosomal recombination, number reduction, and fertilization—are bypassed, resulting in offspring genetically identical to the mother plant, to one another, and to future generations.

Approximately half of all angiosperms can self-fertilize, while the other half depend on cross-fertilization. In contrast, only an estimated 1 in 1000 flowering plant species reproduce via apomixis involving parthenogenesis.

Apomixis has historically received relatively limited scientific attention. This may stem from its rarity, biological complexity, its taxonomic challenges, and its frequent association with polyploidy and hybridization. For a long time, apomixis was considered an eccentric research subject, despite the fact that it touched on major fundamental and applied biological questions.

The apomixis research was technically challenging, and thorough studies required considerable time and effort. Moreover, the terminology was (unnecessarily) complex. Gian Nogler, to whom we dedicate this book, wrote in 1984 an illuminating review article, in which he simplified the terminology, enabling many apomixis researchers to enter this complex field of study.

Apomixis continues to captivate researchers. Solving the challenging questions behind it demands a multidisciplinary approach, combining transmission genetics, cytogenetics, cell and developmental biology, and, more recently, molecular biology and genomics.

This "Apomixis Handbook" combines a comprehensive set of modern methods and techniques for research, as well as reviews on apomixis model species, all organized into six thematic parts.

The chapter in Part I provides an introduction to gametophytic apomixis. Thereafter, the chapters in Part II (Chaps. 2, 3, 4, 5, and 6) focus on the detection and characterization of apomixis. The first method chapter in this part (Chap. 2) describes a protocol for flow cytometric seed screening combined with seed genotyping, which can be used to determine the reproductive mode of plants from dry seeds. The subsequent chapters in Part II center on the study of the ovule, the organ where the developmental processes occur that distinguish apomixis from sexual reproduction. Due to their deep embedding in the surrounding ovular tissue, the ovaries are difficult to study. These chapters present advanced techniques that enable access to these processes for further phenotypic analysis and imaging. This includes confocal microscopy (Chap. 3), 3D imaging, cell segmentation, and annotation (Chap. 4), callose imaging (Chap. 5), and clearing techniques (Chap. 6).

In the third part of the book (Chaps. 7, 8, and 9), emphasis is laid on transcriptomic profiling of apomictic and sexual species. It includes protocols that deal with the isolation of single cells (Chaps. 7 and 8) or tiny amounts of tissues (Chap. 9), combined with RNA-seq and data analysis (Chaps. 8 and 9). These protocols should enable the reader to gain insights into the molecular mechanisms underlying apomixis, but can also be used to answer other scientific questions.

Genetic analysis of apomixis is addressed in Part IV. Polyploidy, segregation distortion, suppression of recombination, and related factors complicate the mapping of apomixis genes. Chromosomal mapping is essential for studying the genetics of apomixis and, when combined with transcriptomics and targeted mutagenesis, offers the possibility to identify the causal genes responsible for apomixis (Chap. 12). A detailed overview of the different approaches, including also unconventional ones such as mapping radiation-induced deletions, is given in Chap. 10.

Choosing the most optimal research strategy often depends on the species involved, and moreover, the application of methods and technologies is best understood in context. Part V is thus dedicated to research on model apomictic systems such as dandelion (*Taraxacum*, Chap. 13), *Pilosella* (Chap. 14), *Boechera* (Chap. 15), and *Cenchrus* (*Pennisetum*, Chap. 16). These case studies demonstrate that scientific discovery is rarely linear and that technological advances often lead to breakthroughs.

Engineering apomixis has long been considered a highly desirable trait in plant breeding. Following recent advances in synthetic apomixis, the final part of the book (Part VI) introduces protocols for engineering either individual components of apomixis, such as apomeiosis (Chap. 17), or a complete synthetic apomixis system (Chap. 18). These practical examples suggest that the field is currently thriving, and the implementation of synthetic apomixis systems in agriculture may soon become a reality.

We hope that this curated collection of methods and overview chapters will be helpful and will serve not only as a reference for studying apomixis but also as an inspiration and foundation for designing new approaches and achieving significant breakthroughs in the field. There has probably never been a more exciting time to study apomixis.

We thank the authors of Chap. 5, Krystyna Musiał and Agnieszka Springer, for kindly providing the image found in the front matter of the book, and all the authors for their outstanding contributions to this book.

Wageningen, The Netherlands

Tatyana Radoeva
Peter J. van Dijk

Acknowledgement

I would like to give a big, special thank you to my husband, Anton Koruev, for his enormous support and patience throughout my work and the entire process of making this book a reality.

Wageningen, The Netherlands *Tatyana Radoeva*

Contents

Part IV Genetic Analysis of Apomixis

Part V Apomixis Model Species

Part VI Engineering Apomixis

Contributors

DAPHNÉ AUTRAN • *DIADE, University of Montpellier, IRD, CIRAD, Montpellier, France*
CAMILLA BANFI • *Department of Biosciences, Università degli Studi di Milano, Milan, Italy*
PHILIPPA JANE BARRELL • *New Zealand Institute for Plant and Food Research Ltd, Christchurch, New Zealand*
ROSS BICKNELL • *New Zealand Institute of Plant and Food Research Ltd, Christchurch, New Zealand*
JOSÉ CARBALLO • *Centro de Recursos Naturales Renovables de la Zona Semiárida (CCT CONICET Bahía Blanca) and Departamento de Agronomía, Universidad Nacional del Sur, Buenos Bahía Blanca, Buenos Aires, Argentina*
ANDREW CATANACH • *New Zealand Institute of Plant and Food Research Ltd., Christchurch, New Zealand*
JOANN A. CONNER • *Department of Horticulture and Institute of Plant Breeding, Genetics and Genomics, University of Georgia, Tifton, GA, USA*
LUCIANA DELGADO • *Instituto de Investigaciones en Ciencias Agrarias de Rosario (IICAR), CONICET, Facultad de Ciencias Agrarias, Universidad Nacional de Rosario, Zavalla, Argentina*
THOMAS DRESSELHAUS • *Plant Cell Biology, Biochemistry, and Biotechnology, Institute of Plant Sciences, University of Regensburg, Regensburg, Germany*
VIVIANA ECHENIQUE • *Centro de Recursos Naturales Renovables de la Zona Semiárida (CCT CONICET Bahía Blanca) and Departamento de Agronomía, Universidad Nacional del Sur, Buenos Bahía Blanca, Buenos Aires, Argentina*
MARÍA FLORES-TORNERO • *Departament de Biologia Vegetal, Facultat de Ciències Biològiques , Universitat de València, Valencia, Spain*
ROVEN ROMMEL FUENTES • *Department of Chromosome Biology, Max Planck Institute for Plant Breeding Research, Cologne, Germany*
LAURINE GILLES • *CIRAD, UMR AGAP Institut, Montpellier, France; University of Montpellier, CIRAD-INRAE-Institut Agro, Montpellier, France*
EMMANUEL GUIDERDONI • *CIRAD, UMR AGAP Institut, Montpellier, France; University of Montpellier, CIRAD-INRAE-Institut Agro, Montpellier, France*
BAS HEIDEMANN • *Department of Plant & Animal Biology, Radboud Institute for Biological and Environmental Sciences, Radboud University, Nijmegen, The Netherlands*
YUJI KE • *Ghent University, Department of Plant Biotechnology and Bioinformatics, Ghent, Belgium; VIB Center for Plant Systems Biology, Ghent, Belgium*
IMTIYAZ KHANDAY • *Department of Plant Sciences, University of California, Davis, CA, USA; Innovative Genomics Institute, University of California, Berkeley, CA, USA*
ANNA M. G. KOLTUNOW • *Centre for Crop Sciences (QAAFI), The University of Queensland, Brisbane, Australia*
ĽUBOŠ MAJESKÝ • *Faculty of Science, Department of Botany, Palacký University in Olomouc, Olomouc-Holice, Czech Republic*
TEREZIE MANDÁKOVÁ • *Central European Institute of Technology, Masaryk University, Brno, Czech Republic; Department of Experimental Biology, Faculty of Science, Masaryk University, Brno, Czech Republic*
RAPHAËL MERCIER • *Max Planck Institute for Plant Breeding Research, Cologne, Germany*

DONALDO MEYNARD • *CIRAD, UMR AGAP Institut, Montpellier, France; University of Montpellier, CIRAD-INRAE-Institut Agro, Montpellier, France*
KRYSTYNA MUSIAŁ • *Department of Plant Cytology and Embryology, Institute of Botany, Faculty of Biology, Jagiellonian University, Cracow, Poland*
RIK H. M. OP DEN CAMP • *Keygene N.V, Wageningen, The Netherlands*
PEGGY OZIAS-AKINS • *Department of Horticulture and Institute of Plant Breeding, Genetics and Genomics, University of Georgia, Tifton, GA, USA*
JURAJ PAULE • *Freie Universität Berlin, ZE Botanischer Garten und Botanisches Museum Berlin, Berlin, Germany*
TATYANA RADOEVA • *Keygene N.V, Wageningen, The Netherlands*
DIANA RIGOLA • *Keygene N.V, Wageningen, The Netherlands*
PETRA ŠARHANOVÁ • *Faculty of Science, Department of Botany and Zoology, Masaryk University, Brno, Czech Republic*
ANJA SCHMIDT • *Institute of Biology, Plant Evolutionary Biology, University of Hohenheim, Stuttgart, Germany*
JUAN PABLO SELVA • *Centro de Recursos Naturales Renovables de la Zona Semiárida (CCT CONICET Bahía Blanca) and Departamento de Agronomía, Universidad Nacional del Sur, Buenos Bahía Blanca, Buenos Aires, Argentina*
AGNIESZKA B. SPRINGER • *The Franciszek Górski Institute of Plant Physiology, Polish Academy of Sciences, Cracow, Poland*
VENKATESAN SUNDARESAN • *Department of Plant Sciences, University of California, Davis, CA, USA; Innovative Genomics Institute, University of California, Berkeley, CA, USA; Department of Plant Biology, University of California, Davis, CA, USA*
CHARLES J. UNDERWOOD • *Keygene N.V., Wageningen, The Netherlands; Department of Chromosome Biology, Max Planck Institute for Plant Breeding Research, Cologne, Germany; Department of Plant & Animal Biology, Radboud Institute for Biological and Environmental Sciences, Radboud University, Nijmegen, The Netherlands*
PETER J. VAN DIJK • *Keygene N.V, Wageningen, The Netherlands*
AURORE VERNET • *CIRAD, UMR AGAP Institut, Montpellier, France; University of Montpellier, CIRAD-INRAE-Institut Agro, Montpellier, France*
YAZHONG WANG • *Department of Chromosome Biology, Max Planck Institute for Plant Breeding Research, Cologne, Germany; Center for Molecular Agrobiology, Institute of Genetics and Developmental Biology, Chinese Academy of Sciences, Beijing, China*
CHRIS WINEFIELD • *Department of Wine, Food and Molecular Biosciences, Lincoln University, Canterbury, New Zealand*
IRIS E. ZAHN • *Department of Plant & Animal Biology, Radboud Institute for Biological and Environmental Sciences, Radboud University, Nijmegen, The Netherlands*
DIEGO ZAPPACOSTA • *Centro de Recursos Naturales Renovables de la Zona Semiárida (CCT CONICET Bahía Blanca) and Departamento de Agronomía, Universidad Nacional del Sur, Buenos Bahía Blanca, Buenos Aires, Argentina*
XIXI ZHENG • *Plant Cell Biology, Biochemistry, and Biotechnology, Institute of Plant Sciences, University of Regensburg, Regensburg, Germany*

Part I

Introduction

Chapter 1

Gametophytic Apomixis: An Introduction

Peter J. van Dijk and Tatyana Radoeva

Abstract

Apomixis, clonal reproduction through seed, is divided into sporophytic (nucellar embryony) and gametophytic apomixis. This book focuses on the latter type. Gametophytic apomixis, while rare, is taxonomically widespread, notably among grasses and *Asteraceae*. Detection methods have advanced significantly, employing flow cytometry and advanced microscopy techniques. The genetic, molecular, and developmental mechanisms behind apomixis involve modifications preventing meiosis and enabling parthenogenesis. Understanding these mechanisms and applying approaches such as forward genetics and mutagenesis holds an important potential for agricultural innovation. Numerous genetic, ecological, and evolutionary questions remain open, highlighting apomixis as an important field for ongoing scientific research.

Key words Apomixis, Apospory, Diplospory, Parthenogenesis, Pseudogamy, Autonomous endosperm

1 What Is Apomixis?

The term *apomixis*, or clonal reproduction through seeds, was originally defined by Hans Winkler in 1908 as "the substitution of sexual reproduction with another, asexual propagation process, that is not associated with nuclear or cellular fusion"[1] [1].

When apomixis occurs in successive generations, we speak of recurrent apomixis. In rare instances, apomixis-like phenomena may occur in sexual species, but these do not result in apomixis in subsequent generations. When this book refers to apomixis, it specifically means recurrent apomixis.

The word *apomixis* is derived from the Greek *Apo* (away from) and *Mixis* (mixing). Its counterpart is *amphimixis*, derived from *Amphi* (both sides) and *Mixis* (mixing). *Amphimixis* refers to sexual reproduction, which involves the fusion of female and male gametes, either from the same plant (in self-fertilization of a

[1] *den Ersatz der geschlechtlichen Fortpflanzung durch einen anderen, ungeschlechtlichen, nicht mit Kern- und Zellverschmeltzung verbundenen Vermehrungsprozeß.*

Tatyana Radoeva and Peter J. van Dijk (eds.), *Apomixis: Methods and Protocols*, Methods in Molecular Biology, vol. 2987, https://doi.org/10.1007/978-1-0716-5001-1_1,

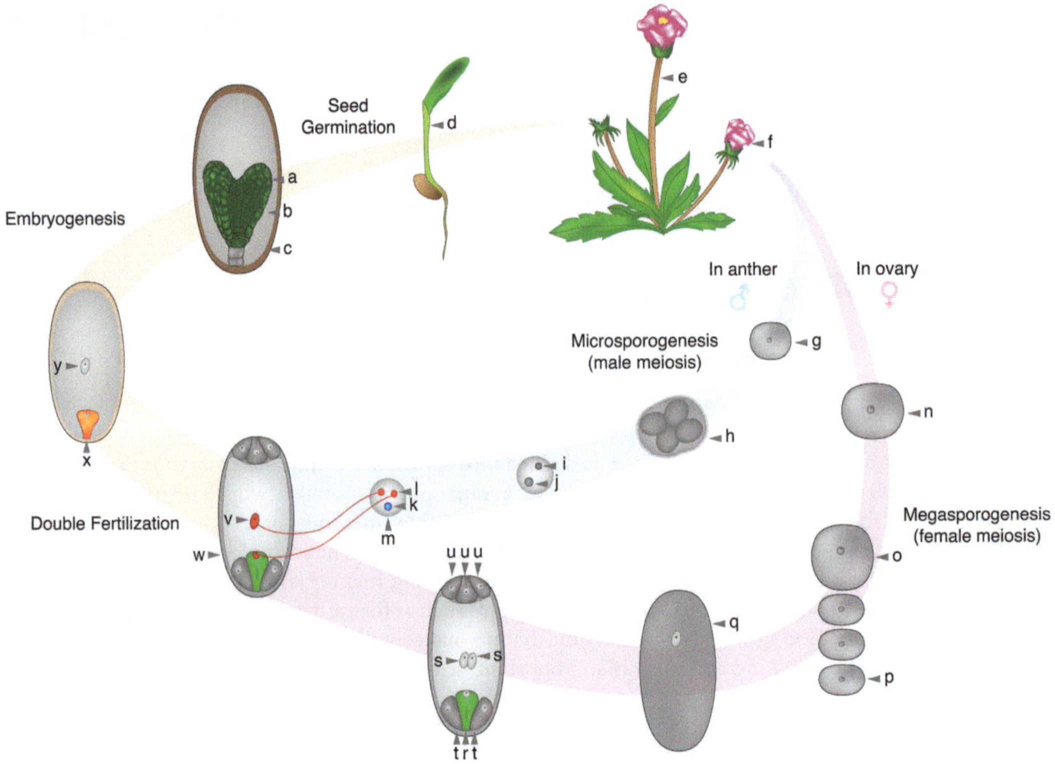

Fig. 1 The life cycle of a sexually reproducing flowering plant. (**a**) mature embryo, (**b**) endosperm, (**c**) seed coat, (**d**) young seedling, (**e**) mature plant, (**f**) inflorescence, (**g**) microsporocyte (male meiocyte), (**h**) tetrad of microspores (male meiotic products), (**i**) generative cell, (**j**) vegetative cell, (**k**) mature pollen grain, (**l**) one of two sperm cells, (**m**) vegetative nucleus, (**n**) megaspore mother cell (MMC) or megasporocyte (female meiocyte), (**o**) functional megaspore, (**p**) one of three nonfunctional megaspores, (**q**) megagametophyte, (**r**) egg cell, (**s**) polar nuclei, (**t**) synergid cells, (**u**) antipodal cells, (**v**) central cell, (**w**) mature megagametophyte or embryosac, (**x**) zygote, and (**y**) endosperm nucleus. In the mature gametophytes, generative nuclei (nuclei that contribute genetic information to the seed) are depicted in red and vegetative nuclei (nuclei that do not contribute genetic information to the seed) are shown in blue. (Modified from Underwood et al. [28]. Originally published in Elsevier's Reference Module in Biomedical Sciences (2024))

hermaphrodite) or from different plants (in cross-fertilization). The three major breeding systems in flowering plants are thus cross-fertilization, self-fertilization, and apomixis. Genotypically speaking, self-fertilization breaks down heterozygosity, cross-fertilization restores it, while apomixis maintains it.

Figure 1 illustrates the normal sexual life cycle in flowering plants. The sporophytic and gametophytic generations are separated by meiosis and fertilization. In the case of gametophytic apomixis, both these hallmarks are bypassed or modified.

2 A Brief History of Apomixis Research

Seed formation without fertilization was first described in 1839 by Smith in the Australian dioecious plant species *Cœlobogyne ilicifolia* (now *Alchornia ilicifolia*) [2]. Brought to Europe, female specimens of this species produced seeds in the absence of male plants. Therefore, seed production had to have occurred without fertilization.

However, embryological research by Strassburger [3] revealed that the embryos were not the result of true parthenogenesis of the egg cell but had derived from somatic cells, the so-called adventitious embryony [3]. Shortly before this, Kerner von Marilaun [4] had suspected the presence of parthenogenesis in *Antennaria alpina*, as female plants in much of northern Europe produced seeds despite the absence of male plants [4].

It was not until 1898 that Juel confirmed, through cytological and embryological studies, that *Antennaria alpina* indeed exhibited parthenogenetic development of the egg cell [5]. This was the first verified case of parthenogenesis in flowering plants.

Demonstrating apomixis in dioecious species is relatively straightforward. However, in hermaphroditic plants, where individuals produce both egg cells and pollen grains, proving apomixis is much more challenging because seed production in isolated plants could also result from self-pollination. Both self-fertilization and apomixis are forms of uniparental inheritance that can produce uniform offspring.

Gregor Mendel was unaware of the existence of apomixis and believed that his apomictic *Hieracium* (now *Pilosella*) plants were self-pollinators. In the second half of the 1860s, Mendel crossed sexual and apomictic *Pilosella* plants and created the first de novo apomictic clones. Through decapitation experiments, in which both the style and anthers were removed (*see* below), Ostenfeld and Raunkiær (1903) demonstrated that Mendel's *Pilosella* species were apomictic [6]. Embryological studies by Murbeck (1904) confirmed this [7]. For a detailed account of the early history of apomixis research, *see* the study cited herein [8].

3 Main Types of Apomixis

Apomixis can be divided into two main types: sporophytic (Fig. 2b) and gametophytic apomixis (Fig. 2c, d). In the case of sporophytic apomixis, one or more clonal sporophytic embryos are formed in the nucellus or integument in addition to a sexual embryo, as was the case in *Alchornia*. The progeny is thus a mixture of sexual and apomictic offspring. This process is also known as adventitious or nucellar embryony. Sporophytic apomixis is common in orchids

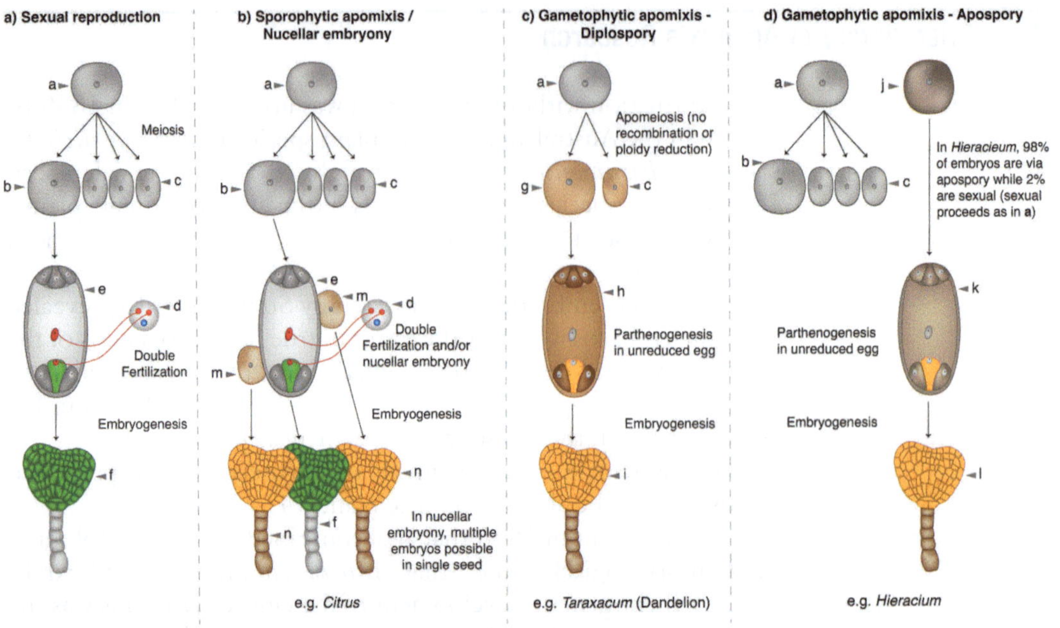

Fig. 2 Sexual (**a**) and asexual (**b**–**d**) reproduction in flowering plants. (**a**) megaspore mother cell (MMC) or megasporocyte (female meiocyte), (**b**) functional megaspore, (**c**) nonfunctional megaspore, (**d**) mature pollen grain, (**e**) mature megagametophyte, (**f**) mature sexual embryo, (**g**) nonrecombined and unreduced functional megaspore, (**h**) unreduced mature megagametophyte, (**i**) mature clonal embryo, (**j**) aposporous initial cell (sporophytic origin), (**k**) aposporous mature megagametophyte, (**l**) mature aposporous embryo, (**m**) nucellar cell (or other cell of sporophytic origin), (**n**) nucellar embryo (nonsexual embryo). (Modified from Underwood et al. [28]. Originally published in Elsevier's Reference Module in Biomedical Sciences (2024))

and occurs in fruit trees such as citrus and mangosteen. In the case of gametophytic apomixis, an embryo sac or megagametophyte is formed, and the clonal embryo develops through parthenogenesis—without fertilization—from an egg cell. This chapter serves as an introduction to gametophytic apomixis and focuses on what is known about the causes of it; sporophytic apomixis is not addressed further in the chapters of this book. Thus, when apomixis is mentioned, gametophytic apomixis is meant.

Approximately 50% of angiosperms are obligate outcrossers, while the other half can self-fertilize [9–11]. Mogie [12] estimated that the percentage of plant species reproducing via gametophytic apomixis is about 0.1%, or 1 in 1000 species [12]. However, it is challenging to provide an accurate estimate of the number of species with apomixis because apomixis affects species concepts, is often associated with hybridization, and requires multiple lines of evidence to confirm with certainty. On one hand, many cases of apomixis may remain undiscovered; on the other, earlier claims of apomixis in certain species have not held up to further scrutiny (*see* below).

Although gametophytic apomixis is rare at the species level, it is taxonomically widespread [13, 14]. Gametophytic apomixis is particularly common in grasses and the *Asteraceae* family. Notably, as far as we know, no cases of gametophytic apomixis have been reported in the large family of *Fabaceae*.

Apomictic taxa are typically located on the outermost twigs of phylogenetic trees. There appear to be no genera consisting exclusively of apomictic species, with the monospecific genus *Houttuynia* as a possible exception [15]. In this genus, no sexual reproduction has been observed so far, although *Houttuynia cordata* shows a considerable amount of genetic variation [16, 17]. The general phylogenetic distribution of gametophytic apomixis suggests that it evolved many times, but that apomictic species are relatively young in evolutionary terms, have limited evolutionary potential, and a short lifespan.

4 Different Types and Mechanisms of Gametophytic Apomixis

For apomixis, at least two modifications of sexual reproduction are necessary: (1) Apomeiosis (the prevention of meiosis, i.e., recombination and halving of the chromosome number) and (2) Parthenogenesis (the development of the egg cell into an embryo without fertilization). In general, apomeiosis only applies to female meiosis and not to pollen meiosis. During pollen meiosis in apomicts, recombination usually occurs, making it possible to use genetic markers in crosses with sexual relatives to map apomixis loci. However, pollen fertility in apomicts is often low, probably due to, amongst others, polyploidy and relaxation of selection against chromosomal rearrangement during female meiosis. Therefore, many crosses may have to be made to produce a sufficiently large mapping population.

In amphimixis, double fertilization occurs: in addition to the egg cell, the central cell of the embryo sac (megagametophyte) is also fertilized (Fig. 1). This results in the formation of endosperm, a tissue that nourishes the developing embryo. In many apomictic plant species, the endosperm also develops through fertilization (pseudogamy). Since endosperm is not part of the next generation, it has no genetic effect on the progeny. In a minority of apomicts, endosperm develops autonomously from the unfertilized central cell. Autonomous endosperm is particularly common among apomictic *Asteraceae*. In these *Asteraceae*, apomixis thus differs from amphimixis in three developmental elements: (1) Apomeiosis, (2) Parthenogenesis, and (3) Autonomous endosperm development.

Cytological studies have demonstrated that there are several mechanisms of apomeiosis. Based on the origin of the megagametophyte (embryo sac), gametophytic apomixis can be further subdivided into the main types: apospory and diplospory (Fig. 2). In diplospory (Fig. 2c), female meiosis (megasporogenesis) is replaced either by a mitotic division (mitotic diplospory—Antennaria type) or by a modified meiosis (Taraxacum type). In this modified meiosis, no recombination occurs during meiosis I, and an unreduced restitution nucleus is formed, after which a normal meiosis II follows. Two megaspores are produced instead of four, one of which degenerates (instead of three degenerating as in amphimixis). The unreduced megaspore undergoes three mitotic divisions, resulting in an unreduced, nonrecombined megagametophyte containing eight nuclei: the unreduced egg cell nucleus with two synergids, the two polar nuclei, and three antipodals. Intermediate forms between mitotic and meiotic diplospory are possible [18]. Diplospory was recently reviewed in depth [19].

In apospory (Fig. 2d), a somatic aposporous initial arises near the megaspore mother cell (MMC), which, through three mitotic divisions, forms an unreduced, nonrecombined megagametophyte containing an unreduced, nonrecombined egg cell. This aposporous egg cell can develop parthenogenetically into an embryo. In most cases, the aposporous embryo suppresses the development of the sexual embryo, although, occasionally, a sexual seed may still be produced. In such cases, apomixis is partial, often called facultative apomixis.

According to the most recent review [14], gametophytic apomixis has been reported in 161 genera, which represents 1.24% of the 13,000 angiosperm genera. Of these, 110 genera exhibit apospory, 68 diplospory, and 17 both. Since apomixis is not easily detected and proven, these figures should be treated with caution.

As explained above, in apomictic seed formation, both apomeiosis and parthenogenesis occur. However, it is also possible for one element to occur independently of the other. Harlan and DeWet [20] introduced useful terminology, where "n" denotes reduced gametes and "$n + n$" represents the fertilized amphimictic zygote (BII hybrids in Rutishauser's [21] terminology). The term "$2n$" refers to an unreduced gamete. Apomictic seeds are indicated as "$2n + 0$" (an unreduced egg cell developing without fertilization). When apomeiosis and parthenogenesis are decoupled, offspring are either "$n + 0$" (polyhaploid progeny) or "$2n + n$" (BIII hybrids according to Rutishauser; hyperpolyploid hybrids).

Apomixis is rarely obligate. Residual sexuality can persist, particularly in apospory. These facultative apomicts produce a mixture of apomictic and sexual offspring. A mixture of sexual, apomictic, and intermediate forms may also arise when the penetrance of the apomixis elements is incomplete. When apomeiosis and parthenogenesis are controlled by separate, unlinked, or partly linked loci, recombinant progeny plants of the "$n + 0$" and "$2n + n$" types can appear.

5 How to Detect Gametophytic Apomixis?

Apomixis is not easy to detect, and several lines of evidence are often needed. Even quite recently, there have been examples of confusion between selfing and apomixis. Claimed facultative apomictic seed set in diploid *Potentilla argentata* was found upon analysis of allozyme markers actually to be the result of selfing [22].

Noyes [23] critically reviewed the evidence for apomixis in the *Asteraceae* family as reported in the literature. In this family, where apomixis is relatively common, apomixis had been reported in 68 genera [24]. Of these, the evidence for apomixis was robust in 22 genera, better documentation was needed in 16 genera, and in 30 genera, there was evidence against apomixis. Similar overestimations likely occur in other families as well.

There are different ways to detect apomixis in hermaphroditic plants. The occurrence of triploids over a large geographical area, without the presence of diploids and tetraploids, is a strong indication of (autonomous) apomixis because sexual triploids typically exhibit severely disturbed meiosis and very low fertility (e.g., in *Taraxacum*) and will therefore not persist in nature. Seed set following bud decapitation, in which styles and anthers are removed before flowering, is also a strong indication of autonomous apomixis. Nonsegregating heterozygosity for multiple genetic markers or higher levels of heterozygosity than expected under Hardy–Weinberg equilibrium are additional indicators of the presence of apomixis.

In 2000, Matzk et al. published a method called flow cytometric seed screen or FCSS (*see* also Chapter 2), a powerful method for determining the reproductive pathway of a plant using flow cytometry of dry seeds [24]. Flow cytometry has made time-consuming and labor-intensive chromosome counts redundant. Over the past 25 years, flow cytometers have become significantly more affordable, allowing more laboratories to utilize this method. Additionally, the protocols have become considerably more robust and generic.

Cytological/embryological observations of apomeiosis and parthenogenesis are still the gold standard. Ovary/ovule clearing methods made the time-consuming preparation and embedding of histological sections unnecessary (*see* Chapters 3, 4, 5, and 6). New microscopic techniques, such as Differential Interference Contrast (DIC)/Nomarski, fluorescence microscopy, and confocal microscopy, often in combination with novel staining methods, have greatly improved resolution.

6 Phenomena Associated with Apomixis

Nearly all gametophytic apomicts found in nature, including diplospory and apospory, are polyploid. A very rare exception to this is diploid apomixis in the genus *Boechera* (*see* Chapter 15). As described above, presumed apomixis in diploid *Potentilla* was in fact shown to be selfing. However, while nearly all gametophytic apomicts are polyploid, the great majority of polyploid plants are actually sexual. In sporophytic apomixis, there is no such correlation with polyploidy.

However, dihaploid plants ($n + 0$) obtained from experimental crosses of, for example, *Ranunculus* [18], *Pilosella* [25], and *Erigeron* [26] can indeed be apomictic. It has been suggested that apomixis genes cannot be transmitted through haploid pollen. In *Taraxacum*, codominant Short Sequence Repeat (SSR) markers linked to the diplospory locus were transmitted via diploid but not via haploid pollen [27]. This is true for various species and different types of gametophytic apomixis. This suggests that haploid nontransmission of apomixis may not be a pleiotropic effect from the apomixis genes themselves. Instead, it could be a secondary effect of apomictic reproduction over long evolutionary timescales, manifested as a linked genetic load of accumulated deleterious mutations (*see* also Chapter 14 on *Pilosella*). Although the structures of only a few apomixis loci have been studied so far, there is clear evidence for recombination suppression during pollen meiosis, hemizygosity, and a high density of transposons [28–30].

Apomixis is also frequently associated with hybridization; this is particularly evident in the genus *Boechera* (*see* Chapter 15). Ernst [31], in his book *Hybridization as the Cause of Apogamy in the Plant Kingdom: A Hypothesis on Experimental Heredity and Evolutionary Theory*, proposed that apomixis could arise directly as a consequence of hybridization [31]. Like polyploidy, the association between gametophytic apomixis and hybridization is one-sided: while many apomicts are hybrids, the vast majority of hybrids reproduce sexually.

Carman [32] proposed that apomixis results from the asynchronous expression of duplicated genes controlling sexual development, a condition often arising from allopolyploidy [32]. Comparative transcriptomic studies between apomictic and related sexual plants reveal substantial differences in gene expression within female reproductive tissues, as well as shifts in gene ontology categories (e.g., [33–37]). This could not only indicate the genome-specific expression of many genes but also reflect the downstream effects of a few major master genes.

7 Genetics and Molecular Basis of Apomixis

Several studies have involved crosses between apomictic pollen donors and sexual relatives (e.g., *Erigeron* [26, 38](*Taraxacum*, Chapter 13, *Pilosella*, Chapter 14, *Pennisetum*, Chapter 16). Although polyploidy, hybridization, and segregation distortion frequently complicate matters, apomixis often segregates within progenies following a simplex genotypic model (e.g., *Aaaa*) for one or a few dominant apomixis loci. Linked genetic markers confirm this. Thus, apomixis appears dominant over amphimixis in these species.

Some studies have demonstrated separate loci for apomeiosis and parthenogenesis. Noyes and Wagner [26] crossed diplosporous dihaploids (*Dd/pp*) with parthenogenetic plants (*dd/Pp*), resulting in diploid apomictic progeny [26]. Apomixis elements initially separated could later be combined, restoring apomixis even at the diploid level.

An important task is the identification of natural apomixis genes. If apomixis genes exist, are they variants of sexual genes or entirely different genes? In insects, for example, in *Hymenoptera*, parthenogenesis can be caused not by the genes of the organism, but by *Wolbachia* bacterial symbionts [39]. In facultatively parthenogenetic *Apis mellifera capensis* and *Drosophila mercuratum*, the first parthenogenesis genes in insects were recently identified [40, 41]. In animals, apomixis always seems directly associated with a disrupted meiosis, whereas in seed plants, apomeiosis and parthenogenesis can often be decoupled [42].

When it comes to the identification of apomixis genes, it presents a challenging problem: all genetic complexities appear to converge here, including polyploidy, suppression of recombination, variation in penetrance, segregation distortion, accumulation of transposons, linkage drag of a mutation load, etc. [43].

Hemizygous Apomixis-Specific Genomic Regions (ASGR) have likely spread across species boundaries via introgression in grass genera *Cenchrus* and *Pennisetum* [44]. Sequencing of Bacterial Artificial Chromosomes (BACs) of the ASGR in *Pennisetum* revealed a *BabyBooM-Like* (*BBML*) gene [45], a strong candidate gene for parthenogenesis because *BBM* induces somatic embryogenesis in *Brassica* [46].

When all three haplotypes of the *PARTHENOGENESIS* (*PAR*) locus in *Taraxacum* were gap-free sequenced, none of the approximately 30 genes stood out as a clear functional candidate parthenogenesis gene based on annotation. However, targeted mutagenesis with Clustered Regularly Interspaced Short Palindromic Repeats/CRISPR-associated protein 9 (CRISPR-Cas9) identified one gene as the *PAR* gene (*see* Chapters 12 and 13). Laser Capture Microscopy showed that this gene is expressed in apomicts already within the egg cell, whereas in sexual plants, the expressed gene enters the egg cell only through pollen during fertilization [47]. A transposon in the promoter of the *PAR* gene is likely responsible for this ectopic egg cell expression.

The *BBML* gene from *Pennisetum* and the *PAR* gene from *Taraxacum* have sexual homologs. The *PAR* gene is also likely responsible for parthenogenesis in the *Asteraceae* genera *Pilosella* and *Hieracium* [48]. These genera also exhibit a transposon insertion in the *PAR* allele promoter at nearly the same position, though *Hieracium* contains a different type of transposon. In all three genera, the same gene is responsible for parthenogenesis, but its evolution occurred independently.

This research demonstrates that amphimixis is the original reproductive mode, suppressed by dominant apomixis. Loss of apomictic genome segments, such as through irradiation-induced mutation experiments, results in the restoration of sexual reproduction.

8 Engineering Apomixis in Sexual Crops

In the 1930s in the Soviet Union, it was already recognized by the famous plant geneticists M. S. Navashin and G. D. Karpechenko that apomixis could offer significant advantages for agriculture [49]. Efforts were made to introduce apomixis into sexual crops through introgression from related wild apomictic species. A notable example is the attempt to introgress apomixis from *Tripsacum* into maize, but these extensive efforts had limited success [50].

Another approach is to induce apomixis through mutagenesis. This can be through forward genetics (from the mutant (loss of apomixis) phenotype to the gene) or reverse genetics (from mutating a gene with a known function to the gain of apomixis phenotype). Given that natural apomixis genes appear to be modified variants of genes in sexual relatives, these approaches seem sensible. Both Singh et al. [51] and Fox et al. [52] followed the forward genetics approach by creating *Mutator* transposable element (*Mu*) insertion populations in maize. Both studies were successful in cloning apomeiosis genes through phenotyping (test crosses) and isolation of insertion flanking sequences. The *Dominant non-reduction4* (*Dnr4*; [51]) gene belongs to the *ARGONAUTE* gene family and influences both female and male meiosis. The recessive non-reduction *nrf4* is female meiosis-specific, but the gene is not annotated [53].

The reverse genetics approach for engineering apomixis has been remarkably successful. In 2009, d'Erfurt et al. succeeded in transforming meiosis into mitosis in *Arabidopsis* by combining three recessive meiotic mutants (*Mitosis instead of Meiosis* [*MiMe*] triple knockout mutant, *see* Chapters 17 and 18) [54]. Khanday et al. [55] successfully synthesized apomictic rice by combining

MiMe with ectopic egg cell-specific expression of a rice ortholog of *BBM* [55]. Ectopic expression was achieved by placing *OsBBM* under the control of the *Arabidopsis* egg cell-specific *EC1* promoter. This is consistent with ectopic expression of *BBML* and *PAR* in natural apomicts, such as *Pennisetum* and *Taraxacum*. Targeted mutagenesis using CRISPR-Cas9 facilitates the generation of *MiMe* knockouts. Wang and colleagues created *MiMe* triple mutants in tomato (*see* Chapter 17). Vernet et al. [56] increased the penetrance of synthetic apomixis by combining transgenes into a single transformation cassette (*see* Chapter 18) [56]. Moreover, it was shown that heterosis was maintained over successive generations.

For practical applications, natural apomicts can provide essential insights into engineering apomixis into crop plants, identifying critical genes and functional polymorphisms. Currently, it remains unclear to what extent apomixis genes are broadly applicable across phylogenetic groups or whether each group requires its own apomixis gene. Goeckeritz et al. [57] have advocated identifying natural apomixis genes in phylogenetically diverse apomicts, as these would function optimally within related agricultural crops [57].

9 Causes of Apomixis Research: What Is Next?

The methods described in this handbook will remain important in future research unraveling the cause of apomixis. History has shown that breakthroughs in apomixis research often follow the introduction of new technologies. For example, comparative RNA-seq of macro-dissected tissues and long-read sequencing are already proving very valuable. Techniques such as Laser Capture Microscopy (LCM) and CRISPR/Cas9 have so far only seen limited application in apomixis studies. These emerging technologies are no longer confined to model species like *Arabidopsis*, maize, or rice. While LCM is currently applied to pools of a few gametophytic cells, single-cell sequencing is on the horizon. Moreover, gene expression in MMC-neighboring cells is also important—making spatial transcriptomics a potentially powerful tool for uncovering mechanisms underlying apomictic reproduction.

To date, no apomict has a fully phased, gap-free whole-genome sequence, but this appears to be only a matter of time. Looking further ahead, pan-genomes of apomicts and their sexual relatives will become available. Key questions are whether these (pan)-genomes contain evolutionary signatures of apomictic reproduction and whether comparative genomics can reveal the causal loci and genes responsible for apomixis. This also presents a major bioinformatics challenge.

In addition to these technological advances, the quality of the biological material remains essential. It is crucial that apomixis in defined sexual lines is entirely absent. Artificial crosses and mutant populations remain of great importance, and for this, high-throughput phenotyping is indispensable—this includes test crosses with sexuals and ploidy progeny analysis, among others. Since apomixis is a trait at the end of the life cycle, developing rapid cyclers can speed up research. For example, in dandelion, for crosses, both apomicts and sexuals were selected that did not need a long vernalization period for flowering. Furthermore, developing robust transformation systems is vital for the functional validation of candidate apomixis genes. In conclusion, combining advanced genomic tools, precise phenotyping, and improved biological materials will greatly accelerate the discovery and validation of genes controlling natural apomixis.

10 Open Questions

Apomixis is a fascinating phenomenon in many respects, including its causes, consequences, and applications in fields such as genetics, developmental biology, taxonomy, evolutionary biology, ecology, and agriculture. It is relevant to empirical research addressing major evolutionary questions, such as the maintenance of sex [58, 59] and hypotheses like the Frozen Niche and the Red Queen. Below are several research questions that largely remain unanswered:

Causes: What causes apomixis? Is it controlled by genes? By a few major genes or by many minor ones? Are there sexual homologs, and how do they differ from apomictic homologs? What explains the functional differences between sexual and apomictic homologs? Are genes for diplospory different from those controlling apospory? Did apomixis arise once within a genus and spread via hybridization, or did it arise independently multiple times? What explains why apomixis is so rare in diploids and why almost all gametophytic apomicts are polyploids? Are apomictic polyploids autopolyploids or allopolyploids?

Consequences: What are the consequences for genome/locus structure over long evolutionary timescales? How does apomixis influence species taxonomy? Are apomictic populations monoclonal or polyclonal, and in the case of the latter, what is the frequency distribution of these clones? What determines clonal diversity? Do mixed populations of apomicts and sexual relatives occur, and how do they influence each other? How frequently do new clones arise over time? Are such mixed populations stable over time? To what extent does epigenetic variation occur within clones, and does this variation have adaptive significance? Are there ecological differences between coexisting clones?

Applications: What are the potential advantages and disadvantages of apomixis in crops regarding breeding, seed production, and cultivation? Can knowledge of natural apomixis genes be utilized to induce apomixis in sexual crops? Are synthetic apomicts stable in their traits across apomictic generations? How can fertility and the penetrance of apomixis be enhanced? What are the consequences of apomictic hybrid production for current F1 hybrid seed production? Should and could alternative business models replace the loss of F1 segregation protection?

These largely open questions highlight how much remains to be discovered about apomixis, making it a rich and promising field for future research across fundamental and applied sciences. This introductory overview focused on the causes of apomixis. It is necessarily brief, with some topics touched on only superficially and many relevant publications not cited. We, therefore, apologize to those authors whose relevant contributions to the field could not be cited due to space constrains.

Conflicts of Interest

The authors have no conflicts of interest to declare that are relevant to the content of this chapter.

References

1. Winkler H (1908) Ueber parthenogenesis und apogamie im pflanzenreich. Gustav Fischer, Jena
2. Smith J (1841) Notice of a plant which produces seeds without any apparent action of pollen. Trans Linn Soc 18:509–512
3. Strassburger E (1878) Ueber polyembryonie. Jenaische Zeitschrift für Naturw 12:647–670
4. Kerner von Marilaun A (1876) Parthenogenesis bei einer angiospermen Pflanze. Sitzb Math Nat Klasse Akad Wiss Wien, Abt I 54:469–476
5. Juel HO (1898) Parthenogenesis bei Antennaria alpina (L.) R. Br Bot Zb 74:369–372
6. Ostenfeld CH, Raunkiaer C (1903) Kastreringsforsøg med Hieracium og andre Cichorieae. Bot Tidsskr 25:409–413
7. Murbeck S (1904) Parthenogenese bei den Gattungen Taraxacum und Hieracium. Bot Not 1904:285–296
8. (2018) Die Anfänge der Apomixisforschung, 1. Auflage. Éditions universitaires européennes, Saarbrücken
9. Darlington CD, Mather K (1949) The elements of genetics. Allen and Unwin, London
10. Brewbaker JL (1959) Biology of the angiosperm pollen grain. Ind J Genet Plant Breed 19:121–133
11. Igic B, Kohn JR (2006) The distribution of plant mating systems: study bias against obligately outcrossing species. Evolution 60:1098–1103. https://doi.org/10.1111/j.0014-3820.2006.tb01186.x
12. Mogie M (1992) The evolution of asexual reproduction in plants, 1st edn. Chapman & Hall, London
13. Hojsgaard D, Klatt S, Baier R et al (2014) Taxonomy and biogeography of Apomixis in angiosperms and associated biodiversity characteristics. Crit Rev Plant Sci 33:414–427. https://doi.org/10.1080/07352689.2014.898488
14. Hörandl E (2024) Apomixis and the paradox of sex in plants. Ann Bot 134:1–18. https://doi.org/10.1093/aob/mcae044
15. Stebbins GL (1950) Variation and evolution in plants. Columbia University Press, New York
16. Wei W, Youliang Z, Li C et al (2006) PCR-RFLP analysis of cpDNA and mtDNA in the genus Houttuynia in some areas of China:

PCR-RFLP analysis of cpDNA and mtDNA in Houttuynia. Hereditas 142:24–32. https://doi.org/10.1111/j.1601-5223.2005.01704.x

17. Wei L, Wu X-J (2012) Genetic variation and population differentiation in a medical herb Houttuynia cordata in China revealed by inter-simple sequence repeats (ISSRs). Int J Mol Sci 13:8159–8170. https://doi.org/10.3390/ijms13078159
18. Nogler GA (1984) Gametophytic Apomixis. In: Johri BM (ed) Embryology of angiosperms. Springer, Berlin Heidelberg, pp 475–518
19. Cornaro L, Banfi C, Cucinotta M et al (2023) Asexual reproduction through seeds: the complex case of diplosporous apomixis. J Exp Bot 74:2462–2478. https://doi.org/10.1093/jxb/erad054
20. Harlan JR, deWet JM (1975) On Ö Winge and a prayer: the origins of polyploidy. Bot Rev 41: 361–390
21. Rutishauser A (1948) Pseudogamie und polymorphie in der gattung Potentilla. Arch Julius Klaus-Stift Vererbungsforsch 23:267–424
22. Holm S, Ghatnekar L, Bengtsson BO (1997) Selfing and outcrossing but no apomixis in two natural populations of diploid *Potentilla argentea*. J Evol Biol 10:343–352. https://doi.org/10.1046/j.1420-9101.1997.10030343.x
23. Noyes RD (2007) Apomixis in the Asteraceae: diamonds in the rough. Func Plant Sci Biotechnol 1:207–222
24. Matzk F, Meister A, Schubert I (2000) An efficient screen for reproductive pathways using mature seeds of monocots and dicots. Plant J 21:97–108. https://doi.org/10.1046/j.1365-313x.2000.00647.x
25. Bicknell RA, Borst NK (1996) Isolation of reduced genotypes of Hieracium pilosella using anther culture. Plant Cell Tissue Organ Cult 45:37–41. https://doi.org/10.1007/BF00043426
26. Noyes RD, Wagner JD (2014) Dihaploidy yields diploid apomicts and parthenogens in erigeron (Asteraceae). Am J Bot 101:865–874
27. Van Dijk P, De Jong H, Vijverberg K, Biere A (2009) An apomixis-gene's view on dandelions. In: Schön I, Martens K, Dijk P (eds) Lost sex. Springer, Dordrecht, pp 475–493
28. Underwood CJ, Van Dijk PJ, Radoeva T (2024) Sexual reproduction versus unisexual; asexual reproduction, plants. In: Reference module in biomedical sciences. Elsevier, p B978044321477600167X
29. Ozias-Akins P, Roche D, Hanna WW (1998) Tight clustering and hemizygosity of apomixis-linked molecular markers in *Pennisetum squamulatum* implies genetic control of apospory by a divergent locus that may have no allelic form in sexual genotypes. Proc Natl Acad Sci 95:5127–5132. https://doi.org/10.1073/pnas.95.9.5127
30. Okada T, Ito K, Johnson SD et al (2011) Chromosomes carrying meiotic avoidance loci in three apomictic eudicot *Hieracium* subgenus *Pilosella* species share structural features with two monocot apomicts. Plant Physiol 157:1327–1341. https://doi.org/10.1104/pp.111.181164
31. Ernst A (1918) Bastardierung als Ursache der Apogamie im pflanzenreich: Eine Hypothese zur experimentellen Vererbung und Abstammungslehre. Gustaf Fischer, Jena
32. Carman JG (1997) Asynchronous expression of duplicate genes in angiosperms may cause apomixis, bispory, tetraspory, and polyembryony. Biol J Linn Soc 61:51–94. https://doi.org/10.1111/j.1095-8312.1997.tb01778.x
33. Sharbel TF, Voigt M-L, Corral JM et al (2010) Apomictic and sexual ovules of *Boechera* display heterochronic global gene expression patterns. Plant Cell 22:655–671. https://doi.org/10.1105/tpc.109.072223
34. Garbus I, Romero JR, Selva JP et al (2017) De novo transcriptome sequencing and assembly from apomictic and sexual Eragrostis curvula genotypes. PLoS One 12:e0185595. https://doi.org/10.1371/journal.pone.0185595
35. Bräuning S, Catanach A, Lord JM et al (2018) Comparative transcriptome analysis of the wild-type model apomict Hieracium praealtum and its loss of parthenogenesis (lop) mutant. BMC Plant Biol 18:206. https://doi.org/10.1186/s12870-018-1423-1
36. Galla G, Basso A, Grisan S et al (2019) Ovule gene expression analysis in sexual and aposporous apomictic Hypericum perforatum L. (Hypericaceae) accessions. Front. Plant Sci 10:654. https://doi.org/10.3389/fpls.2019.00654
37. Pellino M, Hojsgaard D, Hörandl E, Sharbel TF (2020) Chasing the apomictic factors in the Ranunculus auricomus complex: exploring gene expression patterns in microdissected sexual and apomictic ovules. Genes 11:728. https://doi.org/10.3390/genes11070728
38. Noyes RD (2022) Mendelian segregation for parthenogenetic embryo development at the diploid level in the flowering plant *Erigeron*. Am J Bot 109:1641–1651. https://doi.org/10.1002/ajb2.16071

39. Fricke LC, Lindsey ARI (2024) Identification of parthenogenesis-inducing effector proteins in *Wolbachia*. Genome Biol Evol 16:evae036. https://doi.org/10.1093/gbe/evae036
40. Yagound B, Dogantzis KA, Zayed A et al (2020) A single gene causes thelytokous parthenogenesis, the defining feature of the cape honeybee Apis mellifera capensis. Curr Biol 30: 2248–2259.e6. https://doi.org/10.1016/j.cub.2020.04.033
41. Sperling AL, Fabian DK, Garrison E, Glover DM (2023) A genetic basis for facultative parthenogenesis in drosophila. Curr Biol 33: 3545–3560.e13. https://doi.org/10.1016/j.cub.2023.07.006
42. Van Dijk P (2009) Apomixis: basics for non-botanists. In: Schön I, Martens K, Dijk P (eds) Lost Sex. Springer, Dordrecht, pp 47–62
43. Ozias-Akins P, Van Dijk PJ (2007) Mendelian genetics of apomixis in plants. Annu Rev Genet 41:509–537. https://doi.org/10.1146/annurev.genet.40.110405.090511
44. Akiyama Y, Goel S, Conner JA et al (2011) Evolution of the apomixis transmitting chromosome in Pennisetum. BMC Evol Biol 11: 289. https://doi.org/10.1186/1471-2148-11-289
45. Conner JA, Mookkan M, Huo H et al (2015) A parthenogenesis gene of apomict origin elicits embryo formation from unfertilized eggs in a sexual plant. Proc Natl Acad Sci 112:11205–11210. https://doi.org/10.1073/pnas.1505856112
46. Boutilier K, Offringa R, Sharma VK et al (2002) Ectopic expression of BABY BOOM triggers a conversion from vegetative to embryonic growth. Plant Cell 14:1737–1749. https://doi.org/10.1105/tpc.001941
47. Underwood CJ, Vijverberg K, Rigola D et al (2022) A PARTHENOGENESIS allele from apomictic dandelion can induce egg cell division without fertilization in lettuce. Nat Genet 54:84–93. https://doi.org/10.1038/s41588-021-00984-y
48. Bicknell R, Gaillard M, Catanach A et al (2023) Genetic mapping of the LOSS OF PARTHENOGENESIS locus in Pilosella piloselloides and the evolution of apomixis in the Lactuceae. Front Plant Sci 14:1239191. https://doi.org/10.3389/fpls.2023.1239191
49. Соколов ВА (2003) Будет ли следующая «зеленая революция»? Наука и жизнь. 3. С. 25–31
50. Leblanc O, Grimanelli D, Hernandez-Rodriguez M et al (2009) Seed development and inheritance studies in apomictic maize-Tripsacum hybrids reveal barriers for the transfer of apomixis into sexual crops. Int J Dev Biol 53:585–596. https://doi.org/10.1387/ijdb.082813ol
51. Singh M, Goel S, Meeley RB et al (2011) Production of viable gametes without meiosis in maize deficient for an ARGONAUTE protein. Plant Cell 23:443–458. https://doi.org/10.1105/tpc.110.079020
52. Fox TW, Albertsen MC, Williams ME, Lawit SJ, Chamberlin MA, Grossniklaus U, Brunner GA, Chumak N, De Asis JB, Pasquer F. (2016) Methods and compositions for the production of unreduced, non-recombined gametes and clonal offspring. Patent WO2016179522A1
53. Grossniklaus U (2019) The quest for clonal seeds: towards engineering Apomixis in maize. Abstract of a seminar at Nara Institute of Science and Technology Division of Biological Science, 11 June 2019
54. d'Erfurth I, Jolivet S, Froger N et al (2009) Turning Meiosis into Mitosis. PLoS Biol 7: e1000124. https://doi.org/10.1371/journal.pbio.1000124
55. Khanday I, Skinner D, Yang B et al (2019) A male-expressed rice embryogenic trigger redirected for asexual propagation through seeds. Nature 565:91–95. https://doi.org/10.1038/s41586-018-0785-8
56. Vernet A, Meynard D, Lian Q et al (2022) High-frequency synthetic apomixis in hybrid rice. Nat Commun 13:7963. https://doi.org/10.1038/s41467-022-35679-3
57. Goeckeritz CZ, Zheng X, Harkess A, Dresselhaus T (2024) Widespread application of apomixis in agriculture requires further study of natural apomicts. iScience 27:110720. https://doi.org/10.1016/j.isci.2024.110720
58. Maynard Smith J (1978) The evolution of sex. Cambridge University Press, New York
59. Bell G (2019) The masterpiece of nature: the evolution and genetics of sexuality, 1st edn. Routledge

Part II

Detection and Characterization of Apomixis

Chapter 2

Flow Cytometric Seed Screening and Seed Genotyping: Methods for Detecting Reproductive Modes in Flowering Plants

Petra Šarhanová, Ľuboš Majeský, and Juraj Paule

Abstract

Flow cytometry is widely used to estimate ploidy levels and genome sizes in plants and animals. It is also a common tool for determining the reproductive mode in flowering plants through flow cytometric seed screening (FCSS), which can differentiate sexual and apomictic pathways. However, FCSS has limitations, as it is not universally applicable to all apomictic taxa and cannot resolve certain reproductive pathways. These challenges can be addressed by germinating seeds and genotyping seedlings and their parents. However, this approach can be time- and space-intensive and carry other drawbacks (e.g., the ploidy level of the endosperm remains undetected). Here, we present a streamlined method that combines flow cytometry with genotyping of the same seeds. Applying both methods to a single seed enables comprehensive insights into reproductive modes, detecting various forms of apomixis and automixis, assessment of the level of selfing, regardless of the mode of reproduction mode, and identifying pollen donors in the endosperm of apomictic seeds. Each protocol—FCSS and seed genotyping—can also be used independently to determine basic reproductive modes.

Key words FCSS, SSR-seq, Seed genotyping, Mode of reproduction, Apomixis

1 Introduction

Flow cytometric seed screening (FCSS) [1] is a commonly employed technique to discern various reproductive modes in flowering plants. The method is based on the determination of embryo and endosperm ploidy levels in seeds. In most sexually reproducing angiosperms, double fertilization results in an embryo:endosperm ploidy ratio of 2C:3C, where C is defined as the "holoploid genome" [2]. In contrast, apomictic reproduction involves more complex ploidy patterns. Typically, the embryo retains the maternal ploidy level, while the endosperm exhibits increased ploidy due to the development of an unreduced embryo sac. This results in an

Tatyana Radoeva and Peter J. van Dijk (eds.), *Apomixis: Methods and Protocols*, Methods in Molecular Biology, vol. 2987,
https://doi.org/10.1007/978-1-0716-5001-1_2,

embryo:endosperm ratio of 2C:4C/1C:2C (i.e., autonomous endosperm development) or 2C:5C after fertilization (i.e., pseudogamy). These ratios are not exhaustive in apomictic angiosperms. The FCSS can reveal up to ten distinct reproductive pathways, depending on the presence of reduced/unreduced embryo sacs, fertilized/parthenogenetic embryos, and pseudogamous/autonomous endosperms [1]. Nonetheless, the number of potential embryo:endosperm ratios can be even higher, influenced by factors such as the number of polar nuclei, sperm cells, or the reduced/unreduced nature of the sperm cells [3–5]. Thus, despite its versatility, the FCSS has limitations. For example, it cannot distinguish between apomixis and sexuality in taxa with a *Panicum*-type embryo sac, where only one unreduced polar nucleus contributes to endosperm development, resulting in the same embryo:endosperm ratio as in sexual reproduction. Additionally, it is ineffective for seeds with very low amounts of endosperm or when the endosperm has been depleted or degraded over time. In many cases, the G2 phase peak of the embryo may be misinterpreted as apomixis with autonomous endosperm development. These factors, along with automixis, polytubey, and polyspermy, must be carefully considered when interpreting FCSS data.

Genotyping maternal individuals and their offspring by amplicon sequencing—simple sequence repeats sequencing (SSR-seq) [6]—provides an alternative approach for the determination of reproductive modes. This technique detects sequence variation across multiple loci, including single-nucleotide polymorphisms (SNPs), deletions, insertions, and allele dosage, regardless of ploidy level. While traditionally used for plant tissue, amplicon sequencing can also be applied directly to seeds [7].

A more comprehensive characterization of reproduction can be achieved by combining FCSS and genotyping at the single-seed level. Using this approach, the seed is disrupted in a pH-neutral nuclei extraction buffer, and the resulting suspension is divided. One portion is used for ploidy determination by flow cytometry, while the other portion is reserved for DNA extraction, PCR enrichment, and next-generation sequencing to analyze sequence variation. This method is particularly valuable for small seeds, such as those of *Potentilla* or *Boechera*. Alternatively, the seed can be dissected prior to selecting a specific procedure (FCSS or genotyping), allowing the use of FCSS and DNA extraction buffers tailored to the specific laboratory protocols and materials. Variations of this protocol are also feasible, and in some cases, genotyping alone may be sufficient, particularly when the reproductive mode cannot be reliably determined by embryo and endosperm ploidy analysis.

Before analyzing the mode of reproduction, it is essential to first determine the ploidy level and/or genotype of the maternal individual. Additionally, when genotyping the endosperm of apomictic seeds, it is highly recommended to genotype potential pollen donors. This step helps to establish a database of possible paternal alleles and provides a more comprehensive understanding of the reproductive process.

2 Materials

Store all buffers at 4 °C, fluorescent dyes at −20 °C, and commercial kits based on the manufacturer's instructions. Prepare all solutions using ddH_2O.

2.1 Flow Cytometric Seed Screening

1. Galbraith buffer [8] with modifications: 45 mM $MgCl_2$, 20 mM MOPS, 30 mM sodium citrate, 0.1% Triton X-100, 0.1% PVP (*see* **Note 1**).
2. 0.1 mg/mL DAPI (4′,6-diamidino-2-phenylindole) or 1 mg/mL PI (propidium iodide) based on the used flow-cytometric system.
3. Petri dishes (60 mm), razor blades, 1.5-mL microcentrifuge tubes (e.g., Eppendorf), 20-μm nylon mesh filter, cytometric tubes—based on the used flow-cytometric system, e.g., 3.5-mL polystyrene tubes (55 × 12 mm; Sarstedt).
4. Flow cytometer with UV-lamp or laser detection system, e.g., CyFlow ML (Partec, Germany).
5. Ultrasonic bath (optional equipment).

2.2 Genotyping of Seeds

1. HotSHOT buffer [9]: 125 mM NaOH, 1 mM EDTA, 0.1% Tween 20.
2. Neutralizing solution [9]: 125 mM HCl, 10 mM Tris–HCl.

   or
3. Commercial DNA Extraction Kit—e.g., NucleoSpin Plant II (Macherey-Nagel) (*see* **Note 2**).
4. Microcentrifuge tubes (1.5-mL and 2-mL, e.g., Eppendorf).
5. Micropestle (optional equipment).
6. Vortex.
7. Ice.
8. Thermoblock.
9. Centrifuge.
10. PCR primers for loci of interest.
11. PCR tubes, strips, or PCR plates.
12. Multiplex PCR kit (e.g., QIAGEN Multiplex PCR Kit).
13. SPRI or AMPure XP beads (Beckman Coulter).

3 Methods

The protocols for FCSS, DNA extraction, and data interpretation are provided in detail. The PCR amplification, library preparation, and data analyses are to be chosen according to the possibilities and established practice of each laboratory. Here, several recommendations are given for each of these steps. Immediately after seed collection, store the seeds at −20 °C to avoid endosperm consumption.

3.1 Flow Cytometric Seed Screening

1. Chop a single seed (*see* **Note 3**) using a razor blade in a Petri dish containing 10–50 μL of extraction buffer (*see* **Note 1**). Take half of the suspension volume back to the pipette tip, transfer it to a separate microcentrifuge tube, and store it at −20 °C until DNA extraction is performed. For certain taxa, removing the seed coat prior to nuclei extraction can be beneficial. We do not recommend the use of internal standards with the seed material (*see* **Note 4**).
2. Add 700–1000 μL of additional extraction buffer (used for FCSS analysis, not DNA extraction) to the nuclei suspension in the Petri dish. Filter the suspension through a 20-μm nylon mesh filter into a cytometric tube.

   Optional step to enhance nuclei extraction: After allowing the suspension to rest for approximately 30 min, place it in an ultrasonic bath and sonicate for 1 min.
3. Add 50 μL of PI or DAPI stain, mix thoroughly by pipetting or vortexing, and incubate in the dark until flow cytometric measurement. An incubation time of 5–10 min is sufficient for most taxa (*see* **Note 5**).
4. Calculate the embryo and endosperm peak indexes using the mean fluorescence values of the histograms. The ploidy of the embryo and endosperm should be inferred based on the karyologically characterized maternal plants.
5. Interpret the embryo:endosperm ratio to determine the reproductive mode that produced the observed ratio (*see* Fig. 1 for examples).

3.2 DNA Extraction

In the case of small seeds, where the material is limited, we recommend chopping the seed in Galbraith buffer and immediately following with DNA extraction (follow Subheading 3.2.1, **step 1a**). If the volume of the seed allows, divide the seed prior to any analyses and use half for FCSS, and the other half store at −20 °C until DNA extraction (follow Subheading 3.2.1, **step 1b**).

We recommend two protocols based on cost and time considerations. The first, a low-cost option, involves performing PCR amplification shortly after DNA extraction to preserve DNA quality

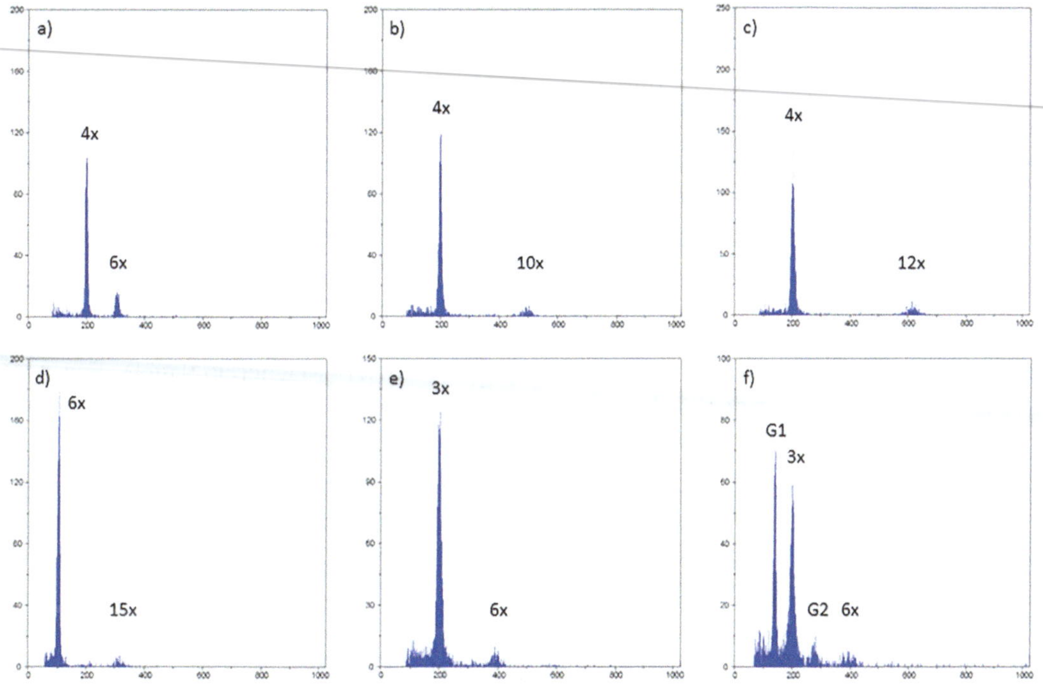

Fig. 1 Examples of histograms from flow cytometric seed screening (*x*-axis: relative fluorescence; *y*-axis: number of particles) in tetraploid *Rubus* (**a**–**c**), hexaploid *Potentilla* (**d**), and triploid *Taraxacum* (**e**, **f**) indicating ploidy level of embryo (first peak) and endposperm (second peak). (**a**) Double fertilization of reduced egg cell and reduced central cell (embryo: $n + n$, endosperm: $2n + n$). (**b**, **d**) Parthenogenetic development of unreduced egg cell ($2n$) and fertilization of unreduced central cell by reduced sperm (pseudogamy; $4n + n$). (**c**) Parthenogenetic development of unreduced egg cell ($2n$) and fertilization of unreduced central cell by either two reduced sperms or one unreduced sperm ($4n + 2n$). (**e**) Parthenogenetic development of unreduced egg cell ($2n$) and autonomous development of unreduced central cell ($4n$). (**f**) Same as (**e**), with a standard included (G1 and G2 peaks) to highlight potential risks when analyzing unknown material or samples with variable reproductive modes, i.e., endosperm ploidy levels

and is particularly effective for larger seeds. The second protocol uses commercial extraction kits, which are ideal for long-term storage and high-quality DNA and work well for smaller seeds. The provided protocol is optimized for the commercial DNA extraction kit listed in Subheading 2, but other kits suitable for your material may also be used.

3.2.1 Low-Cost DNA Extraction Protocol

1. (a) Add 50 μL of HotSHOT buffer to the part of the nuclei suspension taken from FCSS sample preparation and mix by pipetting or vortexing.
   (b) Add 50 μL of HotSHOT buffer to the seed stored in the 1.5-mL microcentrifuge tube and shred the seed in the buffer with micropestle (*see* **Note 6**).
2. Heat the suspension at 65 °C for 30 min.

3. Add 50 μL of neutralizing solution and mix by pipetting or vortexing.
4. Let the suspension cool down in the fridge for 60 min to sediment or centrifuge for 5 min/8000 × g.
5. Transfer the supernatant into a new microcentrifuge tube and dilute it in a 1:10 ratio DNA: ddH2O.
6. The dilution is now ready to be used as a PCR template. Store it at −20 °C and use it within a few days to avoid DNA degradation.

3.2.2 Commercial Kit Protocol

Optimized for NucleoSpin Plant II (Macherey-Nagel). For additional information, follow the manufacturer's protocol. Similar results can be achieved with other commercial kits of choice.

1. Grind the seed with micropestle in the 1.5-mL microcentrifuge tube (*see* **Note 6**) with 50 μL of PL2 buffer, add 1.5 μL of RNase A, and mix thoroughly. Incubate suspension for 30 min at 65 °C.
2. Add 12.5 μL of buffer PL3, mix thoroughly, and incubate for 5 min on ice to precipitate SDS. Centrifuge the lysate for 5 min at 11,000 × g and transfer the supernatant to a new tube (*see* **Note 7**).
3. Add 75 μL of buffer PC and mix thoroughly by pipetting or vortexing.
4. Place the NucleoSpin® Plant II Column with the green ring into a new Collection Tube and load the full volume of the sample into the column. Centrifuge 1 min at 11,000 × g, and discard flowthrough.
5. Load 200 μL of buffer PW1 to wash the column, centrifuge 1 min at 11,000 × g, and discard flowthrough.
6. Load 200 μL of buffer PW2 to wash the column, centrifuge 1 min at 11,000 × g, and discard flowthrough. Repeat this step.
7. Centrifuge 2 min at 11,000 × g to dry silica membrane.
8. Preheat the buffer PE at 65 °C, place NucleoSpin® Plant II Column into a new microcentrifuge tube, and pipette 30 μL buffer PE to the center of the filter to elute the DNA from the silica membrane. Centrifuge 1 min at 11,000 × g. Add another 30 μL of preheated buffer PE and repeat centrifugation in the same tube (*see* **Note 8**).
9. Use the eluted DNA as a template for PCR amplification. For extended storage, freeze at −20 °C.

3.3 Genotyping

1. Use a genotyping method suitable for the species of interest (e.g., microsatellites, gene markers, SNPs). When designing new primers, account for the limitations of the chosen

sequencing platform. For example, avoid designing loci exceeding 250 bp if using a platform with a 300 bp read length.

2. For PCR amplification, use a polymerase suitable for the material. We recommend using multiplex PCR kits to amplify multiple loci in a single PCR reaction.
3. Evaluate the risk of primer-dimer formation and optimize the amplification conditions using available tools. Follow the manufacturer's protocol for PCR enrichment. In general, PCR reactions with a total volume of 10 μL are sufficient to yield an adequate amount of amplicon DNA. If PCR enrichment varies across multiplexed loci, adjust the primer amounts accordingly.
4. If intended for sequencing with multiplexed samples, we recommend using primers with appended barcodes to facilitate higher multiplexing levels in the sequencing library [6].
5. After pooling all loci and/or samples with appended barcodes, clean the products using SPRI or AMPure XP beads to obtain pure amplicons for sequencing library preparation.

3.4 Library Preparation and Sequencing

1. Using a sequencing library kit that does not include PCR steps is highly recommended to minimize the risk of generating chimeric PCR products. Always follow the manufacturer's protocol for the specific kit in use. To reduce the potential for index hopping, use unique dual indexes. For cost-effective optimization, the amounts of all reagents, including indexes, can be halved to reduce library preparation costs.
2. Sequence the amplicon libraries on the available platform or sequencing facility. When calculating the sequencing coverage for the libraries, consider the ploidy level and the number of pooled loci. Coverage of several hundred reads per locus is generally sufficient for the genotyping of the embryos. If the detection of paternal alleles in the endosperm is required, we recommend a sequencing coverage of at least 5000 reads per locus and ploidy level. For example, if the maternal individual or seed is diploid, aim for approx. 10,000 reads per locus; if tetraploid, aim for 20,000 reads per locus.

3.5 Data Analyses and Interpretation

1. Perform analyses of raw reads based on your expertise and the tools available for processing sequencing outputs.
2. A wide range of embryo:endosperm ploidy levels can be observed in variable apomictic taxa, often including rare ploidy level ratios. Several main reproductive modes can be identified by analyzing the most common ratios and correlating them with seed genotyping results (Table 1, *see* **Note 9**). Not all possible outputs can be covered here, depending on the plant system analyzed. Do not forget that nature is full of exceptions.

Table 1
Characterization of potential reproductive modes based on FCSS-derived embryo:endosperm ploidy levels and their corresponding genotypes in comparison to maternal and paternal genotypes

Mode of reproduction	Embryo origin	Endosperm origin	FCSS embryo: endosperm	SSR-seq ♀ alleles in embryo	SSR-seq ♂ alleles in embryo	SSR-seq ♂ alleles in endosperm
Apomixis	Parthenogenetic	Autonomous	2C:4C	All present, no dosage change	None	None
Apomixis	Parthenogenetic	+ 1 reduced sperm	2C:5C	All present, no dosage change	None	½ present
Apomixis	Parthenogenetic	+ 2 reduced sperms	2C:6C	All present, no dosage change	None	½ present
Apomixis	Parthenogenetic	+ 1 unreduced sperm	2C:6C	All present, no dosage change	None	All present
Apomixis *Panicum*-type	Parthenogenetic	Autonomous	2C:2C	All present, no dosage change	None	None
Apomixis *Panicum*-type	Parthenogenetic	+ 1 reduced sperm	2C:3C	All present, no dosage change	None	½ present
Apomixis *Panicum*-type	Parthenogenetic	+ 2 reduced sperms	2C:4C	All present, no dosage change	None	½ present
Apomixis *Panicum*-type	Parthenogenetic	+ 1 unreduced sperm	2C:4C	All present, no dosage change	None	All present
Sexual—outcrossing	Zygotic	Autonomous	2C:2C	½ present	½ present	None
Sexual—outcrossing	Zygotic	+ 1 reduced sperm	2C:3C	½ present	½ present	½ present
Sexual—outcrossing[a]	Zygotic	+ 2 reduced sperms	2C:4C	½ present	½ present	½ present
Sexual—outcrossing	Zygotic	+ 1 unreduced sperm	3C:4C	½ present	All present	All present
Sexual—selfing	Zygotic	Autonomous	2C:2C	Partly absent and/or changed dosage	Same as ♀	Same as ♀

Sexual—selfing	Zygotic	+ 1 reduced sperm	2C:3C	Partly absent and/or changed dosage	Same as ♀	Same as ♀
Sexual—selfing[a]	Zygotic	+ 2 reduced sperms	2C:4C	Partly absent and/or changed dosage	Same as ♀	Same as ♀
Sexual—selfing	Zygotic	+ 1 unreduced sperm	3C:4C	Partly absent and/or changed dosage	Same as ♀	Same as ♀
Automixis	Parthenogenetic	Autonomous	2C:4C	Partly absent and/or changed dosage	None	None
Automixis	Parthenogenetic	+ 1 reduced sperm	2C:5C	Partly absent and/or changed dosage	None	½ present
Automixis	Parthenogenetic	+ 2 reduced sperms	2C:6C	Partly absent and/or changed dosage	None	½ present
Automixis	Parthenogenetic	+ 1 unreduced sperm	2C:6C	Partly absent and/or changed dosage	None	All present
Polyhaploid	Parthenogenetic	Autonomous	1C:2C	½ present	None	None
Polyhaploid	Parthenogenetic	+ 1 reduced sperm	1C:3C	½ present	None	½ present
Polyhaploid	Parthenogenetic	+ 2 reduced sperms	1C:4C	½ present	None	½ present
Polyhaploid	Parthenogenetic	+ 1 unreduced sperm	1C:4C	½ present	None	All present
B_{III} hybrid	Zygotic	Autonomous	3C:4C	All present	½ present	None
B_{III} hybrid	Zygotic	+ 1 reduced sperm	3C:5C	All present	½ present	½ present
B_{III} hybrid[a]	Zygotic	+ 2 reduced sperms	3C:6C	All present	½ present	½ present
B_{III} hybrid	Zygotic	+ 1 unreduced sperm	4C:6C	All present	All present	All present

[a]Results indicate the occurrence of polytubey, where a minimum of three sperm cells are required to achieve this mode of reproduction (*see* **Note 9**)

3. Be aware that if paternal alleles are present only in the endosperm (and not in the embryo), the sequencing coverage of these alleles will typically be an order of magnitude lower than that of alleles from the embryo. This difference depends on the relative sizes of the embryo and endosperm, as well as the number of nuclei, which is species-specific. As a result, PCR and sequencing errors may be misinterpreted as genuine alleles. We strongly recommend genotyping potential pollen donors if endosperm genotyping is required. In some cases, separating the embryo and endosperm prior to genotyping may help clearly distinguish the genotype of each seed compartment.

4 Notes

1. To carry out both FCSS analyses and DNA extraction from the nuclei suspension, it is necessary to disrupt the seed tissue in a neutral buffer, such as Galbraith buffer. However, if the seed is halved before nuclei extraction, buffers compatible with your flow cytometry system can be used independently for DNA extraction. Besides the Galbraith buffer, we had good results with buffers Otto I+II [10] and LB01 [11].
2. A budget-friendly DNA extraction option is to use the Hot-SHOT buffer and Neutralizing Solution [9]. However, this method yields a small amount of DNA that is not suitable for long-term storage, which can be disadvantageous if you need to repeat genotyping on some seeds. If the budget allows, we recommend using commercial extraction kits for consistent results.
3. We do not recommend bulk measurements for FCSS analysis. If the seeds are finely chopped, there should be enough material for both FCSS and genotyping. In some cases, using metal beads and disrupting dry seeds in a 2-mL microcentrifuge tube prior to adding FCSS buffer may help increase the number of nuclei.
4. When working with known material, you can include an appropriate internal standard [12]. However, be aware that the genome size of the internal standard may produce peaks that interfere with the embryo or endosperm peaks, particularly if the endosperm ploidy level varies. To avoid using an internal standard during seed analysis, we recommend calibrating the flow cytometer's gain using either the internal standard or the maternal individual prior to seed measurements.
5. Incubation time may vary for different taxa and should be adjusted experimentally. We did not achieve better results when prolonging the incubation time.

6. Shredding works well in most cases without the use of sand. For tougher seeds, it is advisable to leave the seed in the buffer for several minutes before shredding.
7. The NucleoSpin® Filter (violet ring) columns included in the kit can be used, but they are not essential. Seed debris can be effectively removed through centrifugation. Skipping these columns and reducing the volume of most reagents can help lower costs. In such cases, purchasing additional NucleoSpin® Plant II Columns (green ring) allows you to fully utilize the remaining reagents in the kit.
8. A lower volume of elution buffer may increase the concentration of extracted DNA, which can be particularly advantageous when working with tiny seeds. In such cases, a single elution in 30 μL of elution buffer—or using an extraction kit optimized for low-input samples—may be beneficial.
9. In a single pollen tube, both sperm cells are usually reduced and genetically identical. Thus, genotyping of the endosperm enables differentiation between fertilization by two reduced sperm cells or a single unreduced sperm cell, provided allele differences exist between maternal and paternal individuals. If two reduced sperm cells participate in fertilization, half of the paternal alleles will be absent. However, keep in mind that meiosis in pollen from apomictic species may be altered, and the pollen genotype does not necessarily reflect its ploidy level—i.e., whether it is meiotically reduced or unreduced. Consequently, a robust interpretation of results distinguishing between reduced and unreduced sperm cells fertilizing the central cell of unreduced embryo sacs should be supported by additional methods (e.g., microscopy).

Acknowledgments

This work was supported by the Czech Grant Agency (No. GA23-06726S).

Funding

This study was funded by the Czech Science Foundation (GAČR) project No. 23-06726S, and by Operational Programme Research, Development and Education—Project "Postdoc2MUNI" (No. CZ.02.2.69/0.0/0.0/18_053/0016952).

References

1. Matzk F, Meister A, Schubert I (2000) An efficient screen for reproductive pathways using mature seeds of monocots and dicots. Plant J 21(1):97–108
2. Greilhuber J, Doležel J, Lysák MA, Bennett MD (2005) The origin, evolution and proposed stabilization of the terms 'genome size' and 'C-value' to describe nuclear DNA contents. Ann Bot 95(1):255–260
3. Šarhanová P, Vašut RJ, Dančák M, Bureš P, Trávníček B (2012) New insights into the variability of reproduction modes in European populations of *Rubus* subgen. *Rubus*: how sexual are polyploid brambles? Sex Plant Reprod 25:319–335
4. Dobeš C, Lückl A, Hülber K, Paule J (2013) Prospects and limits of the flow cytometric seed screen–insights from *Potentilla* sensu lato (Potentilleae, Rosaceae). New Phytol 198: 605–616
5. Vašková D, Kolarčik V (2019) Breeding systems in diploid and polyploid hawthorns (*Crataegus*): evidence from experimental pollinations of *C. monogyna*, *C. subsphaerica*, and natural hybrids. Forests 10:1059
6. Šarhanová P, Pfanzelt S, Brandt R, Himmelbach A, Blattner FR (2018) SSR-seq: genotyping of microsatellites using next-generation sequencing reveals higher level of polymorphism as compared to traditional fragment size scoring. Ecol Evol 8:10817–10833
7. Šarhanová P, Majeský Ľ, Sochor M (2024) A novel strategy to study apomixis, automixis, and autogamy in plants. Plant Reprod 37:1–14
8. Galbraith DW, Harkins KR, Maddox JM, Ayres NM, Sharma DP, Firoozabady E (1983) Rapid flow cytometric analysis of the cell cycle in intact plant tissues. Science 220(4601): 1049–1051
9. Brewster JD, Paoli GC (2013) DNA extraction protocol for rapid PCR detection of pathogenic bacteria. Anal Biochem 442(1):107–109
10. Otto F (1990) DAPI staining of fixed cells for high-resolution flow cytometry of nuclear DNA. In: Methods in cell biology, vol 33. Academic Press, pp 105–110
11. Doležel J, Binarová P, Lcretti S (1989) Analysis of nuclear DNA content in plant cells by flow cytometry. Biol Plant 31(2):113–120
12. Doležel J, Greilhuber J, Suda J (2007) Estimation of nuclear DNA content in plants using flow cytometry. Nat Protoc 2:2233–2244

Chapter 3

Analysis of Gametophytic Apomixis Using Confocal Microscopy

Camilla Banfi and Philippa Jane Barrell

Abstract

Apomixis is an asexual reproductive mechanism that takes place deeply inside the female reproductive organs of the plant, in ovules and seeds. In gametophytic apomixis, an unreduced female gametophyte is produced by a modified meiosis of the megaspore mother cell (dipolspory) or from a somatic initial cell (apospory). The unreduced, nonrecombined egg cell develops subsequently into an embryo by parthenogenesis. The cyto-embryological study of apomixis is challenging because of the inaccessibility of these structures. Consequently, images of apomeiosis and parthenogenesis with high definition are limited to a few species. In this chapter, we show the application of a Feulgen staining protocol combined with confocal microscopy for the study of nonreductional megasporogenesis and autonomous embryo formation in diplosporous apomictic *Taraxacum officinale* and aposporous apomictic *Pilosella piloselloides* var. *praealta*. Using a rapid and technically simple method, performed on whole-mount ovaries, we have obtained high-resolution images of the female reproductive cells. Furthermore, we highlight the application of this protocol for the study of *loss-of-diplospory* and *loss-of-parthenogenesis* mutants in the same species.

Key words Confocal microscopy, Schiff's reagent, Apomixis, Apospory, Diplospory, Apomeiosis, Parthenogenesis, Ovule

1 Introduction

Apomixis is a female-specific process that occurs deeply inside the female reproductive organs of the flowers. In this type of asexual reproduction, the ovules develop into seeds bypassing meiosis and fertilization. In gametophytic apomixis, an unreduced female gametophyte is produced either from the megaspore mother cell (MMC) (diplospory) or from a somatic nucellar cell named aposporous initial (AI) (apospory) [1, 2]. Depending on the type of diplospory, the MMC's meiosis is either aberrant or bypassed, forming unreduced megaspores that develop in unreduced gametophytes [3]. By contrast, apospory involves the development of the functional megaspore from the mitosis of the AI, positioned adjacent to the MMC [4]. In either type of apomixis, the resulting

Tatyana Radoeva and Peter J. van Dijk (eds.), *Apomixis: Methods and Protocols*, Methods in Molecular Biology, vol. 2987, https://doi.org/10.1007/978-1-0716-5001-1_3,

unreduced gametes develop autonomously in clonal embryos by parthenogenesis, and the endosperm can form autonomously or by pseudogamy. To date, the detailed description of apomixis at the cellular level has been hampered because the female gametophyte is not easily accessible. Early cyto-embryological studies using tissue embedding and sectioning were conducted in diplosporous and aposporous species, allowing the identification of the main characteristics of two apomictic types [4–6]. However, these techniques were time-consuming, and it was difficult to reconstruct single three-dimensional (3D) images from the individual section planes. Clearing methods are quicker and easier than sectioning and staining because nuclei and walls of reproductive cells can be observed in intact ovules, with differential interference contrast or Nomarski microscopy. These methods have been employed to analyze non-reductional megasporogenesis, aposporous embryo sac formation, and autonomous embryo in many natural apomicts [7–9]. Although chromosomes and the reproductive cellular types were distinguishable, these methods resulted in low-resolution images. Recently, we have cytologically characterized diplospory in *Taraxacum officinale*, *Chondrilla juncea*, and *Erigeron annuus* by applying a previously described protocol for confocal microscopy used for the analysis of female maize meiosis [10]. With this method, we obtained high-resolution images of apomeiosis, and we were able to measure and count chromosomes in 3D reconstruction of MMC nuclei [11].

This chapter includes the protocol for the cyto-embryological analysis of megasporogenesis and megagametogenesis of gametophytic apomixis, which is based on Feulgen staining with Schiff's reagent combined with confocal microscopy. Here, we show the results we obtained using the modified Barrell and Grossniklaus protocol on the diplosporous species *Taraxacum officinale* and the aposporous species *Pilosella piloselloides* var. *praealta*. This method is suitable for the analysis of different developmental stages, from early meiosis to embryo development. Furthermore, we successfully applied it for the characterization of *loss-of-diplospory* (*LOD*) and *loss-of-parthenogenesis* (*LOP*) *Taraxacum* and *Pilosella* mutants, highlighting the differences with respect to the normal apomictic process in wild-type plants.

2 Materials

2.1 Plants

Taraxacum officinale s.l. ((L.) Weber ex F.H. Wigg. 1780) plants are grown in the greenhouse in long-day conditions, with 16 h light at 22 °C and 8 h dark at 16 °C. At these conditions, the plants flower in 6–8 weeks. The apomictic triploid A68 was derived from a natural population near Heteren, The Netherlands; the *loss-of-diplospory* mutant line *LOD* i124 was generated from a triploid apomictic A68 by γ-irradiation, and carries a deletion of the *DIPLOSPOROUS* locus.

Pilosella piloselloides var. *praealta* plants are grown in a glasshouse, with a minimum night temperature of 12 °C, a minimum day temperature of 18 °C, and maximum day temperature of 25 °C. To promote flowering, day-length extension lighting is provided using sodium vapor lamps with a photoperiod of 16 h. The apomict R35 [8] and the γ-irradiation induced *loss-of-parthenogenesis* (*lop*) mutant γ 136 [12] were used in this study.

2.2 Collection and Fixation of Plant Material

1. Plants at the appropriate stage for harvesting flower heads.
2. Fixative solution: mix 90 parts of 50% ethanol (EtOH), 2 parts of formalin, and 5 parts of glacial acetic acid in a fume hood. Prepare fresh fixative solution as much as needed.
3. 70% and 95% EtOH: dilute absolute ethanol with distilled water.
4. Fine forceps for buds collection.
5. 15- or 50-mL tubes, depending on the amount of samples to fix.

2.3 Sample Staining and Slide Preparation

1. Fixed flower heads.
2. 1 M HCl and 5.8 M HCl: prepare by adding the correct amount of 37% HCl to distilled water. To avoid acid drop splashing, add the acid slowly to the water, and not the contrary.
3. Schiff's reagent.
4. Alcohol series made up of absolute ethanol and deionized water: 30%, 50%, 70%, 90%, 95%, and 100% EtOH, used for sample dehydration.
5. Immersion oil.
6. Mixture ethanol: immersion oil. Mix 1 part of ethanol, and 1 part of immersion oil. Vortex well.
7. A couple of syringes with needles or tungsten wire needles for dissection.
8. Fine forceps.
9. Standard glass slides for microscopy.
10. Glass coverslips (24 × 50 mm).
11. 5-mL tubes.
12. Stereomicroscope.

3 Methods

Execute all the procedures at room temperature, unless otherwise specified.

3.1 Collection and Fixation of Plant Material

1. *Day 1: Sample and Fixation*. Sample the flower heads at different stages as close as possible to the receptacle (*see* **Note 1**). Make two perpendicular incisions at the bottom of the receptacle to help fixative penetration; don't cut too deeply to avoid damaging the ovaries (*see* **Note 2**).
2. Place immediately the flower heads in fixative solution for 24 h at room temperature. The fixative volume should be at least three times the sample volume. For *Pilosella*, it is useful for fixative penetration to place the flowers and fixative in a vacuum chamber for 1 hour, with the vacuum released very slowly over approximately 5 min.
3. *Day 2: Storage*. Rinse the buds three to five times in 95% EtOH, 2 min each.
4. Replace with 70% EtOH to completely cover them. At this point, you can directly go to *Day 3* or store the fixed samples in 70% EtOH at 4 °C (*see* **Note 3**).

3.2 Sample Staining and Slide Preparation

1. *Day 3: Acidification and staining*. Transfer the samples to be further processed in 5 mL tube with ddH_2O (*see* **Note 4**). Rinse the buds three times in ddH_2O, 2 min each.
2. Replace water with enough 1 M HCl to cover them. Leave at room temperature for 90 min.
3. Substitute 1 M HCl with 5.8 M HCl and leave at room temperature for 120 min.
4. Replace 5.8 M HCl with 1 M HCl and leave at room temperature for 60 min.
5. Rinse the buds three times in ddH_2O, 3 min each.
6. Exchange water with Schiff's reagent solution to cover the buds. Place the samples in the staining solution at 4 °C, in the dark (*see* **Note 5**).
7. *Day 4: Dehydration*. The samples are dehydrated in EtOH series (30%, 50%, 70%, 90%, 95%, 2×100%) for 30 min each.
8. Replace 100% EtOH with a mixture of ethanol: immersion oil 1:1 (v/v) and leave the samples overnight at 4 °C (*see* **Note 6**).
9. *Day 5: Sample dissection and slide preparation*. The florets are isolated from the flower heads on a glass slide under the stereomicroscope. The anthers are removed with needles, leaving only the ovaries on the slide. There is no need to isolate the single ovules from the ovaries (*see* **Note 7**).

10. Add a few drops of immersion oil on the samples and cover with a coverslip. After the acidification, the tissue is very soft: don't press down on the coverslip to avoid damaging the samples.

3.3 Confocal Microscopy

1. To image *Taraxacum* apomeiosis (MMC to FM stages), we use a Nikon A1+ laser scanning confocal microscope with 60× or 100× oil-immersion objectives. The excitation wavelength is 561 nm, and the signal is detected between 620 and 630 nm. Pictures are recorded at 1024 × 1024 pixels. Single optical sections are acquired for a general description of the stages of diplospory. To analyze detailed chromosome dynamics during meiotic prophase I, the MMC nucleus is imaged with Z-stacks (Z-step = 100 nm). Subsequently, the images are analyzed with the Fiji software [13]. The 3D images presented here have been generated with the Volume Viewer plugin using the Projection display mode and tricubic smooth interpolation. Visualization of the results is performed by applying the thermal LUT with a 2D grad transfer function. The 3D Projection function of Fiji software has been used to reconstruct 3D images as well; in this case, the brightest point projection method is used and interpolation is added. The 3D images of MMC nuclei at the prophase I stage are used for chromosome counting and measurement [11]. The images displayed here are not derived from the deconvolution process (Fig. 1).
2. To image *Pilosella*, a Leica SP5 confocal microscope is used as described previously (Figs. 2 and 3) [14].

4 Notes

1. The stage of gametophyte development with respect to the stage of the flower head highly depends on the species. If you are not sure about which buds to collect, sample a mixture of different stages. Before proceeding with the confocal protocol, you could test some florets with clearing to evaluate the female gametophyte stage.
2. The external bracts can be removed to help the fixative penetration, but this step depends on the species and the number of florets contained in a single flower head. We always remove the bracts from the *Taraxacum* flower buds that contain hundreds of florets. However, if the flower head's florets are a few, as for *Pilosella*, it is better to cut perpendicularly the bracts without removing them. This will avoid the detaching of all the florets from the receptacle during the washing steps and will simplify the changes of the solutions.
3. At this stage, the samples are stable for at least one year.

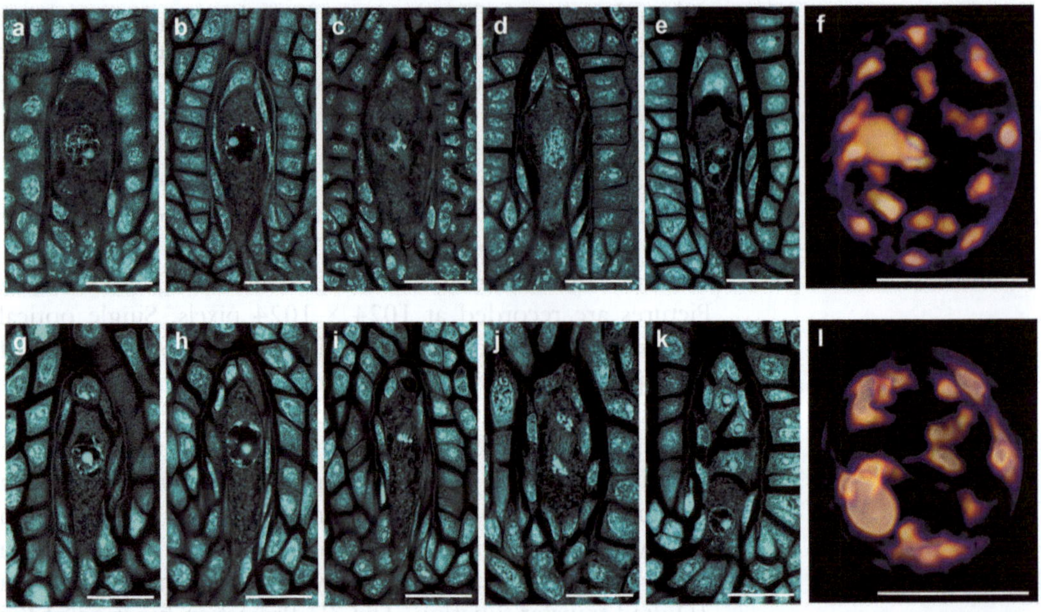

Fig. 1 Apomeiosis in *Taraxacum officinale*, highlighting the main differences between wild-type and *loss-of-diplospory LOD* i124 mutant. (**a**–**e**, **g**–**k**) Single optical sections of ovule's nucellus. In apomictic *Taraxacum* (**a**–**e**), univalent chromosomes are visible as thick threads in the MMC at prophase I stage (**a**) and further condense in diakinesis I (**b**). The first meiotic division is aberrant, with chromosomes scattered along the meiotic spindle at metaphase I stage (**c**). The first meiotic division is avoided by the first division restitution (**d**). A single division follows, generating a dyad of unreduced megaspores (**e**). In the *LOD* i124 mutant, the deletion of the *DIP* locus causes reversion to sexual meiosis (**g**–**k**). During prophase I, the chromosomes are thicker with respect to the wild-type (**g**, **h**). At metaphase I, they align properly at the metaphase plate (**i**) and segregate during anaphase I (**j**). A second division follows, and four reduced spores are produced (**k**). Scale bar is 20 μm. (**f**, **l**) 3D reconstruction of MMC nucleus at diakinesis stage. In wild-type apomictic, a high number of univalent chromosomes is visible (**f**). In *LOD* mutant, the homologous chromosomes are associated into bivalents or multivalents that result larger and in lower number (**l**). Scale bar is 10 μm

4. After *Day 3*, the samples must be analyzed within a few days. If you have many samples, proceed with the acidification step only with the samples you are planning to observe.
5. The Schiff's regent is toxic and irritant; use gloves.
6. The better results are obtained by proceeding directly to the observation of the samples the following day (*Day 5*). However, high-quality images were obtained from samples kept in the mixture of ethanol: immersion oil at 4 °C for a few days.
7. The small florets (comprising the megasporogenesis stages) can be kept entirely, without removing the anthers. This helps in finding the samples once observed at the confocal microscope with high-magnification objectives. For the megagametogenesis stages, it is better to remove the bigger anthers to avoid leaving too much space between the slide and the coverslip because of the anther's volume.

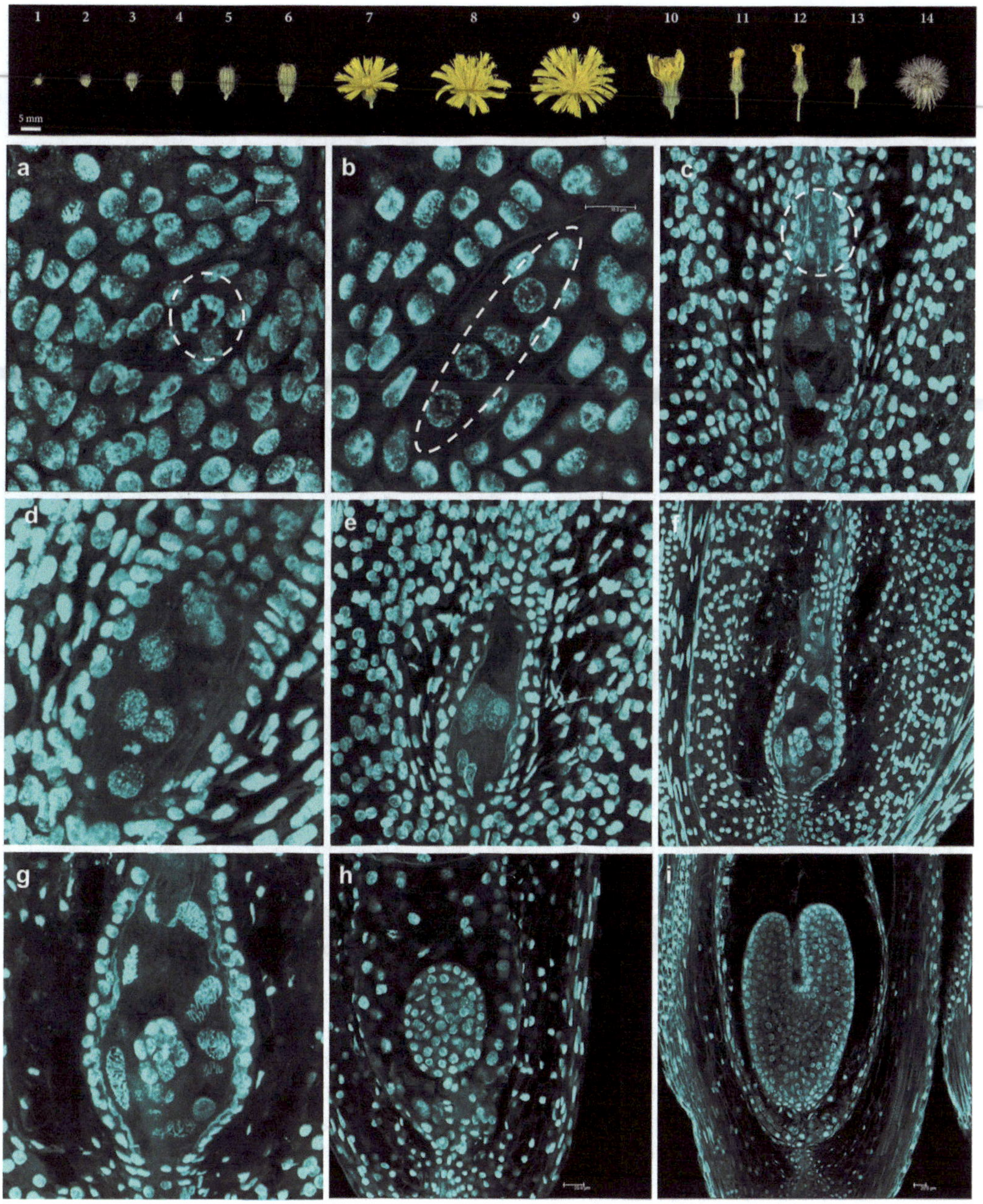

Fig. 2 Panel **a**: Stages of *Pilosella piloselloides* var. *praealta* R35 capitulum development (wild-type). Panel **b** (**a**–**i**): Maximum projections of a series of optical sections from confocal microscopy of *Pilosella* R35 ovules. (**a**, **b**) Capitulum at stage 3: (**a**) MMC with chromosomes condensed (indicated by the dashed circle); (**b**) a tetrad (circled). Scale bar is 10 μm. (**c**, **d**) Stage 4 and (**e**) stage 6: Mitotic divisions of embryo sac development with adjacent AI (circled). (**f**, **g**) Stage 7: Embryo sac with globular embryo and endosperm nuclei visible. (**g**) Magnified view of embryo sac (**f**), with endosperm nuclei synchronously dividing and chromosomes visible in prophase. (**h**) Stage 10: later globular embryo. (**i**) Stage 11: Heart stage embryo

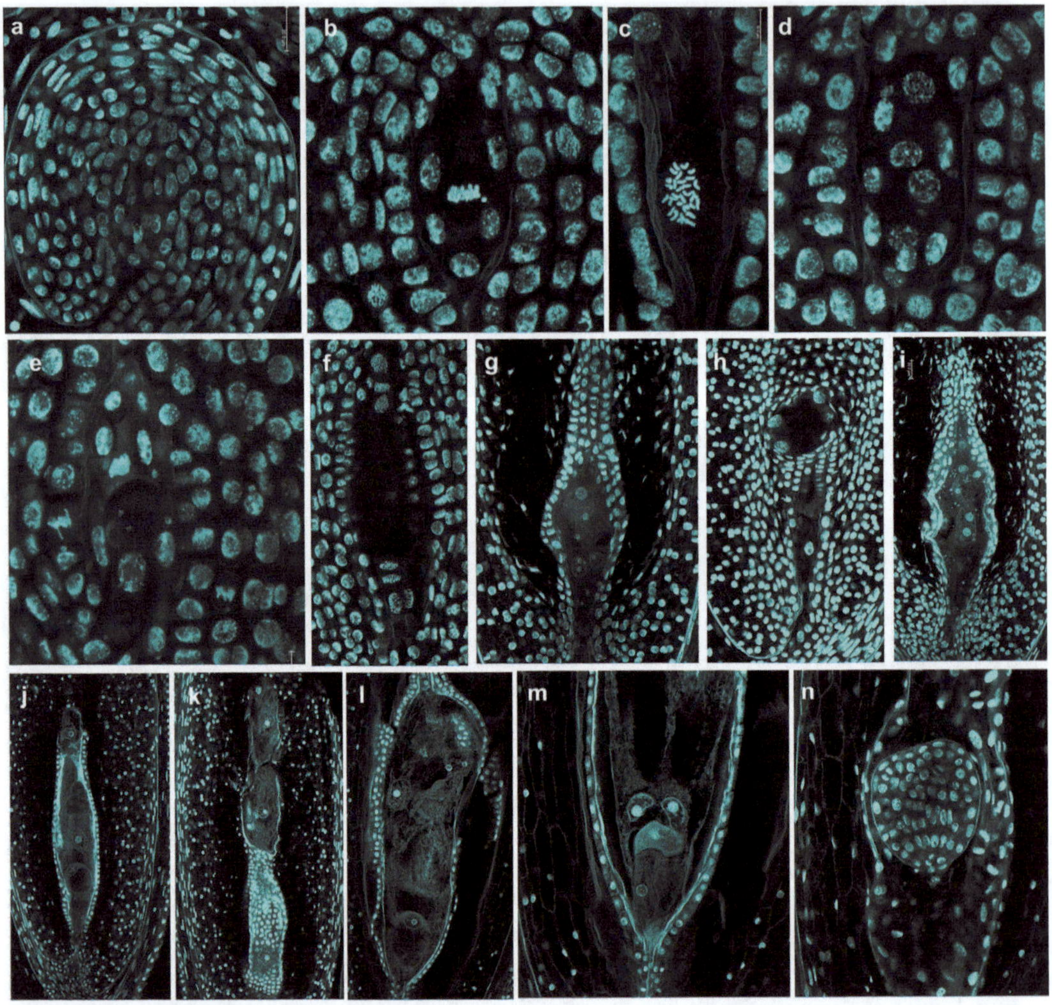

Fig. 3 *Pilosella* deletion mutant γ 136. (**a**–**l**) Unpollinated florets and (**m**–**n**) pollinated florets. (**a**) Capitulum stage 2: Early MMC developing. (**b**–**e**) Stage 3: (**b**) MMC with chromosomes condensed at metaphase and (**d**) tetrad; (**e**) FM. (**f**–**i**) Stage 5: (**f**) third mitotic division of embryo sac formation; (**g**–**i**) embryo sacs, no AI seen. (**i**–**n**) Stage 11: (**i**–**m**) Quiescent embryo sacs with no embryo or endosperm. (**m**, **n**) Pollinated ovules: (**m**) Egg cell, synergid, and unfused central cell halted in development. (**n**) In rare cases, an embryo with endosperm was observed after pollination, indicating that the sexual pathway is still functional in the γ 136 mutant

Acknowledgments

We acknowledge Letizia Cornaro, Alex Cavalleri, Mara Cucinotta, Peter van Dijk, and Lucia Colombo for their thoughtful and valuable suggestions for improving the protocol. *Taraxacum* plants were provided by Keygene. *Pilosella* plants were provided by Ross Bicknell. Part of this work was carried out at UNITECH NOLIMITS, an advanced imaging facility established by the Università degli Studi di Milano.

Conflicts of Interest The authors have no conflicts of interest to declare that are relevant to the content of this chapter.

References

1. Nogler GA (1984) Gametophytic apomixis. In: Johri BM (ed) Embryology of angiosperms. Springer, Berlin, Heidelberg, pp 475–518
2. Koltunow AM (1993) Apomixis: embryo sacs and embryos formed without meiosis or fertilization in ovules. Plant Cell 5:1425. https://doi.org/10.2307/3869793
3. Cornaro L, Banfi C, Cucinotta M et al (2023) Asexual reproduction through seeds: the complex case of diplosporous apomixis. J Exp Bot 74:2462–2478. https://doi.org/10.1093/jxb/erad054
4. Koltunow AM, Johnson SD, Bicknell RA (1998) Sexual and apomictic development in *Hieracium*. Sex Plant Reprod 11:213–230. https://doi.org/10.1007/s004970050144
5. Gustafsson Å (2010) Primary and secondary association in *Taraxacum*. Hereditas 20:1–31. https://doi.org/10.1111/j.1601-5223.1935.tb03176.x
6. Battaglia E (1950) L'Alterazione Della Meiosi Nella Riproduzione Apomittica di *Erigeron Karwinskianus* Dc. var. *Mucronatus* Dc. (Asteraceae). Caryologia 2:165–204. https://doi.org/10.1080/00087114.1950.10797138
7. van Baarlen P, van Dijk PJ, Hoekstra RF, de Jong JH (2000) Meiotic recombination in sexual diploid and apomictic triploid dandelions (*Taraxacum officinale* L.). Genome 43:827–835. https://doi.org/10.1139/gen-43-5-827
8. Koltunow AMG, Johnson SD, Rodrigues JCM et al (2011) Sexual reproduction is the default mode in apomictic *Hieracium* subgenus *Pilosella*, in which two dominant loci function to enable apomixis. Plant J 66:890–902. https://doi.org/10.1111/j.1365-313X.2011.04556.x
9. Janas A, Musiał K, Kościńska-Pająk M, Marciniuk P (2016) Insights into developmental processes in anthers, ovaries, and ovules of *Taraxacum belorussicum* (Asteraceae-Cichorioideae) using DIC optics. Plant Syst Evol 302:617–628. https://doi.org/10.1007/s00606-016-1288-4
10. Barrell PJ, Grossniklaus U (2005) Confocal microscopy of whole ovules for analysis of reproductive development: the *elongate1* mutant affects meiosis II. Plant J 43:309–320. https://doi.org/10.1111/j.1365-313X.2005.02456.x
11. Cornaro L, Banfi C, Cavalleri A et al (2024) Apomixis at high resolution: unravelling diplospory in *Asteraceae*. J Exp Bot 74:2462–2478. https://doi.org/10.1093/jxb/erae477
12. Catanach AS, Erasmuson SK, Podivinsky E et al (2006) Deletion mapping of genetic regions associated with apomixis in *Hieracium*. Proc Natl Acad Sci 103:18650–18655. https://doi.org/10.1073/pnas.0605588103
13. Schindelin J, Arganda-Carreras I, Frise E et al (2012) Fiji: an open-source platform for biological-image analysis. Nat Methods 9:676–682. https://doi.org/10.1038/nmeth.2019
14. Barrell P, Grossniklaus U (2013) Examining female Meiocytes of maize by confocal microscopy. In: Pawlowski WP, Grelon M, Armstrong S (eds) Plant meiosis. Methods in molecular biology. Humana, Totowa, NJ, pp 45–52

Chapter 4

3D Imaging, Segmentation, and Cell Annotation of the Ovule During Megaspore Mother Cell Differentiation in *Paspalum* spp.

Luciana Delgado and Daphné Autran

Abstract

Reproductive development in apomictic plants diverges from the sexual pathway at different key steps. The early steps take place in the ovule, the female organ hosting female sporogenesis and gametogenesis. Cell identities are notably more plastic in the ovule of facultative aposporous plants, where somatic cells can shift to germinal fate. This plasticity likely starts during the early morphogenesis of the ovule, concomitant with gradual differentiation of the sexual megaspore mother cell (MMC). In sexual species, 3D morphogenetic analyses have shown that ovule shape conditions MMC plasticity. However, in aposporous grasses, the morphogenetic events shaping ovule primordia are currently undescribed in 3D and at the cellular level, largely due to the inaccessibility of this organ. To fill this gap, we propose here a comprehensive workflow from ovule sampling to the extraction of 3D cellular quantitative parameters, established for the tropical apomictic grass *Paspalum rufum*. First, this protocol describes 3D imaging of whole-mount ovules at successive developmental stages, covering MMC differentiation, using ClearSee clearing procedure and double cell walls/nuclei staining. Second, it provides a detailed image analysis workflow in the open-source platform, MorphoGraphX. The workflow enables semiautomatic 3D cell segmentation, cell location, and annotation according to tissue layers or adjacency networks, leading to the final extraction of cellular parameters that describe geometry and topology dynamics along with ovule primordia development. This protocol applies to various species of the *Paspalum* genus and is potentially useful for 3D studies of large, curved, and hidden organs in multiple plant species.

Key words Apomixis, *Paspalum*, Ovule, Megaspore mother cell, Morphogenesis, Geometry, 3D imaging, 3D segmentation, Quantitative analysis, MorphoGraphX

1 Introduction

The first steps distinguishing sexual and apomictic development occur in the plant ovule, during the formation of the female germ cell or megaspore mother cell (MMC). Notably, in aposporous species, somatic cells in the ovule can bypass meiosis and produce unreduced gametes [1]. While several genetic and epigenetic candidate control mechanisms have been identified [1], the

Tatyana Radoeva and Peter J. van Dijk (eds.), *Apomixis: Methods and Protocols*, Methods in Molecular Biology, vol. 2987, https://doi.org/10.1007/978-1-0716-5001-1_4,

contribution of ovule morphogenesis itself to developmental plasticity of ovule cells in aposporous species is poorly explored. Toward this aim, the first task is to precisely describe ovule ontology at the cellular level, comparing sexual and aposporous genotypes. Thorough microscopic descriptions have been performed at the 2D level in many apomictic species [2–6]. Yet, the plant ovule is typically a curved 3D object, requiring 3D description to fully understand the dynamics of geometry, topology, and afferent mechanical constraints acting during morphogenesis. This task is challenging due to the relative inaccessibility of young ovules, which are tiny organs located deeply within the ovary, inside the flower. Moreover, quantitative analysis of morphogenetic events requires the digitalization (or segmentation) and annotation of individual cells and/or organs, using dedicated 3D image analysis software tools. Such approaches have been largely developed in the last decades for plant morphogenetic studies [7]. Regarding plant ovules, due to their relative inaccessibility, only recently, full 3D developmental atlas of ovule primordium development at cellular resolution has been established in sexual model species, such as Arabidopsis [8, 9] and maize [10], but not yet in apomictic grasses. In species belonging to the aposporous model *Paspalum*, a single ovule forms within each ovary, a large organ that forms multiple nontransparent layers of teguments, complexifying imaging procedures [2, 11, 12]. Classical 2D approaches use strong clearing procedures [13], often not compatible with 3D structure integrity.

Here, we present an original microscopy protocol and image analysis workflow, allowing 3D cellular reconstruction of early ovules in *Paspalum*; using cell wall and nuclei staining, ClearSee clearing procedure [14], confocal laser scanning microscopy (CLSM) and MorphographX image analysis platform [15, 16] for 3D cell segmentation, annotation, and 3D cell parameters quantification. This protocol is amenable to various ovule developmental stages during megasporogenesis. While we here focus on *Paspalum rufum*, a facultative aposporous species from Argentina, the protocol can be transferred to various *Paspalum* species.

2 Materials

2.1 General Materials and Equipment

1. 2 mL, 1.5 mL, and 0.5 mL microcentrifuge tubes.
2. Distilled water.
3. Syringes and needles (e.g., insulin syringes, 27G–30G).
4. Razor blades.
5. Microscopy glass slides.
6. Coverslips (size: 22 × 22 mm, thickness: #1.5H (0.17 mm ± 0.005 mm).

7. Imaging spacers: 0.12 mm depth, double-side adhesive (e.g., SecureSeal™ Spacer-SS8X9, 9 mm Diameter × 0.12 mm Depth).
8. 10–100 μL pipette and pipette tips.
9. Clear nail polish.
10. Orbital shaker.
11. Dissection stereomicroscope.
12. Confocal laser scanning microscope (CLSM; *see* details of requirements in Subheading 3.5).
13. MorphoGraphX: open-source application available online (https://morphographx.org/software/) that runs on both Linux and Windows (*see* **Note 1**).
14. ITK package (available on the MorphoGraphX web site, (*see* point 13). This package is required in order to use the 3D segmentation and other tools from the Insight Toolkit.

2.2 Reagents and Solutions

1. 10 mM phosphate-buffered saline (1X PBS).
2. Fixative: Freshly made 4% paraformaldehyde in 1X PBS (from powder or from 10% stock solution, prepared under a fume hood). Keep on ice or at 4 °C.
3. Alternative fixative (FAA): 2.5% (v/v) acetic acid, 2.5% (v/v) formaldehyde (37%), and 50% (v/v) ethanol. Keep at room temperature.
4. ClearSee solution: 10% (w/v) xylitol powder, 15% (w/v) sodium deoxycholate, and 25% (w/v) urea in water [14]. Keep at room temperature.
5. SCRI Renaissance 2200, stock solution of the supplier considered 100%. Keep at room temperature, in the dark.
6. Propidium iodide (PI), stock solution of the supplier: 1 mg/mL PI. Keep at 4 °C, in the dark.
7. Staining solution: 1:1000 dilution of SCRI Renaissance 2200 and 1:50 dilution of PI, in ClearSee. Keep at room temperature, in the dark.
8. Mounting solution: 1:1000 dilution of SCRI Renaissance 2200 in ClearSee. Keep at room temperature, in the dark.

3 Methods

3.1 Selection of Ovule Developmental Stages

In *Paspalum* species, since each pistil contains one single ovule, the stage of female reproductive development must be inferred by neither invasive nor destructive strategies. It was previously described that megasporogenesis development and floral organ growth are differentially coordinated processes that vary depending

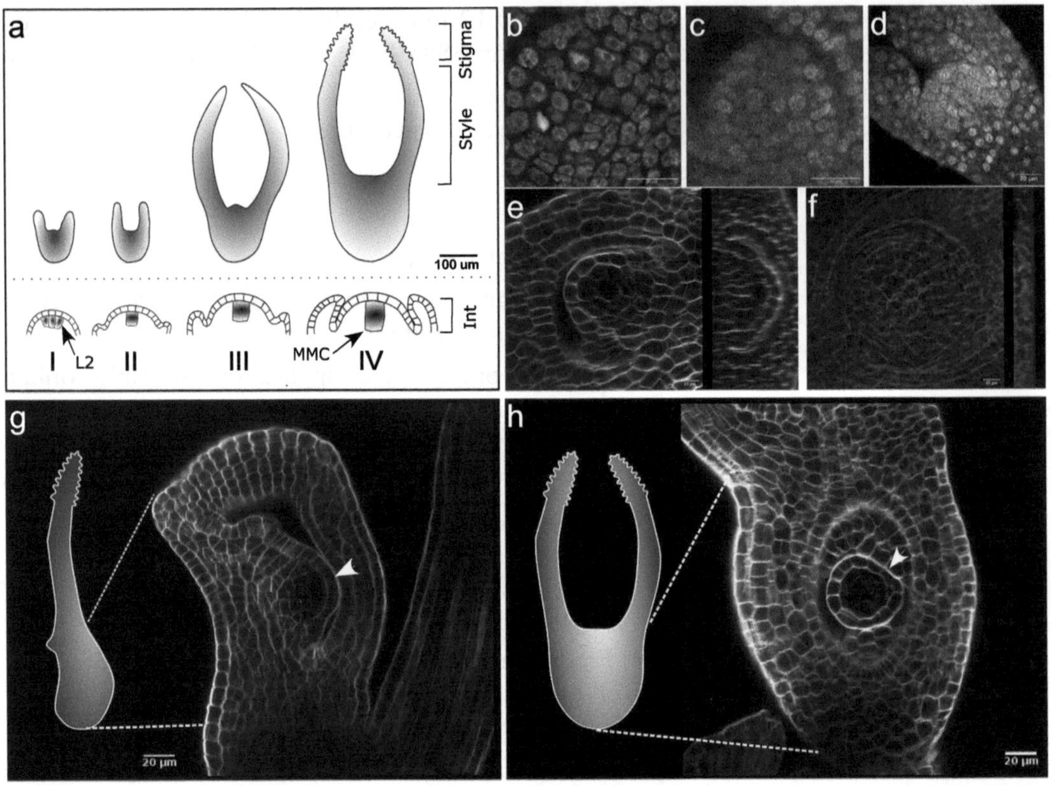

Fig. 1 Sample selection, staining, and orientation for mounting. (**a**) Diagram depicting four different developmental stages of *Paspalum rufum* pistils, showing different shapes and lengths of styles and stigmas, and the corresponding growth stage of the ovule dome and the MMC. (**b**–**d**) Examples of PI staining of different quality: (**b**) optimal, (**c**) low PI signal, (**d**) noisy signal. (**e**–**f**) Examples of orthogonal views showing ovule with (**e**) correct 3D integrity or (**f**) flattened ovule. (**g**–**h**) Examples of images obtained depending on the orientation of the ovary, (**g**) sagittal, (**h**) frontal. White arrow in **g**, **h** indicates the bright signal of the L1 layer of the ovule dome, which should be oriented toward the coverslip when mounting the samples

on the species and the ploidy level [2, 12]. However, it is important to note that young ovule primordia during MMC differentiation were not studied previously. Therefore, in the case of *Paspalum rufum*, to capture the whole process of MMC differentiation, four different stages can be selected considering the shape and length of the style and stigmas (Fig. 1a). The smallest pistils that can be dissected and manipulated, with tiny styles primordia, define Stage I. They contain an ovule primordium with no integuments, and no enlarged cells are seen in ovule layer 2 (L2). At Stage II, the primordia of the styles become elongated and parallel to each other, the inner integuments begin to develop around the ovule dome, and some larger cells appear in the apical region of L2. Throughout Stages III and IV, the integuments gradually increase in size, reaching dome height in stage IV. The MMC is visible as a single, enlarged, and more elongated cell as compared to the surrounding cells (Fig. 1a). Similar features could be considered for other *Paspalum* species, although it is necessary to test them before.

3.2 Pistils Classification and Dissection

1. Under the stereomicroscope, using fine needles, dissect the spikelets and isolate pistils completely. As ovules at young developmental stages are very small (0.1 mm), to visualize them easily, it is better to use only the bottom lighting of the stereomicroscope, while the top lighting is turned off (*see* **Note 2**).

   Classify the developmental stage of the ovule according to the length and shape of the style and the stigma (as described in Subheading 3.1, *see* also Fig. 1a).
2. Sort the pistils by developmental stage and carefully transfer them to separate 0.5 mL microcentrifuge tubes containing approximately 100 μL of distilled water.

3.3 Fixation, Clearing, and Staining

The aim is to preserve organ morphology to obtain images of a suitable quality for 3D segmentation. This method is compatible with deep imaging of whole large organ (0.1–0.2 mm) and with cell wall and nuclei chemical dyes. The ClearSee protocol was adapted from Kurihara et al. [14].

1. Replace water by fixative solution and incubate overnight at 4 ° C (*see* **Note 3**).
2. Rinse twice with 1X PBS.
3. Clear at least 2 days at room temperature in the dark using ClearSee solution (*see* **Notes 4** and **5**).
4. Replace ClearSee by freshly made staining solution, and incubate pistils overnight at room temperature in the dark with orbital stirring (approximately 100 rpm). Fresh staining solution is required to avoid low PI signal (*see* Fig. 1b, c for examples of optimal and low PI signal, respectively).
5. Rinse pistils in ClearSee for at least 3 h or, ideally, overnight to improve signal-to-noise ratio and contrast of nuclear PI staining (*see* Fig. 1d for an example of high PI cytoplasmic noise due to too short rinsing time).

3.4 Mounting

Due to the fragility of cleared samples, to preserve 3D integrity and avoid flattening of the pistils, samples are mounted using 0.12 mm imaging spacers between slide and coverslip (*see* Fig. 1e, f for examples of preserved and flatten samples, respectively).

1. Cut one well of Secure Seal imaging spacers, remove the bottom adhesive liner with forceps, and seal the spacer well in the middle of the glass slide by pressing gently on its surface.
2. Remove the upper adhesive liner of the spacer with forceps.

3. To transfer the pistils from the tubes to the spacer well onto the slide, pipet them with a cut (10–100 μL) tip, with constant monitoring under the stereomicroscope. Keep some ClearSee solution to avoid its crystallization (*see* **Note 6**).
4. Remove all ClearSee solution with a syringe, while monitoring to avoid aspirating the samples.
5. At this step, the samples can be correctly oriented so that the ovule dome is closer to the objective (i.e., toward the coverslip). *Paspalum* pistils have a flat oval shape, and it is desirable that they spontaneously rest on the flatter side, which coincides with the opposite side of the ovule dome (Fig. 1g, h; *see* **Note 7**).
6. Add 15–18 μL of mounting solution into the spacer well.
7. Distribute the liquid with a needle until the entire well area is covered, to avoid air bubbles.
8. Place the coverslip on top of the spacer, and lower it slowly from one side to the other to avoid air bubbles.
9. Seal the edges of the coverslip with clear nail polish.
10. Keep mounted samples refrigerated until imaging to preserve PI signal.

3.5 Microscopy Imaging

Using this protocol, we successfully tested several CLSM systems, such as Leica SP8, Leica Stellaris, or Zeiss LSM880. We present here the key points of the CLSM imaging settings, which can be adapted by the user to their preferred confocal microscope. Detailed settings are provided for Leica SP8 (*see* **Note 8**).

This imaging procedure allowed us to produce image stacks compatible with 3D segmentation, for *Paspalum rufum* ovules from stage I to stage IV (*see* Fig. 2a) as well as for additional *Paspalum* species, i.e., *Paspalum notatum*, *Paspalum chaseanum*, and *Paspalum cromyorrhizon* (*see* Fig. 2b).

1. For Renaissance SR2200, use excitation wavelength at 405 nm, and for PI, use excitation at 561 nm. Use a pinhole of maximum 1 Airy Unit (AU). Adjust laser power and digital/master gain, ensuring both sufficient signal at the bottom of the sample and no overexposure at the upper epidermis. Use laser z correction/compensation along the stack, if required.
2. Set image resolution to 16-bits or 12-bits; both can be open in MGX (*see* Subheading 3.6.1, **step 3**).
3. Adjust stack parameters to ensure cubic voxels, which is required for correct 3D reconstruction. To do so, determine pixel size—depending on the objective used (40× or 63×), the image resolution (512 × 512 or 256 × 256 pixels), and the

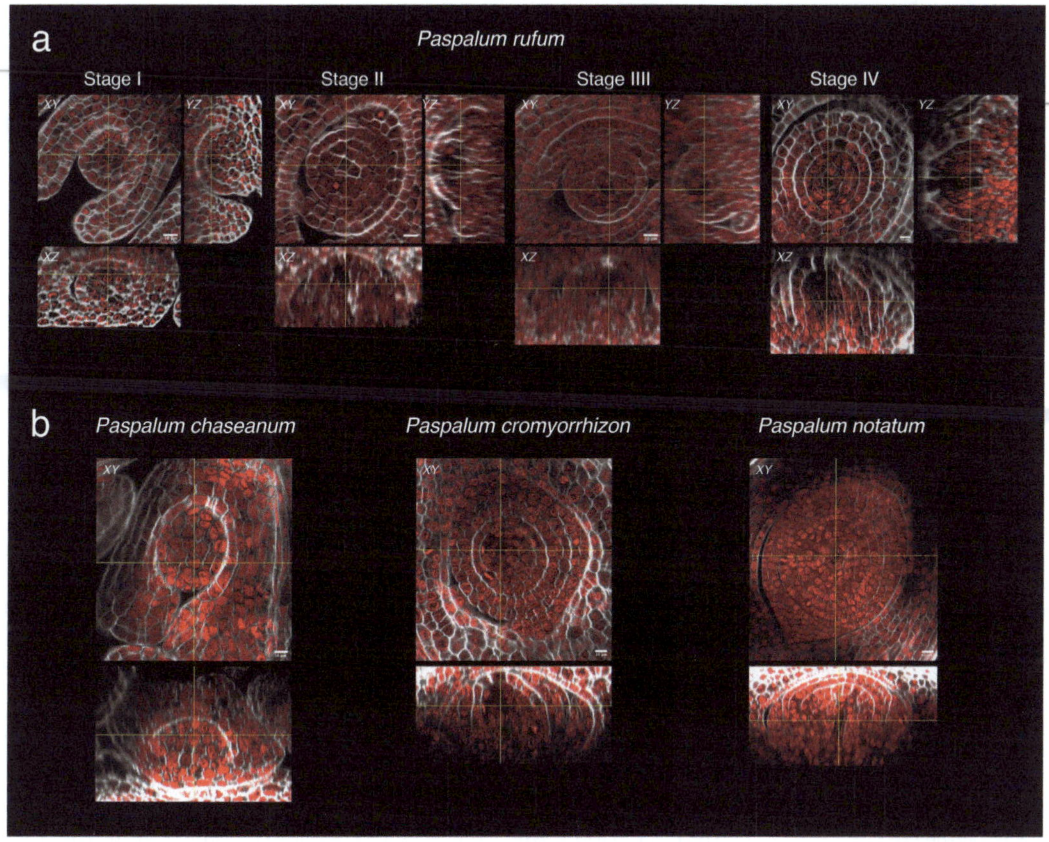

Fig. 2 Representative CSLM raw image stacks of *Paspalum* ovules. (**a**) Representative stacks of ovules at developmental stages I to IV in *Paspalum rufum*. (**b**) Representative stacks of ovules at diverse developmental stages obtained for other Paspalum species (*P. chasesanum, P. cromyorrhizon,* and *P. notatum*). For each panel, image in the plane of the objective (XY) is shown together with orthogonal views (YZ/XZ) of the whole stack. All stacks correspond to a voxel size of approximately 230 nm (*xyz*). White signal: cell walls, stained with Renaissance SR2200. Red signal: nuclei, stained with Propidium Iodide. Scale bars: 10 μm

desired zoom; then adjust the *Z* steps to match pixel size (*see* **Note 9**).

4. Select the initial and final Z-point, ensuring that the stack will contain the full ovule dome, and an additional layer of cells (which is required to ensure correct segmentation of the dome). Note that the upper cell wall of the dome L1 layer presents a very bright Renaissance SR2200 signal that makes it easily recognizable (*see* arrow in Fig. 1g, h). Use orthogonal views to verify dome integrity in the stacks.

3.6 MorphoGraphX Workflow for 3D Cell Segmentation

After acquisition, the image stacks are ready to be used for cell-level 3D reconstruction of the ovule, to quantify cell and organ growth parameters. This implies different steps for cell boundary prediction, segmentation, and curation. All steps are performed in FIJI

Table 1
Glossary of icons for MGX tools

Icon	Volume/mesh tools	Working keys
	Pick label	Alt+left-click
	Fill label	Alt+left-click
	Voxel edit	Alt+left-click
	Delete label	Alt+left-click
	Edit label	Alt+left-click
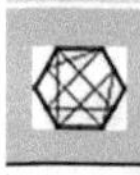	Select connected area	Alt+left-click

and MorphoGraphX (MGX) software (including CNN add-on). Below, each step describes the command line format, detailing the MGX processes used and the values of each parameter, which are given in parentheses. The described protocol is composed from previous related works [9, 15–19], and dedicated MGX video tutorials available online (www.youtube.com/@richardsmithlab6238).

For 3D cell segmentation, only the MGX tools in the upper "Stack Toolbar" are used (Table 1).

3.6.1 Image Pre-processing and Selection of Region of Interest (ROI)

1. In FIJI, open the image stacks as .lif/czi files, split the two channels (Renaissance and PI), and save each channel as TIFF file.
2. Open stack files (TIFF) in Work panel of MGX in Stack 1/Work/Open (Fig. 3a "Stack").
3. As 12 bits TIFF file may result in a very dark image, correct it using: Process/stack/Filters/Brighten Darken (Amount: 16).
4. Due to the large size of the ovule dome, the signal of both dyes decreases significantly along the sample; signal normalization is thus needed: Process/stack/Filters/Normalize Stack (x Radius: 5, z Radius: 5, y Radius: 5, x Sigma: 0.5, y Sigma: 0.5, z Sigma: 0.5, threshold: 5000, blur factor: 0.7) (*see* **Notes 10** and **11**).
5. To select the **Region of Interest** (ROI) containing only the ovule dome, decrease opacity of the working stack and delete manually irrelevant regions using  tool.
6. Save the modified stacks in a new folder: Stack 1/Work /Save. *Optional step*: add "_ROI.tiff" extension to the file name (Fig. 3a "ROI").

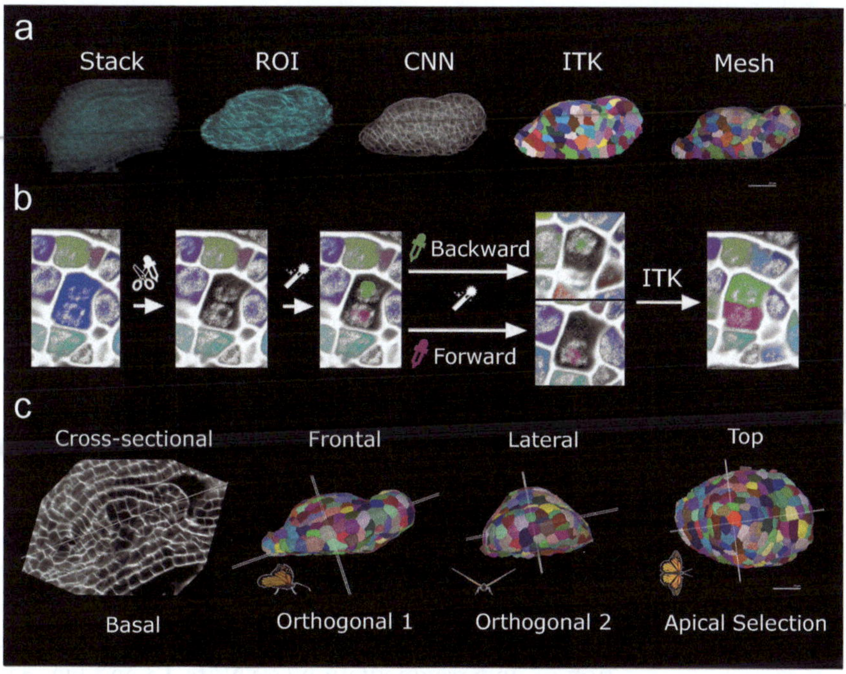

Fig. 3 Graphical description of the MGX processes for segmentation and spatial reference frame definition. (**a**) Overview of the different steps involved, from the stack image to the 3D corrected mesh. (**b**) Details of the steps followed to correct undersegmentation. First remove the mislabeled cell, reseed it according to the number of nuclei observed, optionally move backward and forward to seed the same nuclei in different planes with the same label, continue with the ITK watershed process. (**c**) Graphical description of the steps followed to obtain the spatial reference frame. The first image shows the basal plane on a cross-sectional view of the ovule in the original stack. The following images show the ovule dome in frontal, lateral, and top views and the orthogonal planes are positioned in such a way that they pass through the apical region of the dome. Butterfly is a visual aid of the different orientations of the segmented dome

3.6.2 Automatic CNN Prediction and ITK Auto-Seeded Segmentation

This procedure was adapted from Vijayan et al. [9] and the following MGX video tutorial: Movie 6 "CNN Prediction for 3D Segmentation" (in www.youtube.com/@richardsmithlab6238).

1. Open the ROI image (Stack 1/Work/Open) and blur it using: Process/Stack/Filters/Gaussian Blur (*x* Sigma: 0.3, *y* Sigma: 0.3, *z* Sigma: 0.3).
2. Save blurred stack (Stack 1/Work /Save). *Optional step*: add "_blur.tiff" extension to the original file name.
3. Copy Work to main Stack, for later use, following: Process/Stack/MultiStack /Copy Work to main Stack.
4. Predict the cell wall signal using: Process/Stack/CNN/UNet3D Prediction (Model: insert the path to the network model options, Patch Size128 128 128, Stride: 0.5, Patch 0 0 0, Resample 1.0 1.0 1.0, Standardize No, Use Cuda: Yes), and select the network model *VijayanUNnet*—trained on Arabidopsis ovules Renaissance SR2200 confocal images [9].

5. Save the cell boundary prediction using: Stack 1/Work /Save. *Optional step*: add "_CNN.tiff" extension to the original file name (Fig. 3a "CNN").
6. Blur the predicted signal once using: Process/Stack/Filters/ Gaussian Blur (*x* Sigma: 0.3, *y* Sigma: 0.3, *z* Sigma: 0.3).
7. Save again using: Stack 1/Work/Save. *Optional step*: add "_blur.tiff" extension to the original file name.
8. Display Main and Work in Main panel.
9. Merge blurred raw cell wall signal (ROI) and the blurred cell boundary predicted signal using: Process/Stack/Multistack/ Combine Stacks (add) (*see* **Note 12**).
10. Save it. *Optional step*: add "_comb_add.tiff" extension to the original file name.
11. Proceed with auto-segmentation of the merged image stack using: Process/Stack/ Segmentation/ITK Watershed Auto Seeded (Fully connected: No, Mark watershed line: No, level: 1500). Threshold level value can be adjusted, depending on the signal and image characteristics, to reduce oversegmentation or undersegmentation (*see* **Note 13** and Fig. 3a "ITK").
12. Before saving the segmented stack, remove the outside labels. Select the tool and left-click on the labels outside the sample. As there is no undo function in MGX, before making any additional change, save the segmented volume (in the work store) in the Stack menu.

3.6.3 Manual Curation of Segmentation

Since errors can occur in automatic cell prediction and segmentation, **manual** correction is often required. Both oversegmentation and undersegmentation can happen. For optimal segmentation correction, it is essential to consider nuclei signal, as the cell boundaries may be mispredicted. It is also important to note that oversegmentation is easier to correct. While undersegmentation can also be corrected, it would, however, take longer time.

Start by opening the stacks to consider cell boundary prediction, cell wall raw signal, and nuclei signal images, to perform the curation:

1. Open the merged stack (ended with "_com_add.tif") in the main panel of Stack 1 (Stack1/main/open).
2. Open the Auto-seeded segmented image to be corrected in the work panel (Stack 1/work/open).
3. To take into account the nuclei and cell wall raw signals, open the 16 bit and normalized files, for both PI and Renaissance SR2200 signals, in Stack 2 (Fig. 3b).
4. Before correcting, increase the number of labels from 16 to 64, to make it easier to identify the different labels: enable Stack

1 and select tool from the Volume Toolbar, then change the number of colors to 64 and proceed with Set/Random/Apply.
5. To correct **oversegmentation,** different labels, corresponding in fact to one cell, can be fused together into a single segment. Select one of the labels to fuse with the tool from Volume Toolbar, then, with left-click on the other label to merge cells [18].
6. In order to correct **undersegmentation,** where a single label contains more than one cell, it is necessary to manually seed the cells to be segmented separately (*see* Fig. 3b).
7. Check the "fill" and "seed" checkboxes at the bottom of the view panel.
8. Delete the incorrectly segmented label with tool and add new seeds to each cell with tool, based on the number of nuclei observed.
9. Proceed with segmentation of corrected cells by Process/Stack/Segmentation/ITK Watershed (Fully connected: No, Mark Watershed Line: No).
10. If the problem persists, move Forward and Backward to position the seeds, with the same label (selected by) on the corresponding nuclei in different planes of the 3D ovule (Fig. 3b, for more details, *see* **Note 14**).
11. Use the clipping planes in the View tab to find and then correct errors throughout the sample.

3.7 MGX Workflow to Enable 3D Cell Analyses, Annotation, and Shape Analysis

In this and the following sections, the tools of the MGX lateral "Mesh toolbar" are used.

3.7.1 3D Mesh Extraction

To enable MGX tools to perform cell annotation, 3D measures, shape analysis, and other quantitative characterization, the first step is to **extract a 3D Mesh** (full Mesh) from the corrected stack, using Process/Mesh/Creation/Marching Cube 3D (cube size: 0.6 μm, minimal voxel size: 0; smooth passes: 3 and label: 0) (Fig. 3a "Mesh").

3.7.2 Defining Spatial Reference Frame

The aim here is to define a **spatial reference frame** to obtain the three-dimensional coordinates of each cell with respect to the ovule dome, by placing the three clipping planes at specific positions relative to the dome and also orthogonal to each other (Fig. 3c). It helps to define the apical point of the dome and the underlying candidate MMC, allowing cells to be annotated by cell layer and by their position relative to the MMC (*see* Subheadings 3.8.1, 3.8.2, and 3.8.3).

1. Open the original stack image in the work panel: Stack 1/Work /Open (Original TIFF 16 bit and normalized image).
2. Place a clipping plane at the base of the ovule dome. Playing with decreasing opacity of the full image and moving throughout the volume by enabling clips, locate the basal plane just below the teguments' curvature (*see* Fig. 3c "basal" and **Note 15**).
3. Rotate the other clipping planes to place them perpendicular to the basal plane (orthogonal planes, Fig. 3c).
4. Open the full Mesh in Mesh 1.
5. Looking at the ovule Mesh, move the orthogonal planes to pass through the upper part of the dome (Fig. 3c "Orthogonal 1 and 2").
6. Save the file to keep the position of the spatial reference frame in one file (File/Saved As).
7. Use the intersection between the orthogonal planes to identify the **apical cell** of the ovule dome (Fig. 3c "Apical Selection"). This cell will be later used for layers detection (Subheading 3.8.1).
8. Draw a Bezier grid coincident to each clipping plane to be used later (Subheading 3.9.1): in View panel, check the "CutSurf" box and, below "Cutting Surface" section, tick the checkboxes of Grid and Bezier and then press Reset.
9. Once the grid is visible, move it to place it in parallel to and overlapping with each clipping plane.
10. As one Bezier could be drawn at a time, draw one Bezier grid and Save one File (File/Saved As) for each of the grids defining the reference frame: one **basal** and two orthogonal (**orth1** and **orth2**) MGX files (Fig. 3c).

3.8 MGX Workflow for Cell Type Annotation in 3D

3.8.1 Cell Type Annotation According to Layers

In order to semi-automatically annotate cell types according to their respective cell layers, the previously developed process for the Meristem label was employed in Montenegro-Johnson et al. [20]. This proved to be an effective approach for ovules (Fig. 4a) with satisfactory results obtained by adjusting some parameters.

1. A surface Mesh, covering the ovule dome is used as a reference to detect the different layers [20]. To generate a **tissue surface Mesh,** put ROI image file in the main stack and copy it to Working stack (Process/Stack/Multistack/ Copy main to work). Blur the image using: Process/Stack/Filters/Gaussian Blur, three times (x Sigma: 0.3, y Sigma: 0.3, z Sigma: 0.3).
2. Extract the Mesh using: Process/Mesh/Creation/Marching Cubes Surface (Cube size: 5 μm, Threshold: 1900).

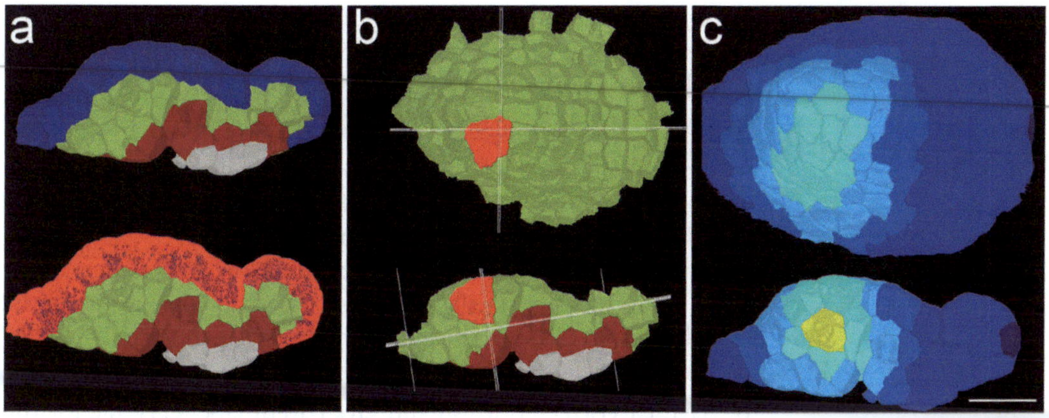

Fig. 4 3D cell type annotation according to cell layers and to candidate MMC. (**a**) Example of the annotated cell layers in a cross-sectional view of the ovule and the selection of L1 cells to be deleted (L1: blue, L2: green, L3 dark-red, and L4: white). (**b**) Orthogonal planes identifying the upper L2 cell, selected as candidate MMC, in a virtually dissected ovule lacking the L1 layer. (**c**) Example of MMC adjacency network from frontal and cross-sectional view. The MMC is shown in yellow, the cells in direct contact with the MMC (contact 1) in light blue, and the cells of contact 2 and 3 in darker blue. The selected areas are highlighted in bright red. Scale bar: 20 μm

3. Shrink the Mesh two or three times at 0.5 μm each pass, using: Process/Mesh/Structure/ **Shrink Mesh** (*see* **Note 16**).
4. Smooth the surface Mesh by three passes, using: Process/Mesh/Structure/ Smooth.
5. Finally, to delete the region of the surface covering the bottom of the ovule dome, select it with a clipping plane using: Process/Mesh/Select/Select clip region; and delete the selections by pressing "supr" key.
6. Open the 3D mesh (generated in Subheading 3.7.1) in Mesh1, and the Surface Mesh in Mesh2, and select the apical cell of the dome previously identified (in Subheading 3.7.2, **step 8**) using: Process/Mesh/Selection/Select labels (Labels: "write here the label of the apical cell previously identified", Replace Selection: No).
7. Detect layers automatically using: Process/Mesh/Cell Atlas 3D/ Meristem/Meristem label (Min. volume: 10, Deph Organizing Center: 2, Radius: 3, Cone angle: 0.65) and save the label in a "CSV" file by: Process/Mesh/Linage Tracking/ Save Parents. *Optional step*: add "Layers_" to the original Mesh file name (*see* **Note 17** and Fig. 4a).

3.8.2 Correction of Layers Detection

1. To correct cell layers, select L1 cells using Mesh/Selection/ parent label (Label: "5").

2. To delete all cells but L1 cells, "Invert Selection" from "Labels & Selection" on the main toolbar, and delete all the selected cells by pressing "supr" key.
3. Save it as L1 Mesh.
4. Correct this L1 Mesh by selecting manually all visible superficial cells, using  tool, and find out those non-superficial cells, unselected, that are badly assigned as L1.
5. Invert selection from Labels & Selection and assign a new parental label to mislabeled L1 cells that correspond to L2 cells using Mesh/Lineage tracking/set parent (Label: "6" for L2).
6. Save L1 corrected Mesh and save parent label CSV file for this Mesh that contains only L1 and corrected L2 cells. *Optional step*: add "L1_" to the original Mesh file name.
7. Correct the original "Layers" CSV file, comparing it with "L1" CSV file and including the modifications of L1 correction directly on the CSV file.
8. Load the full Mesh on Mesh 1.
9. Load the CSV file with the layers corrected by Process/Mesh/ Lineage Tracking/Load Parents.
10. To correct L2, proceed as for L1, but extract only those cells included in L2 layer (labeled by "6").
11. Follow the same for L3 and so on (*see* **Note 18**).
12. Save the corrected CSV file as "**Layers**" using Process/Mesh/ Lineage Tracking/Save Parents.

3.8.3 Radial Cell Type Annotation According to the Candidate MMC

The aim of this procedure is to annotate semi-automatically cell types according to their degree of contact with the candidate MMC in order to define the MMC adjacency network [10] (Fig. 4b, c).

1. Load the spatial reference frame MGX file.
2. Load the L2 Mesh (generated in Subheading 3.8.2, **step 10**) in Mesh 2.
3. To identify the candidate MMC cell in the L2 Mesh, which is defined as the cell at the apical position of the L2 as described previously [8, 10], note the label of the cell tagged by the intersection of the orthogonal clipping planes (Fig. 4b).
4. Load the full Mesh in Mesh 1.
5. Run: Process/Mesh/HeatMap/Analysis/Cell Analysis 3D (*see* **Note 19**).
6. In this Mesh, select the candidate MMC using: Process/ Mesh/Selection/Select Labels (Labels: "label of candidate MMC", Replace Selection: No).

7. Label cells according to candidate MMC contact degrees using: Process/Mesh/Heat Map/Measure 3D/location/Cell Distance (Wall weights: 1, Restrict connectivity to the same cell type: No) [19].
8. Save CSV file as **MMC_adjacency** using: Process/Mesh/Heat Map/Analysis/Heat Map Save (Fig. 4c).
9. This process is quite robust, but if some labeling errors are detected, identify those cells mislabeled in the Mesh and then correct the HeatLabel value directly in the CSV file.

3.9 MGX Workflow for Cell Characterization and Export of Cellular Quantitative Parameters for Statistical Analysis

In addition to cell annotation, MGX offers several processes for cellular quantitative analysis, such as three-dimensional measurements related to cell geometry, relative cell location, and network analysis. In this section, it is shown, with some examples, how parameters obtained from different processes can be integrated into a single table, which can then be used to perform statistical analyses and visualization in other specialized software, such as R.

3.9.1 Cell Location Relative to Basal and Orthogonal Bezier Grids

1. In order to obtain the **3D position** of each cell with respect to the defined Bezier grids, open each of the MGX files previously obtained in Subheading 3.7.2, **step 11** (**basal, orth1,** and **orth2**).
2. For each file, one by one, proceed with: Process/Mesh/Heat-Map/Analysis/Cell Analysis 3D.
3. Then proceed with: Process/Mesh/Heat Map/Location/Distance to Plane from Bezier (Consider Orientation: yes, Reverse Orientation: No).
4. Save separate CSV files for each Bezier (**basal, orth1** and **orth2**), using: Process/Mesh/Heat Map/Analysis/Heat Map Save.

3.9.2 Cell Shape Analysis Process

1. To perform 3D shape analysis, use: Process/Mesh/Cell Axis 3D/Shape analysis/Compute Shape Analysis 3D.
2. To select and save, in Attribute Map, the chosen cell shape parameters included in the process "Shape Analysis", use: Process/Mesh/Cell Axis 3D/Shape analysis/Display Shape Axis 3D (Heatmap: **Select the parameter to measure**, ScaleHeat: Auto, Heat min: 1, Heat max: 3, Show axis/ Axis color/Line Scale/Line Offset/Threshold: as desired, **Create Atrr Maps: Yes**).

3.9.3 Combine Data in a Single Table

Once all the data of cell annotation, cell shape, and location relative to Bezier grids are saved, it is possible to add other parameters from MGX HeatMap processes such as those included in "Geometry" menu (e.g., cell lengths, area, and volume) and combine them all in one single table (*see* Table 2).

Table 2
Example of attribute map including cell annotation and measures exported from MGX

Label[a]	Layers[b]	Bezier Base[c]	Bezier Orth1[c]	Bezier Orth2[c]	Geometry Volume[d]	MMC Adj[e]	3D Shape Max[f]	3D Shape Mid[f]	3D Shape Min[f]
489	6	7.732	5.563	2.940	1111.860	0	19.125	17.700	14.570
16	5	18.873	1.866	6.390	473.878	1	14.473	13.878	10.509
507	6	10.452	−3.427	4.882	582.682	1	16.200	14.206	11.079
77	7	0.840	0.448	10.477	226.535	1	12.836	10.079	7.881
484	7	3.742	−1.869	3.997	559.509	1	16.909	12.826	11.616
521	7	−0.120	1.696	−5.344	542.135	1	15.610	13.489	11.006
7	5	5.279	17.857	0.159	53.851	2	10.854	5.923	3.713
10	5	13.914	12.302	−6.524	359.233	2	14.632	12.700	8.831
148	5	10.585	14.835	−10.838	301.312	2	16.105	10.730	8.156
246	6	−0.016	2.250	17.396	137.932	2	14.480	6.839	6.167
256	6	8.049	3.087	−13.317	359.995	2	13.311	12.764	9.321
290	6	5.040	−1.778	−13.236	514.032	2	18.964	12.203	9.549
179	7	−7.389	5.955	6.135	258.537	2	14.212	10.457	7.893
383	7	−5.272	−2.902	9.138	140.305	2	9.715	9.382	7.140
442	7	−1.575	8.217	−8.419	225.085	2	12.526	9.618	8.588
3	4	−12.127	5.338	−14.554	194.694	3	11.989	8.595	8.118
39	4	−12.272	−1.222	−13.504	195.136	3	10.891	9.637	8.217
111	4	−11.539	2.128	−8.044	46.044	3	7.883	5.929	4.444
341	4	−11.067	−6.260	−7.983	54.658	3	9.410	6.076	4.260
9	5	12.181	−1.284	−19.438	251.789	3	12.402	11.699	7.571
11	5	−0.410	24.424	−13.979	436.023	3	14.021	13.674	10.164
41	5	14.163	7.773	−15.075	259.051	3	13.701	11.718	7.173
55	5	5.548	18.798	−16.502	337.129	3	15.092	13.707	7.181

[a]Label of individual cells, [b]5: L1, 6: L2, 7: L3, [c]Distance from cells to each Bezier grid (μm), [d]cell volume (μm^3), [e]0: MMC (white), 1: contact 1, 2: contact 2, 3: contact 3 (in different tones of gray), [f]length of 3D axes of the cells, maximum, medium, and minimum

1. Open the Mesh in Mesh 1.
2. To enable cell Geometry Analysis, perform Cell Analysis 3D, use: Process/Mesh/HeatMap/Analysis/Cell Analysis 3D (*see* **Note 19**).

3. To combine the data of **Cell Analysis 3D** with cell annotation of Layers, load the CSV file obtained in Subheading 3.8.2, **step 12**, using: Process/Mesh/Lineage Tracking/Load Parents.
4. Export data to Attributes CSV extended file using: Process/Mesh/Heat Map/Operators/Export Heat to Attr Map (Prefix: Empty, Name: **Layers**, Key: label, Heat Map Type: **Parent Heat**, Save to Active Mesh, Clean: No).
5. To combine the data of **MMC_adjacency** with the other attributes, load the corresponding CSV file, use: Process/Mesh/Heat Map/ Heat Map Load.
6. Export data to Attributes CSV extended file using: Process/Mesh/Heat Map/Operators/Export Heat to Attr Map (Prefix: Measure Label Double, Name: **MMC**, Key: label, Heat Map Type: **Label Heat**, Save to Active Mesh, Clean: No).
7. To combine the data of **3D position of each cell to each Bezier grid** with the other attributes, load the corresponding CSV file by Process/Mesh/Heat Map/ Heat Map Load.
8. Export data to Attributes CSV extended file using: Process/Mesh/Heat Map/Operators/Export Heat to Attr Map (Prefix: Measure Label Double, Name: **Bezier_dist_base/_orth1/_orth2**, Key: label, Heat Map Type: **Label Heat**, Save to Active Mesh, Clean: No).
9. To obtain one CSV file including all these data (*see* Table 2), go to: Process/Mesh/Attributes/Save to CSV Extended and select all saved parameters.

4 Notes

1. To process large stacks, a PC with 32–64Gb of RAM and an nVidia graphics card with 4 GB or more dedicated video memory supporting OpenGL3 and Cuda, is recommended. To process stacks by convolutional neural networks (CNN), Cuda is required, and it might be needed to additionally install the libcudnn runtime available from the nVidia website.
2. Please note that a thorough initial dissection is recommended. Since the samples must be cleared, subsequent dissection can be complicated as tissues become transparent.
3. Alternatively, FAA can be used as a fixative. ClearSee clearing has been found to be compatible with this fixative and the final image quality is comparable to that obtained with PFA fixation.
4. Pistils can be stored in the dark in ClearSee at room temperature for at least 1 year, providing parafilm sealing of the tubes to avoid evaporation.

5. To avoid loss or breakage of pistils, change all solutions using a syringe with thin needles. Make sure that the needles reach the bottom of the tube. Monitoring the tubes under the stereomicroscope while changing solutions is advised to ensure that samples are immersed into the solution and to avoid sample loss.
6. If additional dissection is required, place the samples on the slide, outside the spacer well area, and dissect with enough ClearSee solution to avoid crystallization due to evaporation, but as little as possible to avoid slippage of the sample. If crystallization occurs, add ClearSee solution. Once the pistils are isolated, place them one by one into the spacer well area using a needle.
7. If it is difficult to obtain the correct position of the pistils, resulting in apical region of the dome away from the objective, it is possible to mount the sample between two coverslips, allowing flipping the position of the sample if required.
8. We use the following imaging settings for confocal laser scanning microscopy when using a Leica SP8-UV inverted confocal microscope equipped with HyD detectors and a long-distance HC PL-APO CS2 40× 1.1 NA water objective. Bidirectional scan speed used was 8000 Hz using a resonant scanner (pixel dwell time of approximately 72 ns), and signal was averaged by line 16 times (this parameter can be adjusted to find the best compromise between image quality and acquisition time, though we have never used less than 8 times averaging using resonant scanner). Renaissance SR2200 signal emission was collected between 415 nm and 500 nm, with HyD detector (standard mode) gain set from 10 to 15, and 405 nm laser power from 2% to 3%. PI signal emission was collected between 599 nm and 670 nm, with HyD detector gain set from 20 to 40, and 561 nm laser power from 20% to 40%. Using 512 × 512 pixels image resolution, zoom and *z*-step were adjusted to collect stacks with approximate cubic voxels of 0.225 μm (*xyz*). Stack thickness ranged from 60 to 140 μm depending on ovule stage, ploidy level, and species.
9. 512 × 512 pixels images and cubic voxels of approximately 0.225 μm (depending on objective and zoom) allows optimal automatic segmentation. However, 256 × 256 pixels images and approximately 0,530 μm cubic voxels can be segmented successfully by adjusting ITK parameters, which allows reducing image acquisition time.
10. The conditions for both channels (Renaissance SR2200 and PI) were the same for almost all the images, but the threshold may vary depending on the quality of staining and on the sample, note that higher values reduce the noise.

11. When PI signal is noisy, after normalization, perform: Process/stack/Filters/Median of Medians (*X* Radius: 1, *y* Radius: 1, *z* Radius: 1).
12. The "add" option performed best for *P. rufum*; try other alternatives if the segmentation obtained is not optimal (Average, Product, Max, Min, Subtract).
13. If the ovule is oversegmented, higher threshold values would decrease it; and if too many undersegmented cell are observed, lower threshold values will decrease them [16].
14. To correct the segmentation through the ovule dome, first delete the label of mislabeled cells with tool. Move through different clips containing the cells to be corrected, and assign different seeds with tool, in each of the different planes. Then, with tool, copy the label of seeds corresponding to the same cell in the different planes and fill them with the same label using tool. Continue with the segmentation of corrected cells using: Process/Stack/Segmentation/ITK Watershed (No: No) (Fig. 3b).
15. The procedure to move and rotate planes is detailed in the study cited herein [18].
16. Make sure that the surface Mesh is over the L1 cell wall signal and does not invade it.
17. This tool is designed for cellular annotation of shoot apical meristems, L1 to L3 (annotated with "5" to "7" parent labels), and the organizing center (annotated with "4" parent labels), which in this protocol is considered as the L4 in case of larger ovules. Parameters are adapted for the ovule dome by adjusting the Depth of the Organizing Centre in order to place it at L4 and increasing the Radius to span all the cells at this depth, and not only the central cells. The cone angle is adjusted to correctly detect L1 to L3 cells, for details, *see* [20].
18. It is also possible to annotate cell layers manually. Open the full mesh. Select all the surface cells, using: Process/Mesh/Lineage Tracking/Set Parent, assign label "5". Then, save L1 label using: Process/Mesh/Lineage Tracking/Save Parents (as CSV L1 file). To label L2 layer, the same procedure is used, after deletion of the L1 layer: open again the full Mesh, load CSV L1 file using: Process/Mesh/Lineage Tracking/Load Parents, select and delete the L1 cells using: Process/Mesh/Lineage Tracking/Select Parent and "supr" key, repeat the process described above to annotate L2 cells. The next layers are annotated repeating the same procedure. Finally, combine all the CSV files in one single table containing all the layers.
19. As a requisite of MGX process, running the process "Cell analysis 3D" is needed before any HeatMap process is used.

Acknowledgments

This work was funded by EU H2020 Marie Skłodowska-Curie Action Research and Innovation Staff Exchange grant (MAD, Grant agreement ID: 872417) to LD and DA; French National Research Agency grant (LOVE-ME, ANR-24-CE20-2432-02) to DA. Agencia Nacional de Promoción Científica y Tecnológica (ANPCyT), Argentina (PICT-2021-GRF-TII-00316, PICT 2019-03414) to LD; Argentina. Universidad Nacional de Rosario (UNR), Argentina, Project (80020220700076UR, 80020190300021UR) to LD. CONICET, Argentina Project (PIP 11220200101680CO) to LD. LD is a research staff member of Consejo Nacional de Investigaciones Científicas y Técnicas (CONICET), Argentina. We acknowledge the imaging facility MRI, a member of the France-BioImaging national infrastructure supported by the French National Research Agency (ANR-10-INBS-04, "Investments for the future"), in Montpellier; and Pic-Plant platform; funded by DIADE, IRD, University of Montpellier, and the French State—Occitanie Region plan contract CPER PlantEnvi.

Conflicts of Interest The authors have no conflicts of interest to declare that are relevant to the content of this chapter.

References

1. Schmidt A (2020) Controlling apomixis: shared features and distinct characteristics of gene regulation. Genes (Basel) 11:329. https://doi.org/10.3390/genes11030329
2. Soliman M, Espinoza F, Ortiz JPA, Delgado L (2019) Heterochronic reproductive developmental processes between diploid and tetraploid cytotypes of Paspalum rufum. Ann Bot 123:901–915. https://doi.org/10.1093/aob/mcy228
3. Rojek J, Kapusta M, Kozieradzka-Kiszkurno M, Majcher D, Górniak M, Sliwinska E, Sharbel TF, Bohdanowicz J (2018) Establishing the cell biology of apomictic reproduction in diploid Boechera stricta (Brassicaceae). Ann Bot 122:513–539. https://doi.org/10.1093/aob/mcy114
4. Juranić M, Tucker MR, Schultz CJ, Shirley NJ, Taylor JM, Spriggs A, Johnson SD, Bulone V, Koltunow AM (2018) Asexual female gametogenesis involves contact with a sexually fated megaspore in apomictic hieracium. Plant Physiol 177:1027–1049. https://doi.org/10.1104/pp.18.00342
5. Carman JG, Jamison M, Elliott E, Dwivedi KK, Naumova TN (2011) Apospory appears to accelerate onset of meiosis and sexual embryo sac formation in sorghum ovules. BMC Plant Biol 11:9. https://doi.org/10.1186/1471-2229-11-9
6. Grimanelli D, García M, Kaszas E, Perotti E, Leblanc O (2003) Heterochronic expression of sexual reproductive programs during apomictic development in Tripsacum. Genetics 165:1521–1531. https://doi.org/10.1093/genetics/165.3.1521
7. Bassel GW, Smith RS (2016) Quantifying morphogenesis in plants in 4D. Curr Opin Plant Biol 29:87–94. https://doi.org/10.1016/j.pbi.2015.11.005
8. Hernandez-Lagana E, Mosca G, Mendocilla-Sato E, Pires N, Frey A, Giraldo-Fonseca A, Michaud C, Grossniklaus U, Hamant O, Godin C, Boudaoud A, Grimanelli D, Autran D, Baroux C (2021) Organ geometry channels reproductive cell fate in the Arabidopsis ovule primordium. elife 10:e66031. https://doi.org/10.7554/eLife.66031

9. Vijayan A, Tofanelli R, Strauss S, Cerrone L, Wolny A, Strohmeier J, Kreshuk A, Hamprecht FA, Smith RS, Schneitz K (2021) A digital 3D reference atlas reveals cellular growth patterns shaping the arabidopsis ovule. elife 10:1–38. https://doi.org/10.7554/eLife.63262
10. Ouedraogo I, Lartaud M, Baroux C, Mosca G, Delgado L, Leblanc O, Verdeil JL, Conéjéro G, Autran D (2023) 3D cellular morphometrics of ovule primordium development in Zea mays reveal differential division and growth dynamics specifying megaspore mother cell singleness. Front Plant Sci 14:1174171. https://doi.org/10.3389/fpls.2023.1174171
11. Ortiz JPA, Quarin CL, Pessino SC, Acuña C, Martínez EJ, Espinoza F, Hojsgaard DH, Sartor ME, Cáceres ME, Pupilli F (2013) Harnessing apomictic reproduction in grasses: what we have learned from Paspalum. Ann Bot 112:767–787. https://doi.org/10.1093/aob/mct152
12. Laspina NV, Vega T, Seijo JG, González AM, Martelotto LG, Stein J, Podio M, Ortiz JPA, Echenique VC, Quarin CL, Pessino SC (2008) Gene expression analysis at the onset of aposporous apomixis in Paspalum notatum. Plant Mol Biol 67:615–628. https://doi.org/10.1007/s11103-008-9341-5
13. Krysiński S (1887) Beiträge zur histologischen Technik. Arch für Pathol Anat und Physiol und für Klin Med 108:217–219. https://doi.org/10.1007/BF02281361
14. Kurihara D, Mizuta Y, Sato Y, Higashiyama T (2015) ClearSee: a rapid optical clearing reagent for whole-plant fluorescence imaging. Development 142:4168–4179. https://doi.org/10.1242/DEV.127613
15. Barbier de Reuille P, Routier-Kierzkowska AL, Kierzkowski D, Bassel GW, Schüpbach T, Tauriello G, Bajpai N, Strauss S, Weber A, Kiss A, Burian A, Hofhuis H, Sapala A, Lipowczan M, Heimlicher MB, Robinson S, Bayer EM, Basler K, Koumoutsakos P, Roeder AHK, Aegerter-Wilmsen T, Nakayama N, Tsiantis M, Hay A, Kwiatkowska D, Xenarios I, Kuhlemeier C, Smith RS (2015) MorphoGraphX: a platform for quantifying morphogenesis in 4D. elife 4:1–20. https://doi.org/10.7554/eLife.05864
16. Kerstens M, Strauss S, Smith R, Willemsen V (2020) From stained plant tissues to quantitative cell segmentation analysis with MorphoGraphX. Methods Mol Biol 2122:63–83. https://doi.org/10.1007/978-1-0716-0342-0_6
17. Wolny A, Cerrone L, Vijayan A, Tofanelli R, Barro AV, Louveaux M, Wenzl C, Strauss S, Wilson-Sánchez D, Lymbouridou R, Steigleder SS, Pape C, Bailoni A, Duran-Nebreda S, Bassel G, Lohmann JU, Tsiantis M, Hamprecht FA, Schneitz K, Maizel A, Kreshuk A (2020) Accurate and versatile 3D segmentation of plant tissues at cellular resolution. elife 9:1–34. https://doi.org/10.7554/eLife.57613
18. Strauss S, Sapala A, Kierzkowski D, Smith RS (2019) Quantifying plant growth and cell proliferation with MorphoGraphX. Methods Mol Biol 1992:269–290. https://doi.org/10.1007/978-1-4939-9469-4_18
19. Vijayan A, Strauss S, Tofanelli R, Mody TA, Lee K, Tsiantis M, Smith RS, Schneitz K (2022) The annotation and analysis of complex 3D plant organs using 3DCoordX. Plant Physiol 189:1278–1295. https://doi.org/10.1093/plphys/kiac145
20. Montenegro-Johnson T, Strauss S, Jackson MDB, Walker L, Smith RS, Bassel GW (2019) 3DCellAtlas meristem: a tool for the global cellular annotation of shoot apical meristems. Plant Methods 15:1–9. https://doi.org/10.1186/s13007-019-0413-0

Chapter 5

Imaging of Callose in Young Ovules of Diplosporous Species of the Asteraceae Family

Krystyna Musiał and Agnieszka B. Springer

Abstract

Callose (β-1,3-glucan) is one of the cell wall polymers that has multiple biological functions and plays an important role in plant growth and development processes. Callose is the cytological marker of reproductive cells in young ovules of most sexually reproducing angiosperms, whereas in ovules of apomictic plants, callose deposition does not occur or is altered. Aniline blue staining is one of the methods to detect callose. Aniline blue fluorochrome binds preferentially to β-1,3-glucans and is widely used to visualize callose in various plant tissues. Here, we describe an easy-to-perform procedure for fluorescent imaging of callose in whole-mount ovules using aniline blue-stained fixed flowers. This procedure enables rapid and precise detection of callose in ovules in order to determine and compare the pattern of its distribution during early sexual and apomictic reproductive processes.

Key words (1,3)-β-glucan, Aniline blue, Apomixis, Callose, Cell wall, Fluorescence microscopy, Megasporogenesis, Ovule, Plant reproduction

1 Introduction

Callose, a water-insoluble β-1,3-linked homopolymer of glucose with some β-1,6-branches, is one of the plant cell wall components that is produced mainly in specialized cell walls [1, 2]. This amorphous polysaccharide synthesized directly on the plasma membrane performs important regulatory functions in cell-to-cell signaling in the processes of plant growth, development, and reproduction, as well as in plant defense responses to biotic and abiotic stress [3–7]. Despite the importance of callose in intercellular communication during multiple biological processes, our knowledge of the mechanisms governing its biosynthesis and interactions with other biopolymers within the cell wall matrix is still limited [8–10].

In most sexually reproducing angiosperms, callose synthesis and deposition and its subsequent degradation are linked to early reproductive processes in anthers and ovules [2, 11, 12].

Tatyana Radoeva and Peter J. van Dijk (eds.), *Apomixis: Methods and Protocols*, Methods in Molecular Biology, vol. 2987, https://doi.org/10.1007/978-1-0716-5001-1_5,

Deposition of callose around micro- and megasporocytes accompanies the transition from somatic to germline identity during sporogenesis, and therefore callose is a cytological marker that distinguishes young reproductive cells from somatic tissues [12]. In the case of gametophytic apomicts, however, callose deposition in ovules does not occur or is incomplete (it is present as a micropylar cup) [13–15, and references therein]. This may be a consequence of heterochronic expression of genes involved in sexual reproduction that control the activity of factors responsible for callose synthesis and degradation [16, 17]. Comparative studies of callose events in the ovules of other taxa are still necessary to learn more about the mechanisms of cell differentiation that initiate apomictic development. A detailed analysis of the spatial and temporal distribution of callose in the young ovules of various species may be of particular importance for advanced research aimed at understanding how early female reproductive processes are regulated.

Callose deposits in plant tissues can be detected by bright-field microscopy after staining with resorcinol blue or by fluorescence microscopy following immunolabeling with a monoclonal anti-β-1,3-glucan antibody or after a less time-consuming staining with aniline blue [18, 19]. The most widely used method for detecting callose is the relatively simple and quick aniline blue staining. Commercial aniline blue dye is a heterogeneous mixture of various components, including the fluorochrome Sirofluor, which binds specifically to β-1,3-glucans and is responsible for the fluorescence observed during callose detection [20–22]. The aniline blue fluorochrome forms complexes with callose and induces a bright yellow-green fluorescent signal (*see* Fig. 1). The intensity of aniline blue-induced fluorescence is pH dependent, with the fluorescent signal becoming increasingly stronger as pH increases, while callose fluorescence is pale or disappears at lower pH [20–23]. Aniline blue staining is also employed in methods for quantifying callose deposition based on large-scale automated image processing [5, 24, 25].

We have adapted the aniline blue staining protocol that was originally used to detect callose in pollen tubes [26] to devise an easy-to-perform procedure for fluorescent imaging of callose in whole-mount ovules using aniline blue-stained fixed flowers. Figure 2 schematically presents the subsequent steps of this procedure. The great advantage of this method is the ability to quickly and precisely detect callose in ovules without having to perform a series of microtome sections. The aniline dye penetrates the macerated ovular tissue very quickly and the preparation is ready for observation after a short period of time. This procedure has been successfully applied to analyze the pattern of callose deposition in the ovules of sexual and apomictic taxa of the *Asteraceae* family [27–31].

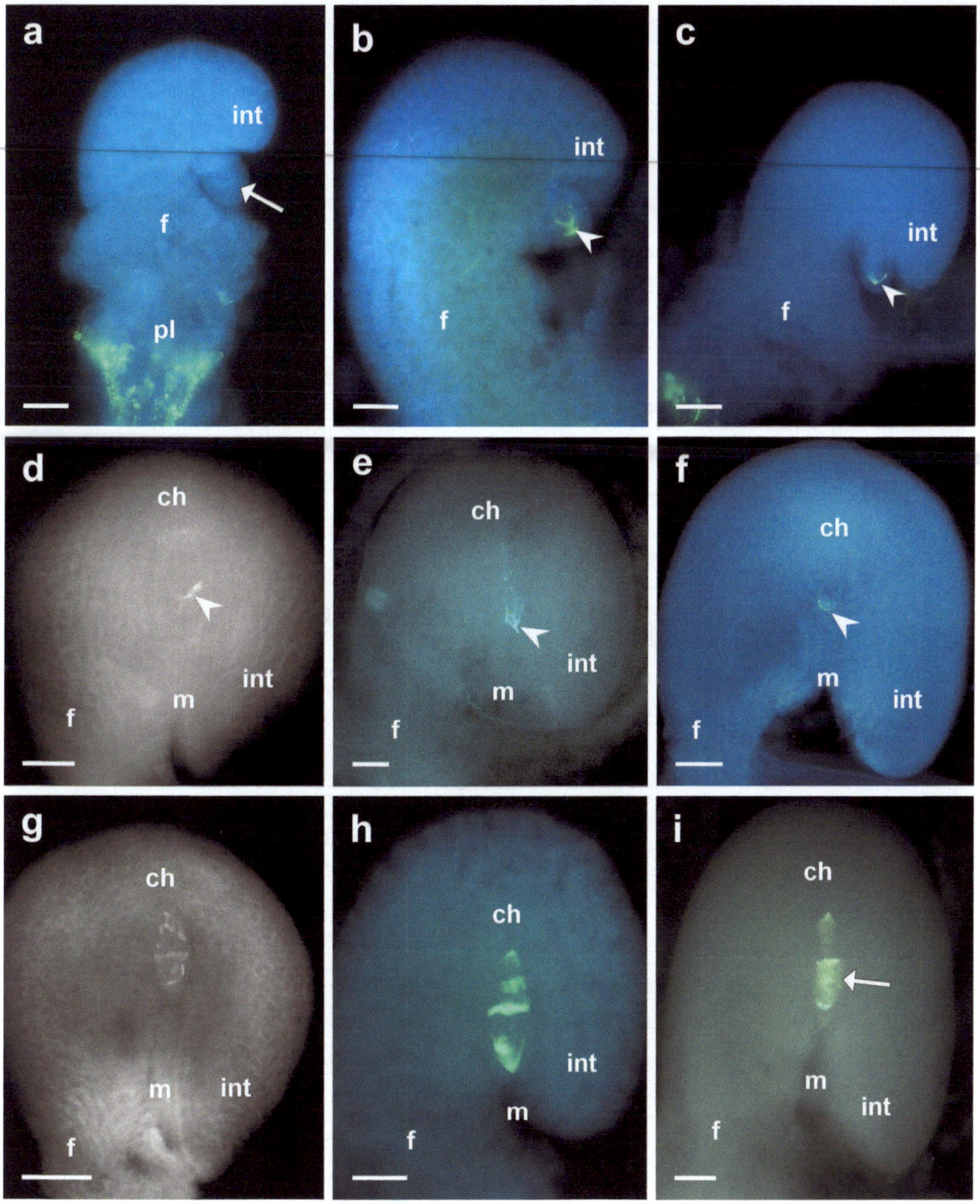

Fig. 1 Aniline blue staining assay to observe callose distribution in young anatropous, tenuinucellate, and unitegmic ovules of diplosporous *Rudbeckia juncea* (**a–c, e, h, i**), *Taraxacum atricapillum* (**d, g**), *Hieracium levicaule* (**f**). Arrowheads point to a fluorescent signal that reveals the presence of callose in the cell wall of the megaspore mother cell. (**a**) Ovule at the stage of archesporial cell (*arrow*). There is no callose in the cell wall of the archesporial cell or in the somatic cell walls of the ovule; a clear fluorescent signal visible in the sieve elements of the vascular bundles in the placental tissue. (**b–d**) Diplosporous megaspore mother cell entering the prophase of the first restitution division; callose present at the micropylar pole of the megasporocyte. **(e)** Diplosporous megaspore mother cell with a restitution nucleus; callose accumulation visible at the micropylar pole of the cell. **(f)** Diplosporous megaspore mother cell at an early stage of mitotic diplospory with callose present at micropylar pole. **(g, h)** Young diplodyad; callose present in the cell walls of megaspores. **(i)** Diplodyad at the stage of functional megaspore; callose accumulation in the cell wall of the micropylar megaspore (*arrow*) that will degenerate. Abbreviations: *ch* chalaza, *f* funicle, *int* integument, *m* micropylar canal, *pl* placenta Scale bars **a–e, h, i** = 25 μm; **f, g** = 50 μm

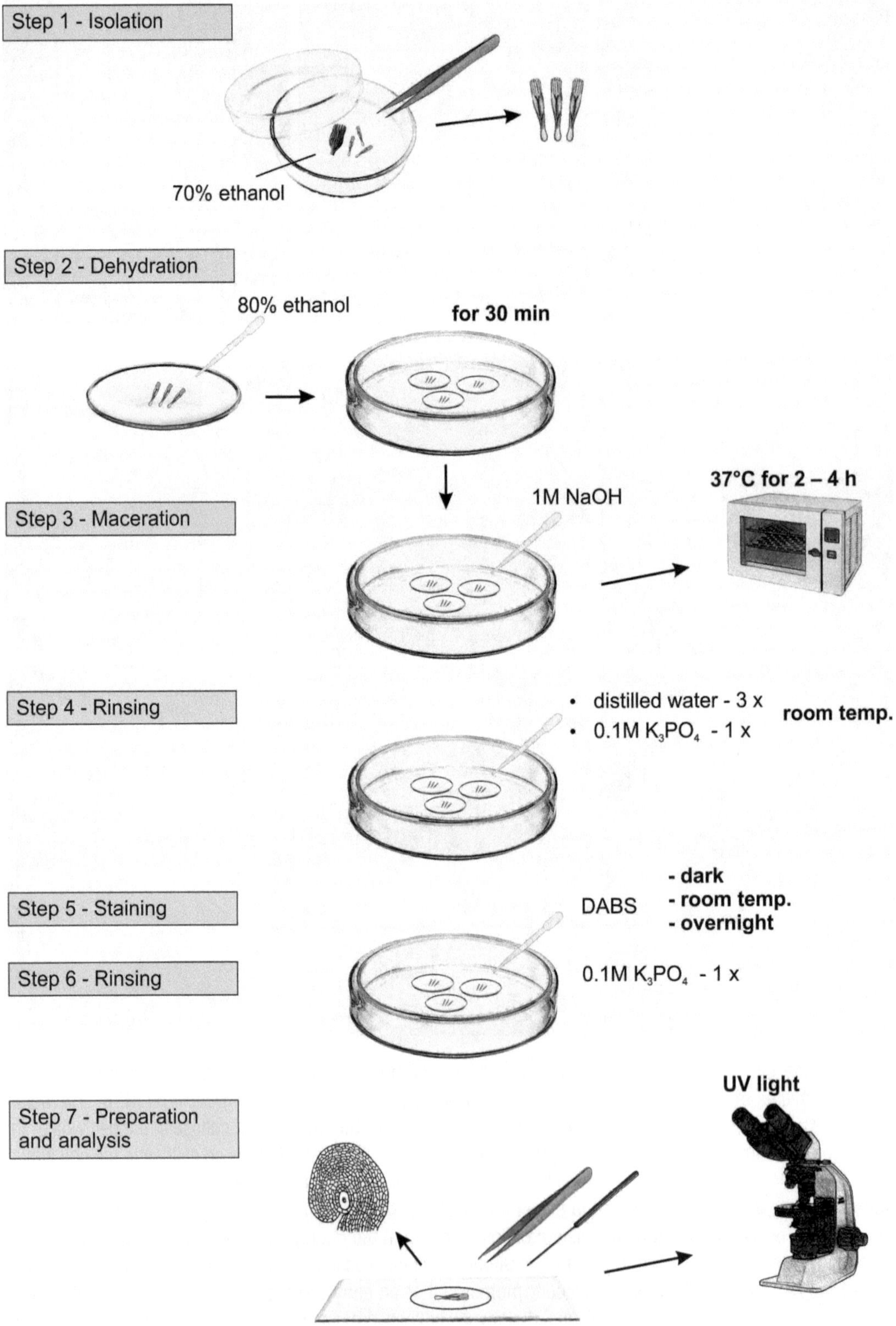

Fig. 2 Procedure for detection of callose in whole-mount ovules

2 Materials

2.1 Plant Material

Inflorescences (capitula) in early stages of development (*see* **Note 1**).

2.2 Solutions

1. Fixation solution: glacial acetic acid and absolute ethanol at a ratio of 1:3 (v/v) (*see* **Note 2**).
2. Ethanol 70%, 80%.
3. Alkaline treatment solution: 1 M NaOH.
4. Decolorized aniline blue solution (DABS): 0.1% (w/v) in 108 mM K_3PO_4, pH ~11 (*see* **Note 3**).
5. Distilled water.
6. 0.1 M K_3PO_4 solution.
7. Mounting solution: 0.1 M K_3PO_4 with glycerol at a ratio of 1:1 (v/v).

2.3 Laboratory Equipment and Supplies

1. Shaker.
2. pH meter.
3. Incubator.
4. Stereo microscope with a bright light source.
5. Microscope equipped with an Epi-Fl Filter Block N UV-2A and a digital camera (here: Nikon Eclipse E400, DS-Fi2 camera, and NIS-Elements software).
6. Sample vials with fitted caps.
7. Fine tweezers, Style #5 or #4.
8. Dissecting needles or fine injection needles.
9. Fine-tip plastic transfer pipettes.
10. Embryo dishes (30 mm dia x 12 mm deep) with lids or watch glasses of 50 mm diameter.
11. Petri dishes 90 mm, 200 mm diameter.
12. Filter papers.
13. Aluminum foil.
14. Microscope slides.
15. Coverslips 24 x 24 mm.

3 Methods

1. Collect whole, young capitula and place them immediately in vials with freshly prepared fixative solution.
2. Fix capitula for at least 24 h at room temperature (*see* **Note 4**), transfer them to 70% ethanol, and store at 4 °C (*see* **Note 5**).

3. Place the selected inflorescence in a small amount of 70% ethanol on a small Petri dish and isolate individual florets with fine tweezers and dissecting needle using a stereomicroscope (*see* **Note 6**).
4. Dispense ~2 mL of 80% ethanol into an embryo dish or watch glass.
5. Transfer the isolated florets with fine tweezers to 80% ethanol on an embryo dish or watch glass for 30 min (*see* **Note** 7).
6. Remove ethanol with a fine-tipped pipette and add ~2 mL of 1 M NaOH, following which incubate the flowers in this solution at 37 °C for 2–4 h (*see* **Note 8**).
7. Remove the NaOH solution with a pipette and rinse the florets with distilled water three times and once with 0.1 M K_3PO_4 solution for 1 min each (*see* **Note 9**).
8. Stain the florets in ~2 mL of decolorized aniline blue solution overnight in the dark for 24 h at room temperature (*see* **Note 10**).
9. Rinse the florets with 0.1 M K_3PO_4 solution.
10. Place a few stained flowers in a drop of mounting solution on a microscope slide and, using a stereomicroscope, isolate the ovules from ovaries with dissecting needles.
11. Gently squash the isolated ovules under a coverslip (*see* **Note 11**).
12. Examine microscope slides under UV light using a microscope equipped with an Epi-F1 Filter Block N UV-2A that consists of excitation filter EX330–380, dichroic mirror DM400, and barrier filter BA420.
13. Observe the bright yellow-green fluorescent signal in the ovule tissues that indicates the location of callose deposits in the cell wall (*see* Fig. 1).
14. Acquire images using a digital camera attached to a microscope.

4 Notes

1. In the case of larger inflorescences, the florets that occupy the outer margin of the capitulum and those located in the center of the inflorescence may differ in their stage of development. In order to determine the stage of ovule development relatively quickly, it is very helpful to use the technique of clearing whole florets with methyl salicylate and analyzing them under a microscope fitted with Nomarski interference contrast optics [27, 32]. Numerous protocols are also available that

demonstrate the successful application of other clearing agents as well as procedures for preparing cleared whole-plant organs for analysis using other microscopy techniques [33–37].

2. The reagents should be mixed directly before use.

3. Store the prepared aniline blue solution at 4 °C overnight and then filter the solution through powdered activated carbon (teaspoon) placed on a filter paper funnel. As a result, the dark blue color of the solution turns slightly yellowish. This step is vital since freshly prepared, blue-colored aniline blue solution will not stain callose [25]. In addition, the pH value of the dye solution is also of key importance. Although the available literature provides various pH values for the aniline blue solution used, generally, the higher the pH (up to 12.0), the more intense the fluorescence signal [20, 23]. Decolorized aniline blue solution should be stored at 4 °C in a dark glass bottle.

4. The volume of the fixative should be at least three to four times greater than the volume of the plant sample being fixed. During fixation, the vial with plant material should be shaken gently or the fixation should be performed in a vacuum desiccator.

5. Fixed capitula can also be stored in a fixative solution at 4 °C, but this may result in excessive hardening of tissues.

6. When isolating individual flowers from inflorescence, 70% ethanol should be successively added to a Petri dish to prevent plant material from drying out.

7. To prevent evaporation of ethanol or other solutions, always cover an embryo dish with a lid. Also, make sure that flowers are completely immersed in any solution used; if they float on the surface of a solution, gently dip them in with tweezers. Instead of embryo dishes with lids, watch glasses can also be used as shown in Fig. 2. In the latter case, watch glasses with samples should be placed on water-moistened filter paper in a large Petri dish to limit evaporation of the solution and prevent plant material from drying out.

8. The optimal time for the incubation of florets in 1 M NaOH should be determined experimentally for each taxon. Incubating flowers in NaOH for too long causes excessive softening of their tissues, which then makes it difficult or even impossible to isolate intact ovules. For dandelion and *Chondrilla juncea* florets, the optimal incubation time in 1 M NaOH is about 4 h at 37 °C. *C. brevirostis* and *Hieracium levicaule* florets, in turn, require a shorter incubation time of 2 and 3 h, respectively.

9. The subsequent steps should be performed with particular care so as not to mechanically damage the ovaries in softened flowers, which are very fragile.
10. To keep samples in the dark, wrap an embryo dish in aluminum foil or place it in a lightproof container. If the procedure is performed using watch glasses, wrap a large Petri dish in aluminum foil and then place watch glasses with samples in it.
11. Before applying a coverslip, it is advisable to remove any remnants of ovary tissue and other flower parts from the mounting solution with fine tweezers. This reduces the formation of undesirable air bubbles in the mounting solution.

Conflicts of Interest

The authors have no conflicts of interest to declare that are relevant to the content of this chapter.

References

1. Chen XY, Kim JY (2009) Callose synthesis in higher plants. Plant Signal Behav 4:489–492. https://doi.org/10.4161/psb.4.6.8359
2. Ünal M, Vardar F, Aytürk Ö (2013) Callose in plant sexual reproduction. In: Silva-Opps M (ed) Current progress in biological research, Rijeka. InTech:319–343. https://doi.org/10.5772/53001
3. Piršelová B, Matušíková I (2013) Callose: the plant cell wall polysaccharide with multiple biological functions. Acta Physiol Plant 35:635–644. https://doi.org/10.1007/s11738-012-1103-y
4. Ellinger E, Voigt CA (2014) Callose biosynthesis in arabidopsis with a focus on pathogen response: what we have learned within the last decade. Ann Bot 114:1349–1358. https://doi.org/10.1093/aob/mcu120
5. O'Lexy R, Kasai K, Clark N, Fujiwara T, Sozzani R, Gallagher KL (2018) Exposure to heavy metal stress triggers changes in plasmodesmatal permeability via deposition and breakdown of callose. J Exp Bot 69(15):3715–3728. https://doi.org/10.1093/jxb/ery171
6. Wang B, Andargie M, Fang R (2022) The function and biosynthesis of callose in high plants. Heliyon 8:e09248. https://doi.org/10.1016/j.heliyon.2022.e09248
7. Ušák D, Haluška S, Pleskot R (2023) Callose synthesis at the center point of plant development–an evolutionary insight. Plant Physiol 193:54–69. https://doi.org/10.1093/plphys/kiad274
8. Schneider R, Hanak T, Persson S, Voight CA (2016) Cellulose and callose synthesis and organization in focus, what's new? Curr Opin Plant Biol 34:9–16. https://doi.org/10.1016/j.pbi.2016.07.007
9. Abou-Saleh RH, Hernandez-Gomez MC, Amsbury S, Paniagua C, Bourdon M, Miyashima S, Helariutta Y, Fuller M, Budtova T, Connell SD, Ries ME, Benitez-Alfonso Y (2018) Interactions between callose and cellulose revealed through the analysis of biopolymer mixtures. Nat Commun 9:4538. https://doi.org/10.1038/s41467-018-06820-y
10. Hsieh YSY, Kao M-R, Tucker MR (2024) The knowns and unknowns of callose biosynthesis in terrestrial plants. Carbohydr Res 538:109103. https://doi.org/10.1016/j.carres.2024.109103
11. Rodkiewicz B (1970) Callose in cell wall during megasporogenesis in angiosperms. Planta 93:39–47. https://doi.org/10.1007/BF003
12. Tucker MR, Koltunow AMG (2014) Traffic monitors at the cell periphery: the role of cell walls during female reproductive cell differentiation in plants. Curr Opin Plant Biol 17:137–145. https://doi.org/10.1016/j.pbi.2013.11.015
13. Tucker MR, Paech NA, Willemse MTM, Koltunow AM (2001) Dynamics of callose

deposition and ß-1, 3-glucanase expression during reproductive events in sexual and apomictic *Hieracium*. Planta 212:487–498. https://doi.org/10.1007/s00425000044

14. Rojek J, Kapusta M, Kozieradzka-Kiszkurno M, Majcher D, Górniak M, Sliwinska E, Sharbel TF, Bohdanowicz J (2018) Establishing the cell biology of apomictic reproduction in diploid *Boechera stricta* (Brassicaceae). Ann Bot 122(4):513–539. https://doi.org/10.1093/aob/mcy114
15. Cornaro L, Banfi C, Cucinotta M, Colombo L, van Dijk PJ (2023) Asexual reproduction through seeds: the complex case of diplosporous apomixis. J Exp Bot 74(8):2462–2478. https://doi.org/10.1093/jxb/erad054
16. Peel MD, Carman JG, Leblanc O (1997) Megasporocyte callose in apomictic buffelgrass, Kentucky bluegrass, *Pennisetum squamulatum* Fresen, *Tripsacum* L., and weeping lovegrass. Crop Sci 37:724–732. https://doi.org/10.2135/cropsci1997.0011183X003700030006x
17. Carman JG (1997) Asynchronous expression of duplicate genes in angiosperms may cause apomixis, bispory, tetraspory, and polyembryony. Biol J Linn Soc Lond 61:51–94. https://doi.org/10.1111/j.1095-8312.1997.tb01778.x
18. Eschrich W, Currier HB (1964) Identification of callose by its diachrome and fluorochrome reactions. Stain Technol 39:303–307. https://doi.org/10.3109/10520296409061248
19. Soukup A (2014) Selected simple methods of plant cell wall histochemistry and staining for light microscopy. In: Žárský V, Cvrčková F (eds) Plant cell morphogenesis: methods and protocols, methods in molecular biology, vol 1080. Springer Science+Business Media New York, pp 25–39. https://doi.org/10.1007/978-1-62703-643-6_2
20. Smith MM, McCully ME (1978) A critical evaluation of the specificity of aniline blue induced fluorescence. Protoplasma 95:229–254. https://doi.org/10.1007/BF01294453
21. Evans NA, Hoyne PA (1982) A fluorochrome from aniline blue: structure, synthesis and fluorescence properties. Aust J Chem 35:2571–2575. https://doi.org/10.1071/CH9822571
22. Stone BA, Evans NA, Bonig I, Clark AE (1984) The application of Sirofluor, a chemically defined fluorochrome from aniline blue for the histochemical detection of callose. Protoplasma 122:191–195. https://doi.org/10.1007/BF01281696
23. Śnieżko R (2000) Fluorescence microscopy of aniline blue stained pistils. In: Dashek WV (ed) Methods in plant electron microscopy and cytochemistry. Humana Press, Totowa. https://doi.org/10.1007/978-1-59259-232-6_5
24. Zhou J, Spallek T, Faulkner C, Robatzek S (2012) Callose measurer: a novel software solution to measure callose deposition and recognise spreading callose patterns. Plant Methods 8:49. https://doi.org/10.1186/1746-4811-8-49
25. Zavaliev R, Epel BL (2015) Imaging callose at plasmodesmata using aniline blue: quantitative confocal microscopy. In: Heinlein M (ed) Plasmodesmata. Methods in molecular biology 1217. Humana Press, New York, NY, pp 105–119. https://doi.org/10.1007/978-1-4939-1523-1_7
26. Martin FW (1959) Staining and observing pollen tubes in the style by means of fluorescence. Stain Technol 34:125–128. https://doi.org/10.3109/10520295909114663
27. Musiał K, Kościńska-Pająk M, Antolec R, Joachimiak AJ (2015) Deposition of callose in young ovules of two *Taraxacum* species varying in the mode of reproduction. Protoplasma 252:135–144. https://doi.org/10.1007/s00709-014-0654-8
28. Musiał K, Kościńska-Pająk M (2017) Pattern of callose deposition during the course of meiotic diplospory in *Chondrilla juncea* (Asteraceae, Cichorioideae). Protoplasma 254:1499–1505. https://doi.org/10.1007/s00709-016-1039-y
29. Musiał K, Kościńska-Pająk M (2019) Callose is integral to meiotic diplospory of the Taraxacum type: new evidence from ovules of *Chondrilla brevirostris* (Asteraceae-Cichorioideae). Bot Lett 166:274–282. https://doi.org/10.1080/23818107.2019.1573378
30. Janas AB, Szeląg Z, Musiał K (2021) In search of female sterility causes in the tetraploid and pentaploid cytotype of *Pilosella brzovecensis* (Asteraceae). J Plant Res 134:803–810. https://doi.org/10.1007/s10265-021-01290-8
31. Janas AB, Marciniuk J, Szeląg Z, Musiał K (2022) New facts about callose events in the young ovules of some sexual and apomictic species of the Asteraceae family. Protoplasma 259:1553–1565. https://doi.org/10.1007/s00709-022-01755-0
32. Janas A, Musiał K, Kościńska-Pająk M, Marciniuk P (2016) Insights into developmental processes in anthers, ovaries, and ovules of *Taraxacum belorussicum* (Asteraceae-Cichorioideae) using DIC optics. Plant Syst

Evol 302:617–628. https://doi.org/10.1007/s00606-016-1288-4

33. Herr JM (1993) Clearing techniques for the study of vascular plant tissues in whole structures and thick sections. In: Goldman CA, Hauta PL, O'Donnell MA, Andrews SE, Van der Heiden R (eds) Tested studies for laboratory teaching, vol. 5, pp 63–84. Proceedings of the 5th workshop/conference of the Association for Biology Laboratory Education (ABLE)
34. Villani TS, Koroch AR, Simon JE (2013) An improved clearing and mounting solution to replace chloral hydrate in microscopic applications. Appl Plant Sci 1(5):1300016. https://doi.org/10.3732/apps.1300016
35. Timmers ACJ (2016) Light microscopy of whole plant organs. J Microsc 263(2): 165–170. https://doi.org/10.1111/jmi.12394
36. Hedhly A, Vogler H, Eichenberger C, Grossniklaus U (2018) Whole-mount clearing and staining of *Arabidopsis* flower organs and siliques. J Vis Exp 134:e56441. https://doi.org/10.3791/56441
37. Tofanelli R, Vijayan A, Scholz S, Schneitz K (2019) Protocol for rapid clearing and staining of fixed *Arabidopsis* ovules for improved imaging by confocal laser scanning microscopy. Plant Methods 15:120. https://doi.org/10.1186/s13007-019-0505-x

Chapter 6

A Method for Detecting Parthenogenesis in *Asteraceae* Species

Tatyana Radoeva and Peter J. van Dijk

Abstract

Embryogenesis in plants is a process that not only occurs after fertilization but can also be initiated in its absence. This form of embryogenesis is known as parthenogenesis, and it is an essential part of a process called apomixis. Apomixis has long been seen as the holy grail of plant breeding because it produces clonal seeds. Therefore, engineering (single components of) apomixis remains a highly desired research goal. Over the last few years, significant progress has been made toward understanding and engineering both complete synthetic apomixis system and individual components of apomixis, such as parthenogenesis, in crops. Characterization of candidate parthenogenesis genes involves thorough phenotypic analysis and microscopic observation of developing embryos, which are, however, not readily accessible because they are embedded in several integumental layers comprising the protective seed coat. Here, we provide a detailed approach for detecting parthenogenesis and we further demonstrate its application to a sexual crop species, i.e., lettuce. We describe an emasculation technique by decapitation, embryo sac isolation, imaging of cleared embryos, and ploidy analysis of embryo sacs utilizing flow cytometry.

Key words Apomixis, Parthenogenesis, Lettuce embryos, DIC, Phenotypic analysis, Flow cytometry, *Asteraceae* species

1 Introduction

A hallmark of sexual reproduction in flowering plants is the process of double fertilization, during which two male gametes (sperm cells) fuse with two female gametes (egg cell and central cell). This results in the formation of the precursor cells of the two major seed components, i.e., the embryo and endosperm [1]. After fertilization, a precise pattern of several cell divisions of the fertilized egg cell (zygote) leads to the end point of embryogenesis, namely the mature embryo. The latter is a miniature version of the plant seedling with the precursors of all the major tissue types of the plant body [2].

Initiation of embryogenesis in plants does not strictly depend on fertilization and can begin in the absence of it, thus asexually.

Tatyana Radoeva and Peter J. van Dijk (eds.), *Apomixis: Methods and Protocols*, Methods in Molecular Biology, vol. 2987, https://doi.org/10.1007/978-1-0716-5001-1_6,

Asexual embryogenesis can be initiated in vitro from a variety of different somatic cells [3, reviewed in 4] as well as gametic cells [5–7]. Notably, in some plant species, this is a genetically derived state and occurs naturally in vivo [5, 8]. An example of the latter is parthenogenesis, embryo formation in the absence of fertilization, which is part of a process called apomixis. Apomixis is a complex trait, consisting of three essential components: (i) apomeiosis, which involves the formation of unreduced, unrecombined egg cell via avoidance of meiosis, (ii) parthenogenesis, the initiation of embryogenesis in the absence of fertilization from such unreduced egg cell, and (iii) endosperm development, initiated either autonomously or via fertilization (pseudogamy). Thus, altogether, apomixis depicts the formation of asexual seeds genetically identical to the mother plant [9]. We further refer to other chapters in this book where apomixis is discussed in more detail. In this chapter, we focus on parthenogenesis and provide a detailed approach for detecting this essential component of apomixis in sexual and apomictic *Asteraceae* species.

Over the last few decades, continuous studies have been conducted using natural apomictic species aiming to understand the genetic and molecular basis of apomixis and to then apply this knowledge toward engineering of apomixis in sexual species. This has led to the identification and cloning of two different single causal genes for parthenogenesis, i.e., the *BABYBOOM-like* (*PsBBML*) gene from *Pennisetum squamulatum* [10] and the *PARTHENOGENESIS* (*ToPAR*) gene from *Taraxacum officinale* and *Pilosella piloselloides* [11]. Following their identification, both genes have been successfully applied to transfer parthenogenesis to sexual (crop) species [10–15].

PsBBML gene is a homolog of the well-known somatic embryogenesis inducer *BBM* [16], and its discovery and function are further discussed in *see* Chapter 16. The recently identified *ToPAR* gene [11] encodes a K2–2 zinc finger and ethylene responsive element binding factor-associated amphiphilic repression (EAR) domain protein. An insertion of a miniature inverted-repeat transposable element (MITE) in the promoter of the apomictic PAR gene allele is the causal polymorphism for parthenogenesis in Taraxacum. This insertion accounts for the egg cell-specific PAR expression, which triggers embryogenesis without the need of fertilization. In contrast, the sexual PAR alleles lack such a transposon element (TE) insertion and are highly expressed in pollen. The latter must deliver the gene transcripts at fertilization to initiate embryogenesis [11]. Remarkably, heterologous expression of the Taraxacum *PAR* gene in lettuce (*Lactuca sativa*) egg cells induces the formation of haploid embryo-like structures without fertilization [11]. Similarly, the ectopic expression of the *ToPAR* gene in foxtail millet [13] and rice [12, 14] led to the successful generation of haploid plants, a product of induced parthenogenesis events.

A critical step in the characterization of selected genes during embryogenesis is the phenotypic analysis of their mutant or mis-expression phenotypes. Microscopic observation of developing embryos is, however, not an easy task because they are embedded within the seed, which is enclosed by a fruit. There are many useful techniques allowing visualization of plant embryos, yet most require sectioning and/or extensive fixation and the use of dangerous chemicals to bypass the layers of tissue protecting the embryo [17, 18]. A better alternative to the latter is the use of tissue clearings in combination with differential interference contrast (DIC; also known as Nomarski) microscopy [19–21]. This generally requires less effort, as it allows for whole-mount preparation of isolated (immature) embryo sacs or seeds (depending on the species) without the need for further manipulation or staining. However, the only prerequisite to consider is the choice of clearing method, as each has a specific protocol for clearing the tissue, and some are more elaborate than others [17–19]. Methyl salicylate-based clearing is an example of a widely used clearing method, which requires fixation first, followed by dehydration, and finally, clearing of the tissue. It is often used for visualization of early embryo sac development [17, 19, 20]. However, besides the elaborate procedure, it also has a strong smell. It is thus advisable to work in a well-ventilated area or use a fume hood not only during the clearing process but also during microscopic observations to minimize exposure to fumes.

Here, we provide a complete protocol for a straightforward detection of parthenogenesis events in *Asteraceae* species (e.g., dandelion and lettuce), which can be easily adapted for use in other species. Emasculation by capitula decapitation, embryo sac isolation followed by visualization using a tissue-clearing technique, and embryo ploidy determination using flow cytometry will be discussed.

2 Materials

2.1 Emasculation by Capitula Decapitation

1. Apomictic plants (e.g., dandelion, *Taraxacum officinale* A68) or transgenic sexual plants (e.g., lettuce, *Lactuca sativa*) carrying a genetic construct for induction of parthenogenesis (e.g., *pAtEC1.1::ToPAR* [11]).
2. Castro-Viejo scissors (e.g., 102 mm, Straight tips, Electron Microscopy Sciences, Cat. # 72933).
3. Sewing thread (*see* **Note 1**).
4. Fine tweezers.

2.2 Embryo Sac Isolation and Clearance

1. Seed heads (capitula) 3 days after decapitation.
2. Stereomicroscope with a bright light source.
3. Microscope equipped with DIC (Nomarski) optics and camera (*see* **Note 2**).
4. Microscope slides.
5. Microscope coverslips (20 x 20 mm).
6. Double-sided adhesive tape (*see* **Note 3**).
7. Fine tweezers (e.g., Rubis Switzerland 5-SA).
8. Dissecting needles (e.g., insulin needle 0.30x8mm).
9. Syringe.
10. Pipette (100 μl).
11. Pipette tips.
12. Clearing/mounting solution: Chloral hydrate: water: glycerol (w/v/v) 8:3:1 (*see* **Note 4**).

2.3 Flow Cytometry Analysis of Isolated Embryo Sacs

1. All materials listed in subheading 2.2 (except point 12).
2. 1.5-ml microcentrifuge tubes.
3. Wide-opening pipette tips.
4. 50-μm nonsterile CellTrics® filters.
5. Small-volume sample tube for flow cytometry (e.g., Sysmex).
6. Galbraith's buffer (*see* **Note 5**): 45 mM $MgCl_2$, 20 mM MOPS, 30 mM Na-citrate, 0.1% Triton X-100, 40 μg/ml RNAse A, 15 mM TCEP. Set pH to 7.0 with KOH, and filter sterilize. Keep in fridge.
7. 2 mg/ml propidium iodide (PI) staining stock solution in deionized water (*see* **Note 5**). Store in aliquots at −20 °C.
8. Flow cytometer (e.g., BD Accuri C6 or later version).

3 Methods

3.1 Emasculation by Capitula Decapitation

1. Select young flower buds (capitula) that are ideally 1 day before anthesis (Fig. 1a). Avoid using the first to the third flower buds, as they usually contain a few aborted or abnormal seeds, which may alter the interpretation on the number of parthenogenetic embryos.
2. Clip approximately one-third of the capitulum (Fig. 1b) with fine, sharp scissors (*see* **Notes 6** and 7).
3. Mark the decapitated flower buds using a sewing thread knotted around the peduncle (*see* **Note 8**; Fig. 1c).

Fig. 1 Procedure for detecting parthenogenetic events in *Asteraceae* species. (**a, b, c**) Emasculation by decapitation, (**d**–**h**) embryo sac isolation, and (**i, j**) embryo imaging after clearing in lettuce (*Lactuca sativa*). (**a**) Young capitulum approximately 1 day before anthesis. Dashed line shows where the capitulum should be cut. (**b**) Decapitated capitulum. The arrow points at latex coming out of the cut side. (**c**) Decapitated capitula marked with sewing thread knotted around the peduncle. (**d**) Capitulum 3 days after decapitation (DAC). (**e**) Opened capitulum with exposed developing seeds. (**f**) Developing seed placed on a double-sided tape. The gray dashed line marks the incision point. (**g**) Developing seed with a cut on the side indicated by gray dashed lines. (**h**) Dissected seed with exposed embryo sac, which is marked by gray dashed circle. Nomarksi images of (**i**) wild-type and (**j**) parthenogenetic lettuce embryo(-like structure). Asterisks in (**i**) point at endosperm cells. Note the lack of endosperm in (**j**)

3.2 Embryo Sac Isolation and Tissue Clearing

The combination of tissue clearing and DIC microscopy allows for direct observation of plant embryos. The DIC utilizes polarized light that interacts strongly with the (usually transparent) sample, creating high contrast against the background [22]. To make the tissues transparent and thus suitable for DIC microscopy, we use a clearing/mounting solution containing glycerol and chloral hydrate [21]. This clearing solution creates an excellent refraction index, which causes light to bend at the surface of the object, making it visible. Following the tissue-clearing technique described below, it is possible to use whole-mount preparation of isolated embryo sacs without further manipulation (such as fixation, dehydration) or staining. Thus, after isolation, the embryo sacs can be directly deposited in a drop of the clearing/mounting solution and immediately thereafter observed by DIC microscopy.

1. Collect whole, decapitated capitula from transgenic lettuce carrying the *pAtEC1.1::ToPAR* construct 3 to 5 days after decapitation (DAC; Fig. 1d).
2. Attach a piece of double-sided adhesive tape to a microscope glass slide and place the slide under stereomicroscope. Turn off the bottom-lightning of the stereomicroscope, as this will heat up the sample quickly.
3. Place one of the collected capitula on the slide. Proceed with opening the capitulum, using fine tweezers (Fig. 1e). Take out

all developing seeds and place them on the slide. Alternatively, take and work with one seed at a time.

4. Before you start dissecting the seeds and isolating the embryo sacs, dispense 30–50 μL of clearing solution onto a separate microscope slide (*see* **Notes 9** and **10**).
5. Open the seed using a fine needle to make a longitudinal cut (Fig. 1f). To avoid damaging the fragile embryo sac, make the cut more toward the side rather than toward the middle of the seed (Fig, 1f, g). Expose the embryo sac by gently separating the cut sides of the seed using fine tweezers on one side and the needle on the other (Fig. 1h).
6. Carefully collect the embryo sac with the shaft of the needle (the opening should be oriented away from the embryo sac) and deposit it in the clearing solution. Collect all embryo sacs from one capitulum in one drop. Carefully put a coverslip on top and avoid squashing the sample. Add more clearing solution if necessary or remove any excess solution with a piece of filter paper.
7. Leave the finished slides for about 10 min at room temperature before observation. Alternatively, slides can be kept at 4 °C for several weeks.
8. Observe the cleared samples using a DIC microscope. Locate the embryo sacs under a low magnification (10x) and change to a higher magnification (20x) when taking images (Fig. 1i, j).

3.3 Flow Cytometry Analysis of Isolated Embryo Sacs

1. Collect whole, decapitated capitula from transgenic lettuce carrying the *pAtEC1.1::ToPAR* construct 5 days after decapitation (DAC) and nontransformed (wild-type; sexual) capitula 5 days after self-pollination to serve as a sexual control (*see* **Note 11**).
2. Add 150 μL of cold Galbraith's buffer with TCEP and RNAse A to 1.5-ml microcentrifuge tubes. Place the tubes on ice.
3. Isolate approx. 15 embryo sacs of each genotype (parthenogenetic and sexual) by following the procedure described in subheading 3.2, steps 1–5 and collect them in the prepared buffer. Pool the embryo sacs of each genotype into one tube (*see* **Note 12**).
4. Use a pipette tip (fitting a 100–1000 μL pipette) to homogenize the samples and release nuclei into the buffer by tapping vigorously multiple times until the buffer becomes cloudy.
5. Take the respective number of flow cytometry sample tubes and place 30–50-μm nylon filters (e.g., CellTrics®) on top of them.
6. Use wide tips to collect the homogenate and filter it into the sample tube to remove tissue debris.

7. Add an additional 150–200 μL of cold Galbraith's buffer with TCEP and RNAse A to reach a total sample volume of 300 μL and 7.5 μL (2.5 μL per 100 μL filtrate) PI staining solution to each tube.
8. Briefly tap the tubes to homogenize. Incubate the stained samples on ice in darkness for 30 min prior to analysis (*see* **Note 13**).
9. Measure the samples on a flow cytometer (e.g., BD Accuri C6). Operate the flow cytometer at the Slow Flow Rate setting (*see* **Note 14**).
10. To obtain well-defined peaks (Fig. 2), perform the following steps:
    (a) Plot FL2-A (log) versus FL3-A (log) fluorescence emission (Fig. 2a).
    (b) Select the region containing the nuclei (R1, region-of-interest; *see* Fig. 2a, b).
    (c) Plot FL2-A (log) versus particle count (linear) and apply gate (Fig. 2c).

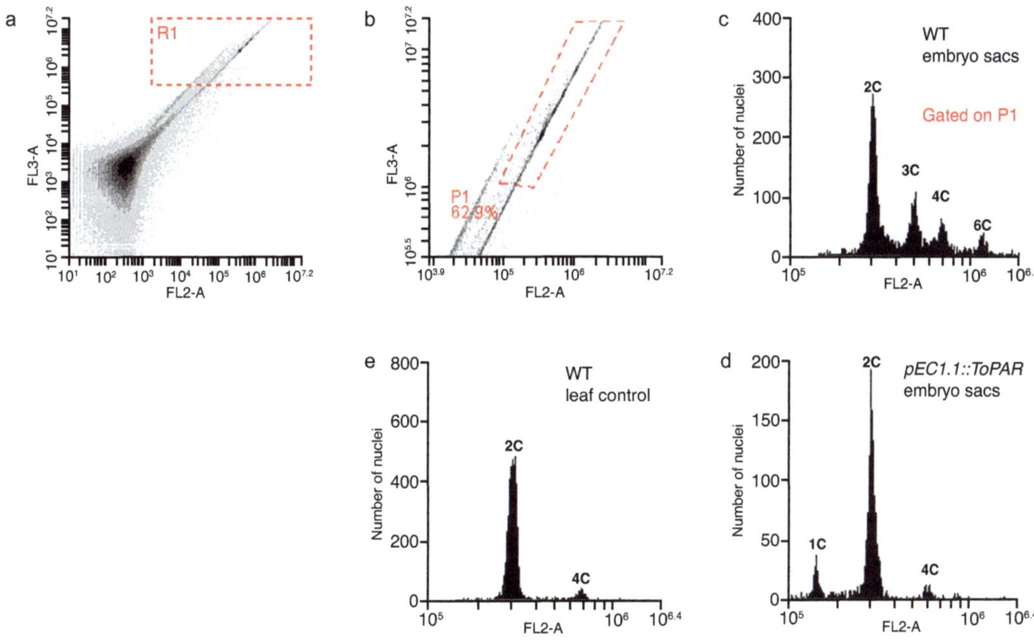

Fig. 2 Flow cytometry analysis of homogenates prepared from lettuce (*Lactuca sativa*) embryo sacs (**a**–**d**) and leaf tissue (**e**). (**a**) Biparametric contour plot of FL2-A versus FL3-A fluorescence emission. (**b**) Enlargement of the region-of-interest (R1) containing the nuclei. The latter are enclosed by polygonal region P1. (**c**) Histogram of F2-A fluorescence, gated on region P1 of panel (**b**). The histogram depicts the ploidy of embryo sacs from control nontransformed (wild-type) lettuce 5 d after self-pollination. (**d**) Histogram depicting the ploidy of embryo sacs from transgenic lettuce carrying the *pAtEC1.1::ToPAR* construct 5 d after decapitation. (**e**) Histogram depicting the ploidy of leaf tissue from control nontransformed (wild-type) lettuce Abbreviations: 1C, 2C, 3C, 4C, and 6C designate the appropriate C-values for the individual peaks

4 Notes

1. Choose a thread color that contrasts well with green, such as red, bright pink, dark purple, and dark blue, to ensure that it can be easily distinguished.
2. In this case, we use a Nikon Eclipse E600, but any alternative is possible.
3. It is important to use transparent double-sided tape because the developing seeds of lettuce and dandelion are very light green to almost white in color, while the embryo sacs are transparent. Using white tape, for instance, would make embryo sac isolation difficult.

   In contrast, white double-sided tape is recommended for species where embryo sac isolation is not needed and young developing seeds are isolated from the (silique or other type of) fruit, e.g., *Arabidopsis thaliana*.
4. We have tested chloral hydrate from various brands, and the one from ROTH (Art. No. K318.3) provides the best and most consistent results as a clearing agent in plants.
5. The described protocol is optimized for the Accuri C6 flow cytometer based on [23, 24]. Depending on the flow cytometer type, different buffers and fluorochromes might be required (such as Otto's buffer [25, 26] and UV-excited fluorochromes DAPI and Hoechst 33257).
6. Lettuce belongs to the latex-bearing species and, upon tissue damage, secretes a milky fluid called lactucarium (latex). Latex becomes sticky and quickly coagulates when exposed to air. It is considered to play an essential role as the first line of defense against herbivores and plant pathogen infection [27, 28].

   A small amount of latex will probably come out of the decapitated capitulum (Fig. 1b). It is advisable to remove it, as it will cause the decapitated capitulum to dry out or stop developing. Younger capitula tend to secrete more latex, which can completely clog the cut site. Therefore, the use of such capitula is not recommended, but they are also often way too young.
7. Depending on the flower morphology of the species, a different emasculation technique might be needed.
8. Marking the peduncle (the main stalk of an inflorescence) with a colorful thread will make it easier to find the decapitated capitula a few days later.
9. Two droplets of clearing solution and thus two samples can easily fit on one slide. Use separate coverslips (20 × 20 mm).

10. It is important to prepare the slide with the clearing solution before the actual isolation of embryo sacs. This will ensure that the isolated embryo sacs can be immediately placed in the solution and will not dry out, which can happen very quickly.
11. We advise taking a young leaf (~1 cm^2 or ~ 10 mg) of the same or other plant with known ploidy (e.g., diploid lettuce) to serve as a control sample tissue next to the sexual embryo sacs sample (Fig. 2e). To prepare this extra control for flow cytometry, place the excised leaf tissue in a plastic 60-mm petri dish standing on ice. Add 800 μl of cold Galbraith's buffer with TCEP and RNAse A and chop the tissue using a razor blade for 1–2 min to release the nuclei into the buffer. Thereafter, follow steps 5–10 in subheading 3.3.
12. If this method is used for species with smaller seeds, pool more embryo sacs or even whole seeds together.
13. PI is light sensitive; therefore, cover the samples either with aluminum foil or with the lid of the ice bucket.
14. Start with measuring the control leaf sample with known ploidy (e.g., a diploid lettuce leaf) to establish baseline fluorescence peaks. Run, thereafter, the embryo sac samples.

Conflicts of Interest

The authors have no conflicts of interest to declare that are relevant to the content of this chapter.

References

1. Bleckmann A, Alter S, Dresselhaus T (2014) The beginning of a seed: regulatory mechanisms of double fertilization. Front Plant Sci 5: 452. https://doi.org/10.3389/fpls.2014.00452
2. Ten Hove CA, Lu K-J, Weijers D (2015) Building a plant: cell fate specification in the early Arabidopsis embryo. Development 142:420–430. https://doi.org/10.1242/dev.111500
3. De Vries SC, Booij H, Meyerink P et al (1988) Acquisition of embryogenic potential in carrot cell-suspension cultures. Planta 176:196–204. https://doi.org/10.1007/BF00392445
4. Radoeva T, Weijers D (2014) A roadmap to embryo identity in plants. Trends Plant Sci 19:709–716. https://doi.org/10.1016/j.tplants.2014.06.009
5. Van Dijk P, De Jong H, Vijverberg K, Biere A (2009) An Apomixis-gene's view on dandelions. In: Schön I, Martens K, van Dijk P (eds) Lost sex. Springer, Dordrecht, pp 475–493
6. Custers JBM (2003) Microspore culture in rapeseed (Brassica napus L.). In: Maluszynski M, Kasha KJ, Forster BP, Szarejko I (eds) Doubled haploid production in crop plants. Springer, Dordrecht, pp 185–193
7. Siemons C, Jonkers S, Vlieg RC et al (2025) Establishment and maintenance of embryogenic cell fate during microspore embryogenesis. Plant J 121:e17243. https://doi.org/10.1111/tpj.17243
8. Garcês H, Sinha N (2009) The 'Mother of Thousands' (Kalanchoë daigremontiana): a plant model for asexual reproduction and CAM studies. Cold Spring Harb Protoc 2009:pdb.emo133. https://doi.org/10.1101/pdb.emo133
9. Van Dijk PJ, Op Den Camp R, Schauer SE (2020) Genetic dissection of Apomixis in dandelions identifies a dominant parthenogenesis

locus and highlights the complexity of autonomous endosperm formation. Genes 11:961. https://doi.org/10.3390/genes11090961

10. Conner JA, Mookkan M, Huo H et al (2015) A parthenogenesis gene of apomict origin elicits embryo formation from unfertilized eggs in a sexual plant. Proc Natl Acad Sci USA 112: 11205–11210. https://doi.org/10.1073/pnas.1505856112
11. Underwood CJ, Vijverberg K, Rigola D et al (2022) A PARTHENOGENESIS allele from apomictic dandelion can induce egg cell division without fertilization in lettuce. Nat Genet 54:84–93. https://doi.org/10.1038/s41588-021-00984-y
12. Song M, Wang W, Ji C et al (2024) Simultaneous production of high-frequency synthetic apomixis with high fertility and improved agronomic traits in hybrid rice. Mol Plant 17:4–7. https://doi.org/10.1016/j.molp.2023.11.007
13. Huang Y, Liang Y, Xie Y et al (2024) Efficient haploid induction via egg cell expression of dandelion PARTHENOGENESIS in foxtail millet (Setaria italica). Plant Biotechnol J 22: 1797–1799. https://doi.org/10.1111/pbi.14302
14. Chahal LS, Conner JA, Ozias-Akins P (2022) Phylogenetically distant BABY BOOM genes from Setaria italica induce parthenogenesis in Rice. Front Plant Sci 13:863908. https://doi.org/10.3389/fpls.2022.863908
15. Jie X, Yajie J, Shenlin Y et al (2025) In vivo haploid induction via parthenogenesis gene ToPAR in rice. Rice Sci S1672630824001094. https://doi.org/10.1016/j.rsci.2024.12.006
16. Boutilier K, Offringa R, Sharma VK et al (2002) Ectopic expression of BABY BOOM triggers a conversion from vegetative to embryonic growth. Plant Cell 14:1737–1749. https://doi.org/10.1105/tpc.001941
17. Lentini Z, Tabares E, Buitrago ME (2020) Vibratome sectioning and clearing for easing studies of cassava embryo formation. Front Plant Sci 11:1180. https://doi.org/10.3389/fpls.2020.01180
18. Feng Y, Wang T, Liu L (2022) An efficient clearing protocol for the study of seed development in tomato (Solanum lycopersicum L.). JoVE 64445. https://doi.org/10.3791/64445
19. Musiał K, Płachno BJ, Świątek P, Marciniuk J (2013) Anatomy of ovary and ovule in dandelions (Taraxacum, Asteraceae). Protoplasma 250:715–722. https://doi.org/10.1007/s00709-012-0455-x
20. Jongedijk E (1987) A rapid methyl salicylate clearing technique for routine phase-contrast observations on female meiosis in Solanum. J Microsc 146:157–162. https://doi.org/10.1111/j.1365-2818.1987.tb01336.x
21. Llavata-Peris C, Lokerse A, Möller B et al (2013) Imaging of phenotypes, gene expression, and protein localization during embryonic root formation in Arabidopsis. In: De Smet I (ed) Plant Organogenesis. Humana Press, Totowa, pp 137–148
22. Horn A (2022) Nomarski microscopy. In: The physics of laser radiation–matter interaction. Springer, Cham, pp 391–395
23. Galbraith DW, Harkins KR, Maddox JM et al (1983) Rapid flow cytometric analysis of the cell cycle in intact plant tissues. Science 220: 1049–1051. https://doi.org/10.1126/science.220.4601.1049
24. Galbraith DW (2009) Simultaneous flow cytometric quantification of plant nuclear DNA contents over the full range of described angiosperm 2C values. Cytom Pt A 75A:692–698. https://doi.org/10.1002/cyto.a.20760
25. Otto FJ (1992) Preparation and staining of cells for high-resolution DNA analysis. In: Radbruch A (ed) Flow cytometry and cell sorting. Springer, Berlin\Heidelberg, pp 65–68
26. Doležel J, Göhde W (1995) Sex determination in dioecious plants Melandrium album and M. Rubrum using high-resolution flow cytometry. Cytometry 19:103–106. https://doi.org/10.1002/cyto.990190203
27. Agrawal AA, Konno K (2009) Latex: a model for understanding mechanisms, ecology, and evolution of plant defense against herbivory. Annu Rev Ecol Evol Syst 40:311–331. https://doi.org/10.1146/annurev.ecolsys.110308.120307
28. Gu C, Zeng B, Wang M et al (2024) Study on active components and mechanism of lettuce latex against Spodoptera Litura. Chem Biodivers 21:e202400993. https://doi.org/10.1002/cbdv.202400993

Part III

Transcriptomics of Apomixis

Chapter 7

Laser-Assisted Microdissection for Cell- and Tissue Type-Specific Transcriptome and RNA-Seq Analysis

Anja Schmidt

Abstract

Monitoring gene activity at high spatial and temporal resolution is key to understanding the regulation of developmental programs. It allows researchers to uncover developmental trajectories in complex organs or tissues and to determine the expression of genes in rare cell or tissue types, such as the cells of the plant female reproductive lineages (germline), which are embedded within the sporophytic tissues of the flower. Investigations into plant germlines are of scientific interest in order to gain an understanding of transcriptional processes governing reproduction. Moreover, in general terms, plant germlines are well suited as models for investigating gene regulatory programs underlying organogenesis, as they are typically composed of only a small number of highly specialized cell types. In recent years, a number of transcriptome analyses have uncovered regulatory programs underlying major steps of germline development during both sexual and asexual reproduction through seeds (apomixis). Apomixis is of great interest, as it leads to a formation of clonal embryos by omission or alteration of meiosis by the cell specified as first cell of the female germline combined with parthenogenesis (absence of fertilization of the egg cell). To overcome the long-faced challenge of the inaccessibility of the cells for cell type-specific investigations, the establishment of laser-assisted microdissection (LAM) has been an important methodological advance. The method has been improved for isolation of high-quality RNA for RNA-Seq analyses for investigations on *Arabidopsis thaliana* and sexual and apomictic accessions from the closely related genus *Boechera*. In this chapter, the method from tissue fixation to RNA isolation is detailed.

Key words Apomixis, *Boechera*, Laser-assisted microdissection, RNA-seq, Transcriptome

1 Introduction

Plant seeds are major sources of energy for animal and human nutrition. In particular, asexual reproduction through seeds (apomixis) has long been recognized for its outstanding potential for harnessing in agriculture [1]. Apart from their importance for applications, due to their small size and composition of a few specialized cell types, plant germlines are attractive as models for developmental processes and organogenesis [2, 3]. Due to their simple organization and short lifespan, especially investigations

Tatyana Radoeva and Peter J. van Dijk (eds.), *Apomixis: Methods and Protocols*, Methods in Molecular Biology, vol. 2987,
https://doi.org/10.1007/978-1-0716-5001-1_7,

into the female germlines are of great interest. Elucidation of transcriptional programs can give important new insights into the gene regulatory programs underlying cell identity and fate, and also on the evolution of these programs in adaptation to environmental conditions.

In sexual plants, female germline development starts with the specification of a megaspore mother cell (MMC) in a specialized domain of the ovule, which is the female reproductive tissue of the flower. The MMC undergoes meiosis to give rise to a tetrad of megaspores, thereby mediating the transition from the sporophytic to the gametophytic generation. Typically, only one of the megaspores survives. This functional megaspore undergoes three rounds of mitosis in a syncytium. Subsequently, after cellularization, the mature female gametophyte is formed, which is also referred to as embryo sac. It is typically composed of four highly specialized cell types, which are the egg cell and central cell as the two female gametes, the synergid cells required for pollen tube attraction and reception upon fertilization, and the antipodal cells [4]. After double fertilization of the female gametes, the egg cell gives rise to the embryo and the central cell to its nourishing tissue, the endosperm. In comparison to sexual reproduction, a few important developmental steps are modified in apomicts (reviewed in [5]). In contrast to the MMC in sexual species, which is undergoing meiosis, an apomictic initial cell forms as the first cell of the apomictic germline that omits or alters meiosis. It gives rise to a female gametophytic lineage in which the genome remains unreduced and unrecombined. Consequently, after gametogenesis, the genome of the egg cell remains identical to that of the sporophytic cells of the mother plant. Importantly, the egg cell autonomously gives rise to an embryo in the absence of fertilization (parthenogenesis). To gain an understanding of the gene regulatory control underlying germline development in sexual plants and apomicts is of high interest from a scientific perspective and as a precondition for applications for crop seed improvement. Moreover, transcriptional investigations not only provide important insights into the gene regulatory programs underlying cell fate specification and cell identity but also offer an understanding of the evolution of gene regulatory programs in adaptation to environmental conditions.

Over recent years, much progress has been made in methodologies allowing to study gene expression at a high spatial and temporal resolution, with a special focus on single cell or cell type specific analyses [6]. In this chapter, laser-assisted microdissection (LAM) to isolate high-quality RNA for RNA-Seq from pools of similar cells of interest is described. Unlike recently emerging methods for single-cell transcriptional analysis of more complex tissues, LAM enables specific identification and selection of individual cells and to isolate them from thin sections of tissues [7, 8]. And even though minor contaminations with surrounding cells cannot

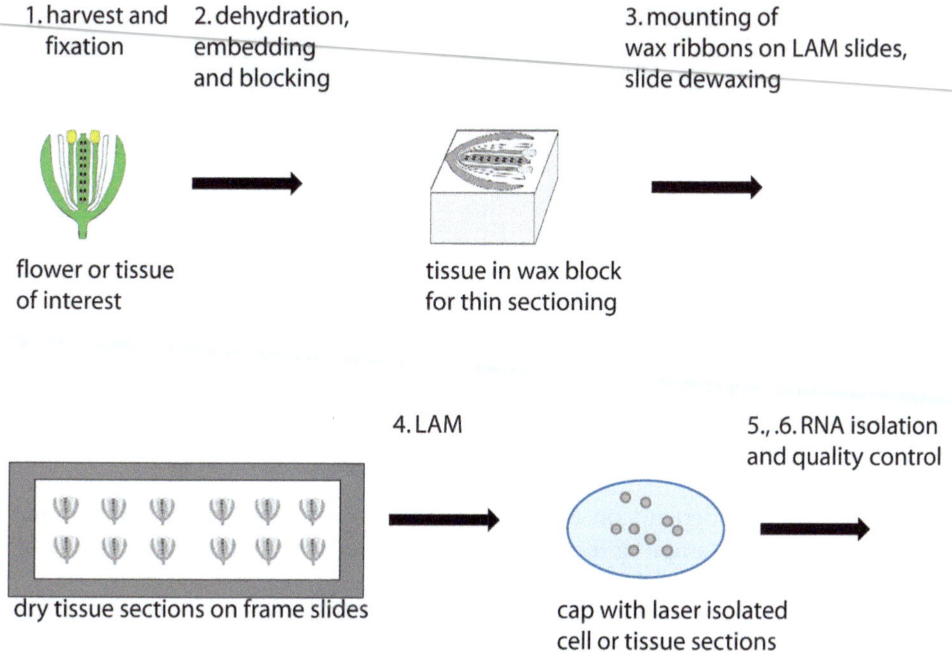

Fig. 1 Schematic overview of subsequent procedures of the method. First the tissue of interest, in this case a flower, is harvested and fixed (1). After dehydration and embedding in paraffin wax, the tissue is embedded in wax blocks and prepared for microdissection using a microtome (2). The thin sections are mounted on LAM slides (typically metal-framed plastic slides) and dewaxed (3). From the dewaxed slides, cells or tissues of interest are isolated with a laser and collected on a suitable cap (4). This is followed by RNA isolation from the cells or tissues on the cap (5 and 6)

be completely avoided, this allows to determine the transcriptomes of a cell type of interest with high specificity and accuracy [2, 7–11].

Briefly, the method requires fixation of freshly harvested tissues with a noncrosslinking fixative, followed by dehydration and embedding in Paraffin wax (for an overview of the procedure, *see* Fig. 1). After mounting blocks of wax-embedded tissues onto a suitable support for fixation in a microtome, thin sections are generated and mounted on special slides for laser microdissection, which are typically metal-framed plastic slides. The tissue on the slides is dewaxed prior to laser-assisted microdissection (LAM). The LAM is done with a microscope equipped with a laser, which allows to microscopically identify and dissect the cells or tissue types of interest. Of great importance for this method to be applicable is the morphology of the tissue and in the dry sections under the microscope (*see* Fig. 2). It is critical that cells of interest can be determined and are visible even at the low contrast of the dried sections, as dyes or cell identity markers cannot be typically used. In our experience, cells should have at least a diameter of about 10 μm to allow their isolation and collection by LAM. Nevertheless, the possibilities and opportunities in using this technology largely

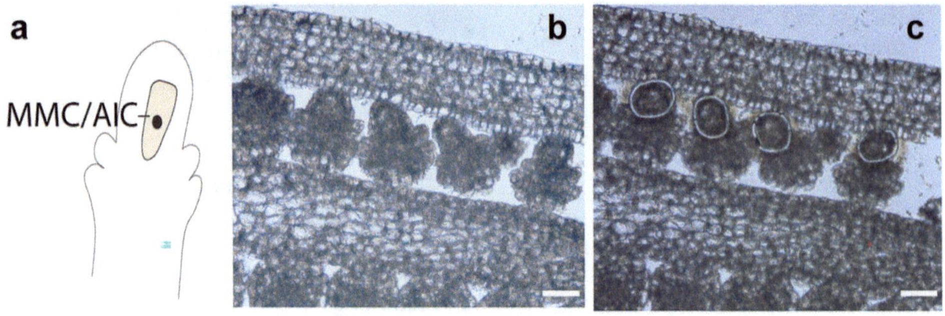

Fig. 2 Example of tissue sections for LAM. (**a**) Schematic presentation of a young ovule harboring the first cell of the female germline, which is the sexual megaspore mother cell (MMC) or an apomictic initial cell (AIC). (**b**, **c**) 6-μm-thick dry thin sections of young flowers of apomictic *Boechera pallidifolia* before (**b**) and after (**c**) applying a laser to dissect part of the ovule harboring the MMC or AIC. Scale bars are 10 μm

depend on the precise instruments used, as the laser sources and sampling technologies differ and have an important impact (also *see* **Note 1**). In addition, with regard to the time and cost efficiency of the sampling, the numbers and frequencies of the cell types of interest in the mounted tissues on a single slide are highly relevant. In order to be applicable, it should be feasible to isolate on average at least 10 of the cell- or tissue types of interest per slide, except for when the cells and tissues have a much larger diameter than the limit mentioned above.

Prior to RNA isolation, enough cells from the same cell type need to be pooled. For isolation of RNA and for generation of high-quality RNA-Seq data, typically collecting of more than 100 sections of the same cell type is required. For downstream transcriptome analysis, obtaining high RNA quality is of crucial importance. Therefore, the method described here has been established and optimized to reproducibly obtain high-quality RNA for downstream transcriptional analyses (at least RNA Integrity Number (RIN) of 7) of *Arabidopsis thaliana* and closely related sexual and apomictic accessions from the genus *Boechera* [2, 7–12]. This method should also be suitable for investigations into additional *Brassicaceae* members and members of other plant families.

From a scientific point of view, the usefulness of the approach to uncover genes and gene regulatory programs underlying germline development has been proven (reviewed in [12]). Over the recent years, these investigations have already allowed gaining novel insights into the transcriptional basis underlying developmental processes and cellular identity in plant germlines. Importantly, the role of members of the ancient RNA helicase gene family in the plasticity of plant reproductive development during sexual and apomictic reproduction has been demonstrated, and several additional new candidate genes for apomixis could be identified [9–11].

In summary, transcriptional profiling using LAM in combination with RNA-Seq is therefore a powerful method to determine genes and gene regulatory programs involved in reproduction, cell-type specification, and developmental processes. This chapter provides a detailed description of the method, from tissue fixation to RNA isolation, and highlights critical steps (*see* Fig. 1). RNA amplification and RNA-Seq are not objectives of this protocol.

2 Materials

2.1 Plant Material and Equipment

1. Flowering plants with buds at the developmental stage of interest.
2. Growth chamber for plant cultivation at controlled conditions.
3. Stereomicroscope.
4. Heating oven.
5. Heating plate.
6. Microtome.
7. Microscope equipped with a laser for microdissection (*see* **Note 1**).
8. Tissue-processing machine (*see* **Note 2**).
9. Embedding machine and blocking station.
10. Bioanalyzer or Tape Station (*see* **Note 3**).
11. RNAse Zap.
12. RNAsc-free water.
13. Gloves.

2.2 Fixation

1. Acetic acid (*see* **Note 4**).
2. Ethanol (*see* **Note 4**).
3. Injection needles.
4. Fine forceps.
5. 2-ml microcentrifuge tubes.
6. 15-ml or 50-ml plastic tubes.
7. Ice box.
8. Ice.
9. Vacuum pump.
10. Exsiccator.

2.3 Embedding Using an Automated Tissue-Processing Machine

1. Ethanol (*see* **Note 4**).
2. Xylol (*see* **Note 4**).
3. Paraffin wax (*see* **Note 5**).

4. RNAse-free water.
5. Tissue-embedding cassettes.
6. Optional: Biopsy-embedding cassettes.
7. Automated tissue-processing machine.
8. Embedding machine and blocking station.
9. Fine forceps.
10. Preparation needles.
11. Balancing trays.

2.4 Embedding Using a Manual Protocol

1. Ethanol (*see* **Note 4**).
2. Xylol (*see* **Note 4**).
3. Paraffin wax (*see* **Note 5**).
4. RNAse-free water.
5. Tissue-embedding cassettes.
6. Pipettes and pipette tips.
7. Fine forceps.
8. Preparation needles.
9. Balancing trays.

2.5 Preparation of Thin Sections and Mounting on Slides

1. One-sided razor blades.
2. Microtome blades.
3. Microtome.
4. Fine forceps.
5. Preparation needle.
6. RNAse-free plastic box.
7. Black cardboard.
8. Special slides for LAM (*see* **Note 6**).

2.6 Laser-Assisted Microdissection

1. Microscope equipped with a laser.
2. Hardware and software to control the laser.
3. Tubes or caps suitable for collection of isolated tissue sections (*see* **Note 7**).

2.7 RNA Isolation and Quality Control

1. RNA isolation kit (*see* **Note 8**).
2. Set of pipettes and tips (*see* **Note 9**).

3 Methods

3.1 General Remarks and Preparation Including Removal of RNAses

All procedures are carried out at room temperature, unless otherwise specified. To allow isolation of high-quality RNA, degradation needs to be prevented at any step of the protocol. It is ideal if a separate room or bench can be used for RNA work only. Throughout performing the protocol, fresh gloves need to be worn (*see* **Note 10**). Glassware should be baked to remove RNAses (*see* Step 4 under this subheading), any plasticware and plastic consumables are to be used brand new without any pretreatment. Certified RNAse-free plasticware should preferentially be used.

Please carefully study and consider safety instructions and safety data sheets for all chemicals used. Keep in mind that xylol is harmful and can penetrate certain gloves and that it can dissolve certain plasticware and that methanol is toxic. All solvents should only be used under a chemical hood or similar device. For any instrument required to perform the procedure, carefully study manuals and safety instructions prior to first use. Wear protective glasses during work and special glasses for protection from laser radiation when performing LAM.

1. Clean working bench to remove RNAses. First, treat the bench with RNAse-Zap (*see* **Note 11**). Remove RNAse-Zap using brand-new paper towels from a tissue box and clean the bench with RNAse-free water (*see* **Note 12**). Treat the bench with pure ethanol and dry afterward using paper towels.
2. Treat metal surfaces like the heating plate following the steps described under Step 1 in this subheading.
3. Regularly clean rotors of centrifuges 1 day prior to work by removing them from the centrifuges and following cleaning sequence described under Step 1 in this subheading, except that brand-new cotton swabs are used for cleaning slots inside the rotors. After cleaning, rotors are left upside down on a fresh paper towel overnight to allow drying to complete. Before use, fix them carefully back in the centrifuge.
4. Wrap any glassware and metal equipment like forceps (without any parts of plastic) in aluminum foil and bake them for 8 hours in a heating oven at 180 °C. Alternatively, baking can be done for 3 hours at 210 °C. Make sure to allow cooling to room temperature before removing from the oven and remove aluminum foil only directly prior to use.

3.2 Tissue Fixation

Fixation of tissues is required to stop metabolic processes, to avoid changes in RNA abundance after harvest, and to conserve RNA quality. To allow isolation of up to full length RNA transcripts after fixation, use of noncrosslinking fixatives is required. Steps 1–7 below in this subheading describe the procedure.

1. Prepare fresh farmer's fixative (3:1, (v/v) ethanol: acetic acid) in 15-ml or 50-ml plastic tubes, depending on the amount of tissue to be fixed. Alternatively, 5:1 (v/v) ethanol: acetic acid can be used. Fixative always needs to be prepared freshly on ice in an ice bucket directly prior to use. For up to 20 buds from *Arabidopsis* or *Boechera,* use 10 ml of fixative.

   For larger buds or tissues, directly before fixation, dissect with a razor blade to generate smaller parts of tissues.
2. Carefully harvest the buds at the developmental stages of interest and immediately submerge them in the ice-cold fixative. For isolation of cells from mature gametophytes, harvest buds 2 days after emasculation of the largest 2–3 buds per inflorescence. Follow this procedure both for *Arabidopsis* and for sexual and apomictic species or accessions of *Boechera*.
3. As soon as possible after harvest of tissues, fill the bottom of an exsiccator with ice and place the tubes open without lids in the ice. Using a vacuum pump connected to the exsiccator, apply vacuum for 15 minutes.
4. Release the vacuum and repeat the vacuum infiltration one more time for 15 minutes.
5. Remove the vacuum, close the lids of the tubes, and incubate the samples on ice overnight (*see* **Notes 13** and **14**).
6. Prior to embedding, exchange the fixative with 70% EtOH prepared from pure EtOH and RNAse-free water (*see* **Note 15**). Avoid drying off the samples at any processing step. Keep the tissues on ice until embedding.
7. Start embedding by following the procedure described in subheadings 3.3. or 3.4, the day after fixation.

3.3 Dehydration and Embedding Using an Automated Tissue-Processing Machine

The application of an automated tissue-processing machine is helpful to save time during processing steps of dehydration and infiltration of the tissues with paraffin wax (*see* **Note 2**). The machine will typically be prepared for automated dehydration in ethanol followed by infiltration with xylol, and thereafter with melted paraffin wax. In case an automated tissue-processing machine is not available, skip any steps described in this subheading and proceed directly to steps described in subheading 3.4. For sample preparation and embedding using an automated tissue-processing machine, follow Steps 1–18 outlined here.

1. The tissue-processing machine needs to be prepared with bottles containing 70% EtOH, 90% EtOH, 100% EtOH, xylol, and with paraffin wax (*see* **Notes 16** and **17**).
2. Shake out the required amount of embedding or biopsy cassettes from the storage bag. Use of embedding or biopsy cassettes depends on the size of the tissues in relation to the

holes in the cassettes. Mark the cassettes using a pencil, ideally on the inclined surfaces.

3. Open the cassettes and place them in a small volume of 70% EtOH, for example, on the bottom of a brand-new petri dish.
4. Using fine forceps, gently transfer the samples to a cassette. One or a few samples (typically up to about 5) can be processed in one cassette. Carefully close the lid of the cassette.
5. Store the cassette loaded with tissues in a glass staining trough filled with 70% EtOH.
6. Repeat Steps 3–5 until all samples are loaded into cassettes.
7. Remove the basket of the tissue-processing machine from the retort. For transferring the cassettes to the basket, fill in an appropriate RNAse-free container with 70% EtOH (*see* **Note 18**). For handling, it is best to position and orient the cassettes, ideally with the inclined labeled surfaces on top, facing the same direction. In case not a sufficient number of samples are processed at once for the basket to be loaded completely, leave empty spaces between the filled basket positions when loading the cassettes into the basket.
8. After loading, quickly put the lid on the basket and directly transfer the basket to the retort of the tissue-processing machine and close the retort.
9. On the tissue-processing machine, immediately start the appropriate program. A recommended standard setting would be to program the machine to infiltrate the samples for 1 hour with 70% EtOH, 3 × 1 hour with 90% EtOH, 3 × 1 hour with 100% EtOH, 2 × 1 hour with 100% xylol, 1 × 15 minutes with 100% xylol at room temperature, and then 2 × 1 h paraffin wax and 1 × 3 h with paraffin wax at 56 °C (*see* **Note 19**).
10. After finalizing the program, the retort remains to be filled with melted wax at 56 °C until the command to drain the retort is approved (*see* **Note 20**). Proceed to blocking of the samples immediately after draining the retort and on the same day that the program in the tissue-processing machine is finalized.
11. Start heating paraffin wax baths in the embedding machine and blocking station to 56 °C several hours before final embedding and blocking. Make sure that the wax is fully melted in this machine before proceeding to the next steps.
12. Open the retort of the tissue-processing machine and quickly transfer the basket with the samples to the molten paraffin wax in the wax bath. Remove the lid from the basket and cover the wax bath with its lid.
13. Pile up as many balancing trays as embedding cassettes or samples to be processed facing bottom up on a surface of the

blocking station heated to 56 °C. For the labeling of the balancing trays, use an ethanol-resistant permanent marker. Do not use acetone-resistant markers.

14. Turn the balancing tray that is topmost in the pile. Using the blocking station, fill a preheated balancing tray half with molten paraffin wax heated to 56 °C. Quickly lift the lid from the wax bath and remove an individual embedding cassette from the basket. The sample can be flushed from the embedding cassette into the balancing tray with liquid paraffin wax from the outlet of the blocking station or be gently transferred using fine forceps. Make sure to avoid squeezing of the tissues. Fill the balancing tray with wax up to about 2–3 mm below the rim. Before the wax starts solidifying, the samples can be oriented using forceps or preparation needle. Typically, individual samples are arranged to be spaced about 1 cm in each direction.
15. Repeat Steps 13 and 14 until all samples and cassettes are processed.
16. Remove the lid from the wax bath, take out the basket of the tissue-processing machine, replace lids on the basket and on the paraffin wax bath, and transfer the basket back to the retort of the tissue-processing machine.
17. Start the appropriate program of the tissue-processing machine to clean retort and baskets from wax.
18. The embedded samples can be stored at room temperature until the wax has hardened. After hardening, transfer embedded tissues in balancing trays to a clean plastic box and store them at 4 °C (*see* **Note 21**).
19. Using the embedding machine and blocking station, prepare supports for the samples to be processed in the microtome by filling open embedding cassettes without lids with melted paraffin wax. Allow the wax to harden and store supports in an RNAse-free box.

3.4 Manual Embedding

For manual embedding described in Steps 1–12 in this subheading, fixed tissues are processed directly continuing from Step 6 described in subheading 3.2. All steps can be performed in 2-ml plastic tubes. Solvents are exchanged by pipetting. Skip this subheadings if performing the steps described in subheading 3.3.

1. After fixation, the tissue is first dehydrated in a series of ethanol on ice. First, remove the fixative, and afterward at each step, the different dilutions of ethanol, carefully by pipetting. Avoid drying off the samples. Incubate samples in a series of 70%, 80%, 90% EtOH (each prepared by diluting with RNAse-free water), each step for 1 hour, followed by an incubation in 100%

EtOH overnight at 4 °C or on ice (*see* **Note 17**), replace 100% EtOH two times for 1 hour on the next day.

2. Similarly, treat samples with a xylol gradient of 25%, 50%, 75%, 100% xylol in EtOH, each step for 1 hour at room temperature, before an incubation in 100% xylol overnight.
3. On the next day, heat samples to 56 °C in an oven and slowly add small pieces or droplets of solid paraffin wax. Allow them to melt or be dissolved before adding the next.
4. Continue adding solid pieces or droplets of paraffin wax over the next 2–3 days at 56 °C, until the solvent is completely replaced by liquid paraffin wax. For each step, carefully remove surplus volume of xylol with wax mixture by pipetting.
5. Several hours before blocking, fill 50 ml plastic tubes with solid paraffin wax. Place the tubes in an incubation oven heated to 56 °C until paraffin wax is completely melted.
6. For final embedding, place brand-new balancing trays on a heating plate set to 57 °C bottom up and allow them to preheat. For the labeling of the balancing trays, use an ethanol-resistant permanent marker. Do not use acetone-resistant permanent markers.
7. Turn one balancing tray and place it bottom down on the heating plate. Fill the balancing tray with melted paraffin from Step 6.
8. Use fine forceps to carefully transfer the sample to the balancing tray.
9. A preparation needle can be used to orient the samples. Typically, if more than one piece of tissue or bud is embedded in the same tray, it should be spaced about 1 cm in each direction.
10. Repeat Steps 7–9 until all samples are processed.
11. Store the embedded samples at room temperature until the wax has solidified and afterward transfer embedded samples in balancing trays to a clean plastic box for storage at 4 °C (*see* **Note 21**).
12. Prepare supports for the samples to be processed in the microtome by filling embedding cassettes without lid with melted paraffin wax. Allow the wax to harden and store supports in an RNAse-free box.

3.5 Preparation of Thin Sections and Mounting of Wax Stripes on Slides

This subheading describes all Steps (1–22) required to generate thin sections and to mount them on special slides for LAM.

1. Pick up balancing trays with embedded samples in paraffin wax from 4 °C.

2. Carefully remove the plastic balancing tray from the wax block and place the latter on a light table illuminating the block with the embedded buds or tissues facing upward.
3. Using an RNAse-free one-sided razor blade (*see* **Note 22**), carefully dissect out a squared wax block surrounding the sample, e.g., bud or flower and orient it as before with the tissue facing up on the light table.
4. Continue to repeat Steps 2 and 3 until all samples are prepared that should be used for LAM at a time.
5. To mount the blocks on prepared supports (embedding cassettes without lids, which are filled with solid paraffin wax), first, flatten the wax surface of the support using a one-sided razor blade. Afterward, use the flame of an ethanol lamp to melt a small piece of paraffin wax on the tip of a spatula. Use the hot paraffin wax to fix the block with the sample on the wax of the support with the tissue facing up and not toward the support.
6. Repeat Step 5 until all samples are processed.
7. Place samples at 4 °C and allow the wax to harden for at least 20 minutes.
8. Take the samples back to the light table and use an RNAse-free one-sided razor blade to remove surplus paraffin wax surrounding the sample by cutting vertically to the support to produce a squared surface with parallel edges.
9. Repeat Step 8 until all samples are processed.
10. Set a heating plate to 42 °C.
11. Open an RNAse-free plastic box and place black cardboard at the bottom.
12. For using the microtome refer to the manual. First, prepare the microtome by inserting an RNAse-free microtome blade.
13. Safely mount the samples that are attached to the support at the appropriate sample holder of the microtome to prepare paraffin sections of 6–10 μm thickness (*see* **Note 23**).
14. During preparation of thin sections using the microtome, grab the first thin section of the forming wax stripe with fine forceps and keep holding it while dissecting. Be careful not to touch the microtome blade with the forceps.
15. Using the forceps and a preparation needle to support the stripe at the other end, transfer wax stripes that are formed onto the cardboard in the plastic box (*see* Step 11) once they have reached a length of between 10 and 15 cm.
16. If possible, preselect stripes and part of stripes containing cell and tissue types of interest under a dissecting scope and discard the rest (*see* **Note 24**).

17. Set a heating plate to 42 °C.
18. Place the required number of slides for LAM, typically metal-framed plastic slides with the flat side facing upward on the heating plate to preheat.
19. Use an RNAse-free razor blade to cut the wax ribbons to the maximal length fitting onto the plastic window of the LAM slide (typically about 4–5 cm).
20. Use an RNAse-free pipet to transfer 1–2 ml of RNAse-free water to the plastic windows of each LAM slide. Alternatively, place the heating plate under a chemical hood and pipet the minimal volume of methanol on the window of the slides that is just sufficient to moisten the surface (*see* **Note 25**).
21. Transfer the short wax ribbons to the windows of the LAM slides using fine forceps and a preparation needle.
22. Cover the heating plate with a lid that does not touch the slides and does not prevent an exchange of air. Allow the slides to dry at 42 °C overnight (12–14 hours). Alternatively, reduce drying time to a minimum of 2 hours until sections are completely dry.

3.6 Laser-Assisted Microdissection

The precise steps and procedures of LAM largely depend on the instrument used. Please refer to the respective user manual. Here, Steps 1–20 are exemplified for the instrument used by us (*see* **Note 1**).

1. Take the slides from the heating plate and dewax in an appropriate slide holder 2× for 10 minutes in 100% xylol under a chemical hood. Make sure to submerge all plastic parts, but to keep the broader rim of the metal frame dry.
2. Take out the slides while only touching the dry part of the metal frame and allow them to dry for about 10 minutes under the chemical hood by inserting them in an appropriate dry slide holder.
3. Close the slide holder with the lid and transfer the slides to the LAM device.
4. Switch on the LAM device (computer, microscope, and laser control box).
5. Start the appropriate program controlling microscope and laser.
6. Place a sampling cap, e.g., 0.5 ml without diffusor, in the cap holder with the tube open and position the cap holder in the instrument.
7. Support the LAM slide with a glass slide (prepared RNAse-free by baking beforehand) so that the plastic and the glass are sandwiching the tissue of the samples.

8. Insert the slide into the appropriate slide holder in the LAM device.
9. In order to make a scan of the slide, select the 4× objective and move the cap to the "up" position. Make a scan of the slide.
10. Use the scan to select positions with cells or tissues of interest and pinpoint them to be able to move back to the position for efficient dissection.
11. Select the appropriate objective (e.g., 40× or 60×).
12. On a part of the slide without tissues, verify laser position. If required, readjust laser position according to the procedures described in the manual of the instrument.
13. According to the requirement, adjust laser power, laser speed, focus, and stage movement following the procedures described in the manual. Verify and calibrate the objective used according to the manual.
14. To start dissecting, move to the first pinpointed position and move the cap "down." With the appropriate hand tool, mark the borders of the cell or tissue to be dissected. Press "cut" for dissection. Inspect that the laser has efficiently cut around the marked cell or tissue. Often, repeating the "cut" action is required. Once the tissue is dissected, move cap "up."
15. Move to the next pinpointed position and repeat Step 14 until all cells or tissues of interest on the slide are dissected. Make sure that the sampling position on the cap is slightly changed for each dissection. This can be set automatically.
16. Before changing to the next slide, it is recommended to switch the sampling cap with a new one so that a fresh cap is used for additional isolation of larger tissue sections from the same slide intended for RNA quality control.
17. Before proceeding to the next slide, switch back to the sampling cap for the cell type of interest. The cap can be used for collecting the same cell- or tissue type for up to 4 hours.
18. After completing the sampling on one cap, visually inspect for contaminations under the 4x objective. If possible, remove contaminations carefully by using a fresh fine 2-μl pipet tip. In case contaminations cannot be removed, the sample needs to be discarded.
19. Close the cap and finalize LAM by switching off the steering program, laser control, microscope, and computer.
20. Store caps at −80 °C until RNA isolation for a maximum of 4 months. It is, however, recommended to proceed with RNA isolation as soon as possible.

3.7 RNA Isolation and Quality Control

For RNA isolation, any isolation method or kit suitable for isolation of RNA amounts in pg-to-ng range can be used (*see* **Note 8**). Please follow the manufacturer's instructions.

1. Proceed to RNA isolation directly after LAM or after removal of the caps from −80 °C. First, allow caps to adjust to room temperature.
2. Carefully open the tubes and for extraction on the caps, pipet 11 μl of extraction buffer to the wall of the tube. Close the tube and shake the buffer down to cover the cap surface. Change tips between samples.
3. Place the tubes with cap down for 30 min at 42 °C to allow extraction and follow further isolation procedure following manufacturer's instructions (*see* **Note 26**).
4. After RNA isolation, proceed to RNA quality control (*see* **Notes 27** and **28**).

4 Notes

1. Instruments for laser-assisted microdissection are offered by different companies, and they differ in features of the laser and in precise sampling technology. Although it is not the aim here to promote any technology, instrument, or supplier, it should be taken into account that sampling technologies based on catapulting of dissected tissues or based on gravity typically require larger sections (in our experience, at least 30 μm in diameter) for efficient sampling. The instruments manufactured by "Molecular Machines & Industries," in contrast, use a technology where a cap is in direct contact to the slides with the mounted tissue during microdissection and the isolated cells or tissues afterward stick to the cap by electrostatic forces. In our hands, this allows to efficiently collect sections down to a size of approximately 10 μm in diameter [10].
2. Machines for automated dehydration and processing of tissues for embedding (e.g., Leica ASP200/300 s) allow fast and automated processing of the tissues overnight. It is recommended not to use crosslinking fixatives, such as formaldehyde, in the embedding machine even for other purposes than LAM, as there is typically carryover between bottles in the machine.
3. For RNA-quality check at high sensitivity, gel electrophoresis using typical laboratory equipment for molecular biology is not suitable. It is highly recommended to use Agilent Bioanalyzer, Tape Station, or similar technology.

4. Any organic solvents or organic acids used should have the highest purities that are commercially available. For xylol, use a mixture of isomers.
5. The precise type of paraffin wax can make a difference for the embedding and for the RNA quality that can be achieved with the procedure. We have been using Paraplast-Xtra for our work with good results.
6. For LAM, typically metal-framed plastic slides are used. Please, refer to the recommendations of the supplier of the used LAM instrument for suitable slides. For best quality, use slides that are certified DNAse- and RNAse-free.
7. It depends on the used LAM instrument, which types of tubes or caps are compatible. Please refer to the recommendations of the supplier. When using silicon-filled caps, the type without diffusor should be used for better visibility of tissue morphology.
8. The isolation kit should be suitable for isolation of RNA from pg-to-ng range. We have been using the Arcturus PicoPure Kit with very good results.
9. For best results, low binding and certified RNAse-free tips should be used brand new.
10. While working with gloves, avoid touching any surfaces that could be contaminated with RNAse, for example, door handles.
11. RNAse Zap is not only harmful to RNAses but also detrimental to RNA. After treatment of the bench or any instruments, it needs to be carefully removed.
12. Water from water purification instruments including ion exchange resins and UV treatment can typically be used for this step.
13. Overnight incubation in fixative should typically not exceed 14–15 hours maximum.
14. To prevent the ice melting overnight, the ice boxes should be closed with lids or covered with aluminum foil, and they should be stored in a cold chamber or fridge at 4 °C.
15. Larger volumes of RNAse-free water can be used from the water purification instruments (also *see* **Note 12**). For smaller volumes, use of commercially available RNAse-free water is recommended. Do not use diethylpyrocarbonate (DEPC) for water treatment or DEPC-treated water.
16. Typically, the reagents in the machine are reused for a certain number of runs or amount of time. Regularly exchange solvents and paraffin wax as recommended by the suppliers of the tissue-processing machines.

17. For 100% EtOH, use highest-purity grade commercially available, typically exceeding 99.5%.
18. During the transfer, it is important that samples do not dry.
19. The tissue-processing machine can be switched on several hours before embedding. This will affect the total time needed until the processing is finished. In case the paraffin wax is not completely melted when the program is started or processing time will be too limited for the wax to melt until the retort needs to be flushed with melted wax, the retort will automatically be flushed with the liquid from the first processing step (typically 70% EtOH) and further processing is halted until the machine is ready to start. In our experience, there is no detectable effect on the quality of embedding due to using one or the other alternative.
20. In our experience, a few hours difference concerning the time of incubation in wax at the last processing step, before draining the retort, does not affect the quality of embedding.
21. At 4 °C, samples can typically be stored for several weeks without significant loss of RNA quality.
22. To ensure that the razor blades are RNAse-free, they can be wrapped in aluminum foil and baked as described in subheading 3.1, Step 4.
23. The precise thickness of the sections needs to be optimized for each cell or tissue type. Thinner sections typically result in better resolution and visibility of cells in the dried sections under the microscope, but at the same time, RNA amount and quality often decrease as a consequence of decreased thickness of sections.
24. Time and cost requirements for LAM can highly increase when cell types of interest cannot be placed on a single slide at sufficient amounts.
25. Handle methanol under a chemical hood. It is recommended to test the impact of using methanol or water for this step on the RNA quality and to only use methanol if a significant improvement is visible.
26. In case it is required to pool harvests on different caps in one sample, combine them in one tube directly after RNA extraction on the cap. Proceed quickly after addition of EtOH to RNA extraction, as RNA starts precipitating.
27. For RNA quality control, use of an Agilent Bioanalyser or Tape station is recommended.
28. RNA quality for amplification and library preparation for RNA-Seq should be about RIN 7 or higher.

Conflicts of Interest

The authors have no conflicts of interest to declare that are relevant to the content of this chapter.

No funding received or to be declared.

References

1. Spillane C, Steimer A, Grossniklaus U (2001) Apomixis in agriculture: the quest for clonal seeds. Sex Plant Reprod 14(4):179–187
2. Wuest SE, Vijverberg K, Schmidt A, Weiss M, Gheyselinck J, Lohr M, Wellmer F, Rahnenführer J, von Mering C, Grossniklaus U (2010) Arabidopsis female gametophyte gene expression map reveals similarities between plant and animal gametes. Curr Biol 20(6):506–512
3. Schmid MW, Schmidt A, Grossniklaus U (2015) The female gametophyte: an emerging model for cell type-specific systems biology in plant development. Front Plant Sci 6:907
4. Sprunck S, Groß-Hardt R (2011) Nuclear behavior, cell polarity, and cell specification in the female gametophyte. Sex Plant Reprod 24(2):123–136
5. Schmidt A, Schmid MW, Grossniklaus U (2015) Plant germline formation: common concepts and developmental flexibility in sexual and asexual reproduction. Development 142(2):229–241
6. Brownfield LR, Figueiredo DD, Borg M, Schmidt A (2023) Editorial: plant development and reproduction at single cell and cell type-specific resolution. Front Plant Sci 14: 1261685
7. Florez Rueda AM, Grossniklaus U, Schmidt A (2016) Laser-assisted microdissection (LAM) as a tool for transcriptional profiling of individual cell types. J Vis Exp 111:53916
8. Schmid MW, Schmidt A, Klostermeier UC, Barann M, Rosenstiel P, Grossniklaus U (2012) A powerful method for transcriptional profiling of specific cell types in eukaryotes: laser-assisted microdissection and RNA sequencing. PLoS One 7(1):e29685
9. Schmidt A, Wuest SE, Vijverberg K, Baroux C, Kleen D, Grossniklaus U (2011) Transcriptome analysis of the Arabidopsis megaspore mother cell uncovers the importance of RNA helicases for plant germline development. PLoS Biol 9(9):e1001155
10. Zühl L, Volkert C, Ibberson D, Schmidt A (2019) Differential activity of F-box genes and E3 ligases distinguishes sexual versus apomictic germline specification in Boechera. J Exp Bot 70(20):5643–5657
11. Binmöller L, Volkert C, Kiefer C, Zühl C, Slawinska MW, Loreth A, Nauerth BH, Ibberson D, Martinez R, Mandakova TM, Zipper R, Schmidt A (2024) Differential expression and evolutionary diversification of RNA helicases in Boechera sexual and apomictic reproduction. J Exp Bot 75(8):2451–2469
12. Schmidt A (2020) Controlling Apomixis: shared features and distinct characteristics of gene regulation. Genes (Basel) 11:3

Chapter 8

Manual Isolation of Parthenogenetic Egg Cells Combined with Comparative Transcriptomics to Identify Parthenogenesis Genes in Tripsacum

Xixi Zheng, María Flores-Tornero, and Thomas Dresselhaus

Abstract

Parthenogenesis, meaning "creation by virgin," is a key component of apomixis (asexual reproduction through seeds). It refers to spontaneous embryogenesis from an unfertilized egg cell resulting in offspring that is genetically identical to the mother plant. The diplosporous apomictic grass *Tripsacum dactyloides* (Tripsacum) represents a valuable natural model system for studying gene expression patterns associated with parthenogenesis. All investigated polyploid Tripsacum accessions reproduce via facultative apomixis with parthenogenetic frequencies of >90%. Here, we report a strategy and method regarding how parthenogenesis genes can be identified from a non-model species whose genome is not yet available. We describe the manual dissection of parthenogenetic and sexual egg cells from polyploid and sexual Tripsacum accessions, respectively, the generation of RNA-seq data from few cells and their mapping to a closely related species (here maize) whose genome is known. We further describe a comparative approach to identify candidate parthenogenesis genes and how their parthenogenetic potential and the reproduction mode can be elucidated in transgenic approaches.

Key words Parthenogenesis, Apomixis, Egg cell, *Tripsacum dactyloides*, Maize, Micromanipulation

1 Introduction

Apomixis, an asexual reproduction process through seeds, is an attractive trait for plant breeding, as it can immediately fix any desirable traits [1]. It is found in 34 orders, 80 families, and 326 genera of angiosperms, predominantly in wild species, but is absent in most major crop species [2]. We focus on diplosporous gametophytic apomixis [3], a complex trait consisting of three main components: (i) apomeiosis, bypassing of meiotic reduction divisions from a megaspore mother cell, resulting in a single (unreduced) embryo sac per ovule followed by (ii) parthenogenesis,

The authors Xixi Zheng and María Flores-Tornero have equally contributed to this chapter

Tatyana Radoeva and Peter J. van Dijk (eds.), *Apomixis: Methods and Protocols*, Methods in Molecular Biology, vol. 2987, https://doi.org/10.1007/978-1-0716-5001-1_8,

autonomous development of the unreduced egg cell into an embryo without fertilization. In some species, (iii) endosperm develops autonomously without fertilization, while in others, it requires fertilization of the central cell by a single sperm cell, a process known as pseudogamy. The complexity of apomixis, its prevalence in polyploid plants, and the lack of reference genomes for such species present significant challenges for the identification of genes associated with apomeiosis and parthenogenesis.

Tripsacum dactyloides (Tripsacum) is a natural diplosporous gametophytic apomictic species exhibiting an apomeiotic–parthenogenetic–pseudogamous pathway of reproduction in all polyploid accessions, such as tetraploids, hexaploids, and octaploids. Studies have reported that the progeny reproduction rate through parthenogenesis in these polyploids is above 90% [4]. In contrast, diploid Tripsacum strictly reproduces sexually. Egg cells of polyploid Tripsacum with unreduced chromosomes can autonomously develop into a zygotic state, expected to display a gene expression pattern distinct from that of sexual egg cells. In sexual reproduction, egg cells remain arrested in the G0 phase without proceeding to cell division [5, 6]. The molecular mechanisms controlling these processes in Tripsacum are unknown, presenting an exciting area for further research. Notably, Tripsacum is a close relative of maize, with syntenic genes that share high sequence identity between the two species [7, 8]. This close genetic relationship allows maize to be used as a reference genome for mapping transcriptomic data obtained from Tripsacum. Moreover, maize can be effectively utilized for verifying candidate apomictic genes identified in Tripsacum.

Here, we report a strategy and method repertoire to identify parthenogenetic genes from Tripsacum as an apomictic grass model. We describe a method for manually isolating egg cells from both, parthenogenetic and sexual Tripsacum accessions incorporating a low-input RNA extraction and library construction protocol. By comparing the transcriptomes of these two egg cell types after mapping to the maize reference genome, we show how the data can be further exploited, for example, with a focus on cell cycle progression and transcription factor-mediated gene regulation. We further suggest using stable maize transformation to either ectopically express or knockout candidate genes to reprogram the gene expression pattern of sexual egg cells to mimic the characteristics of parthenogenetic egg cells. Flow cytometry is then employed to efficiently identify haploid plants generated through parthenogenesis, providing a robust framework for studying and manipulating apomictic traits.

2 Materials

2.1 Plant Material, Emasculation, Spikelet, and Ovule Dissection

1. Diploid and polyploid flowering Tripsacum plants (*see* Fig. 1a).
2. Washing solution: 0.55 M mannitol (Thermo Fisher Scientific), 570 mOsm/L, pH 5.0.
3. Enzymatic solution: 1.5% pectinase, 0.5% pectolyase Y23 (highly purified), 1% hemicellulase, 1% cellulase (chromatographically purified, ≥ 45 units/mg of dry weight), dissolved in 0.55 M mannitol 570 mOsm/L, pH 5.0.

Fig. 1 *Tripsacum dactyloides* (Tripsacum) plant and inflorescence architecture. (**a**) Tripsacum plant showing two immature inflorescences (red arrowheads). (**b**) Male florets (staminate spikelets) are located at the tip of the mature inflorescence and about 12 female florets (pistillate spikelets) at the bottom. Scale bar: 1 cm

4. ≥ 99.8% ethanol absolute.
5. RNase-free water.
6. Liquid nitrogen.

2.2 Egg Cell Isolation

1. 35-mm imaging Petri dishes (*see* Fig 2c).
2. 60-mm glass Petri dishes (*see* Fig 2c).
3. High-precision tweezer (*see* Fig 2c).
4. Curved forceps (*see* Fig 2c).
5. Disposable hypodermic needles (0.4-mm diameter, 13-mm length) (*see* Fig 2c).
6. Scalpel (No. 24 sterile carbon steel surgical blade) (*see* Fig 2c).
7. Disposable 50-micron glass capillary pipettes (Hirschmann™ 9600150; *see* Fig 2c).
8. Borosilicate glass cylinders (1.5-mm diameter, 10-mm length) to generate glass needles.
9. Humid chamber (e.g., a tip box with wet paper on the bottom of the box and on the inner surface of the lid).
10. 0.5-μl RNA LoBind tubes with safe locker.
11. Dewar container for liquid nitrogen.
12. Stereomicroscope with a 5:1 zoom minimum.
13. Inverted microscope (e.g., Eclipse TE200-S, Nikon), equipped with a double Imaging Petri dish stage, and a high-resolution 20x objective (*see* Fig. 2a, b).
14. Manual micromanipulator (e.g., CellTram®, Eppendorf; *see* Fig. 2f).
15. Programmable micropipette puller (e.g., P-97 Flaming/Brown type micropipette puller, Sutter Instruments®; *see* Fig. 2d).

2.3 Library and RNA-seq Preparation

1. RNeasy Plus Micro Kit (e.g., QIAGEN, 74134).
2. SMART-Seq® mRNA LP (with UMIs, Takara, 634762).
3. Unique Dual Index Kit (e.g., Takara Bio).
4. AMPure XP beads (e.g., Beckman Coulter).
5. KAPA Library Quantification Kit (e.g., Roche, 07960140001).
6. Agilent 2100 Bioanalyzer with the RNA 6000 Pico LabChip reagent set.
7. Illumina NextSeq 2000 Sequencing System.

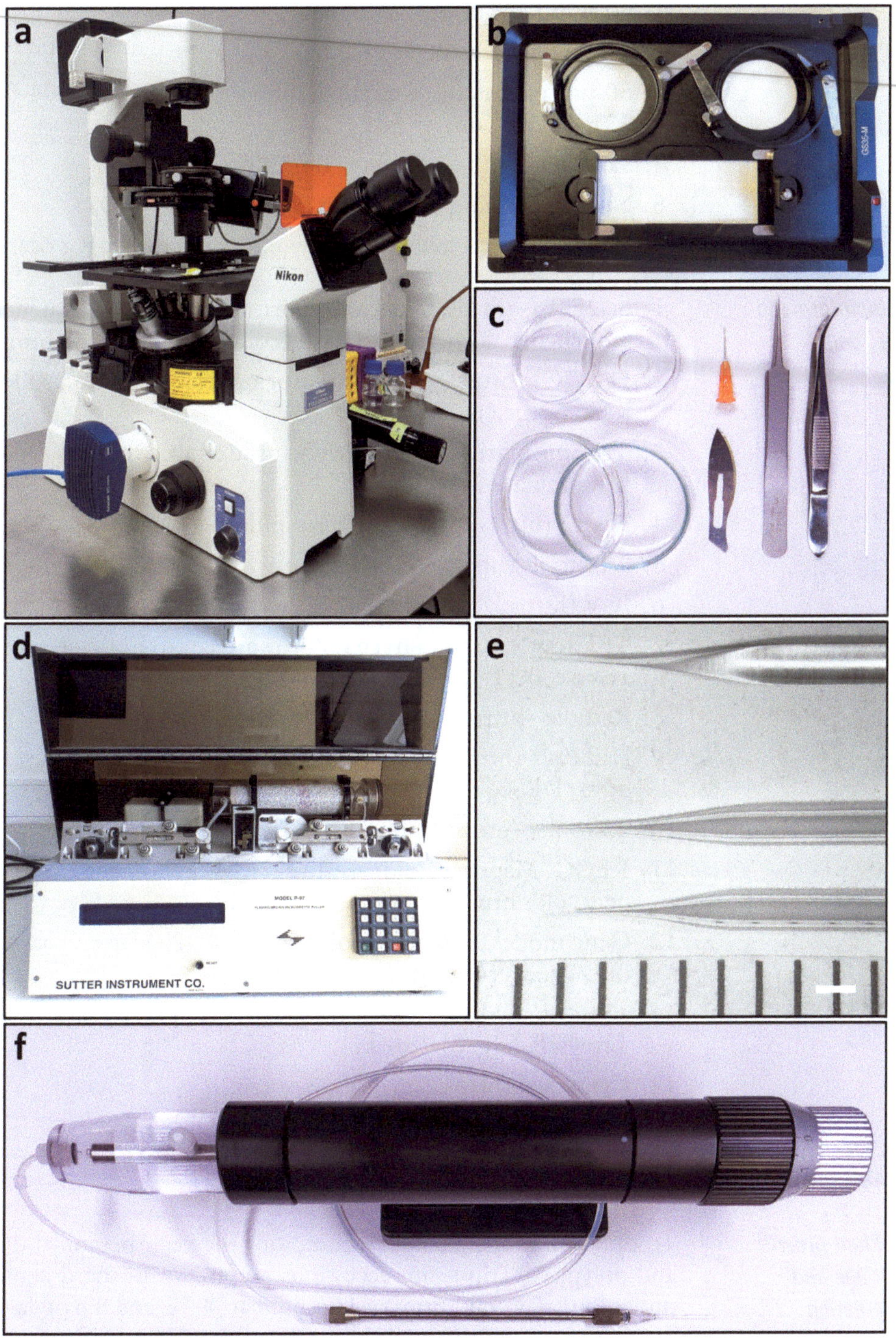

Fig. 2 Materials and equipment for egg cell isolation. (**a**) Inverted microscope. (**b**) Double Imaging Petri dish holder. (**c**) Materials for manipulation, dissection, and observation of egg cells. (**d**) Programmable micropipette puller to create glass capillary and sharp glass needles. (**e**) Sharp glass needle (top), glass needle (middle), and glass capillary (bottom). (**f**) Manual micromanipulator equipped with a glass capillary. Scale bar: 1 mm

2.4 Flow Cytometry

1. 60-mm Petri dishes (*see* Fig 6a).
2. Single-edged blade (*see* Fig 6a).
3. 50-μm non-sterile CellTrics® filters (e.g., Sysmex, 04-0042-2317; *see* Fig 6b).
4. Sample tube (e.g., Sysmex, 04-200; *see* Fig 6b).
5. Nuclear Staining Solution (e.g., Sysmex, 05-5007).
6. CyFlow Space Flow cytometry (e.g., Sysmex; *see* Fig 6c).

2.5 Bioinformatic Online Tools

1. bcl2fastq (https://emea.support.illumina.com/sequencing/sequencing_software/bcl2fastq-conversion-software.html).
2. FastQC (https://www.bioinformatics.babraham.ac.uk/projects/fastqc/).
3. Trimmomatic (http://www.usadellab.org/cms/?page=trimmomatic).
4. Hisat2 (http://daehwankimlab.github.io/hisat2/).
5. SAMtools (http://www.htslib.org/).
6. IGV (https://igv.org/).
7. HTSeq-count (https://htseq.readthedocs.io/en/release_0.11.1/index.html).
8. RStudio (https://posit.co/download/rstudio-desktop/).
9. DESeq2 (https://bioconductor.org/packages/release/bioc/html/DESeq2.html).
10. g:Profiler (https://biit.cs.ut.ee/gprofiler/gost).
11. KEGG Mapper Color (https://www.genome.jp/kegg/mapper/color.html).
12. Gene model translator tool in MaizeGDB (https://maizegdb.org/gene_center/gene).
13. Grassius Maize TF database (https://grassius.org/browsefamily/Maize/TF).
14. Cytoscape (https://cytoscape.org/).

3 Methods

3.1 Plant Growth Conditions and Emasculation

To obtain flowering plants throughout the year, perennial diploid and polyploid Tripsacum accessions are grown in the greenhouse under long day conditions of 16 h light at 28 °C and 8 h of darkness at 21 °C, and a constant air humidity of 60–65% (*see* **Note 1**). The best time to perform egg cell isolations is summer, as plants produce more flowers. When plants start to flower, track the development of florets and remove the whole staminate spikelet (male part of inflorescences; Fig. 1b) the day before the first anthers open to avoid pollen shading on the pistillate spikelet (female part of the inflorescence). Immediately thereafter, protect the remaining spikelet with a small plastic or paper bag to prevent pollination from other neighboring plants.

During the following days, monitor the presence of silks in the protected pistillate spikelet. In the case of diploid plants, the optimal stage to cut the pistillate spikelet and initiate the isolation of sexual cells is when silks are 3–4 cm in length. In the case of polyploid apomictic plants, only those fruitcases in which the silk did not yet emerge are suitable for isolation of parthenogenetic egg cells, as embryo development is already initiated when the silk is visible.

3.2 Spikelet and Ovule Dissection

This procedure can be performed at room temperature.

1. Once the pistillate spikelet contains fruitcases with the desired stage, carefully cut the part of the spikelet to be processed and place it inside a glass Petri dish with wet paper on the bottom to avoid dehydration.
2. Dissect the fruitcases from the spikelet with the help of curved forceps and a scalpel. Pay attention to the dissection angle, using the small hole of the fruitcase as a reference (Fig. 3a). This cut divides the ovule into two parts; the embryo sac will remain in the lower dissected part of the fruitcase (Fig. 3b). This step is crucial to facilitate later steps in the isolation process (*see* **Note 2**).
3. Repeat step 2 with the next fruitcase of the spikelet (*see* **Note 3**).
4. Take all the lower halves of the fruitcases and place them in another glass Petri dish filled with 6 ml of 0.55 M mannitol solution.
5. Place the glass Petri dish under a stereomicroscope and observe immersed cuttings with 5× magnification.
6. Hold the lower half of a fruitcase with curved forceps and place it with the cut section facing up (Fig. 3c). The lower cut part of the ovule containing the embryo sac becomes visible as a circular yellowish structure in the middle of the cut section (*see* **Note 4**).
7. Gently remove the lower part of the ovule with the help of a hypodermic needle (Fig. 3c). This structure detaches smoothly from the rest of the tissue. After releasing it into mannitol solution, it will remain at the bottom of the Petri dish. If the dissection angle was properly performed, it is possible to see a darker area—the embryo sac—inside the lower part of the ovule (Fig. 3d).

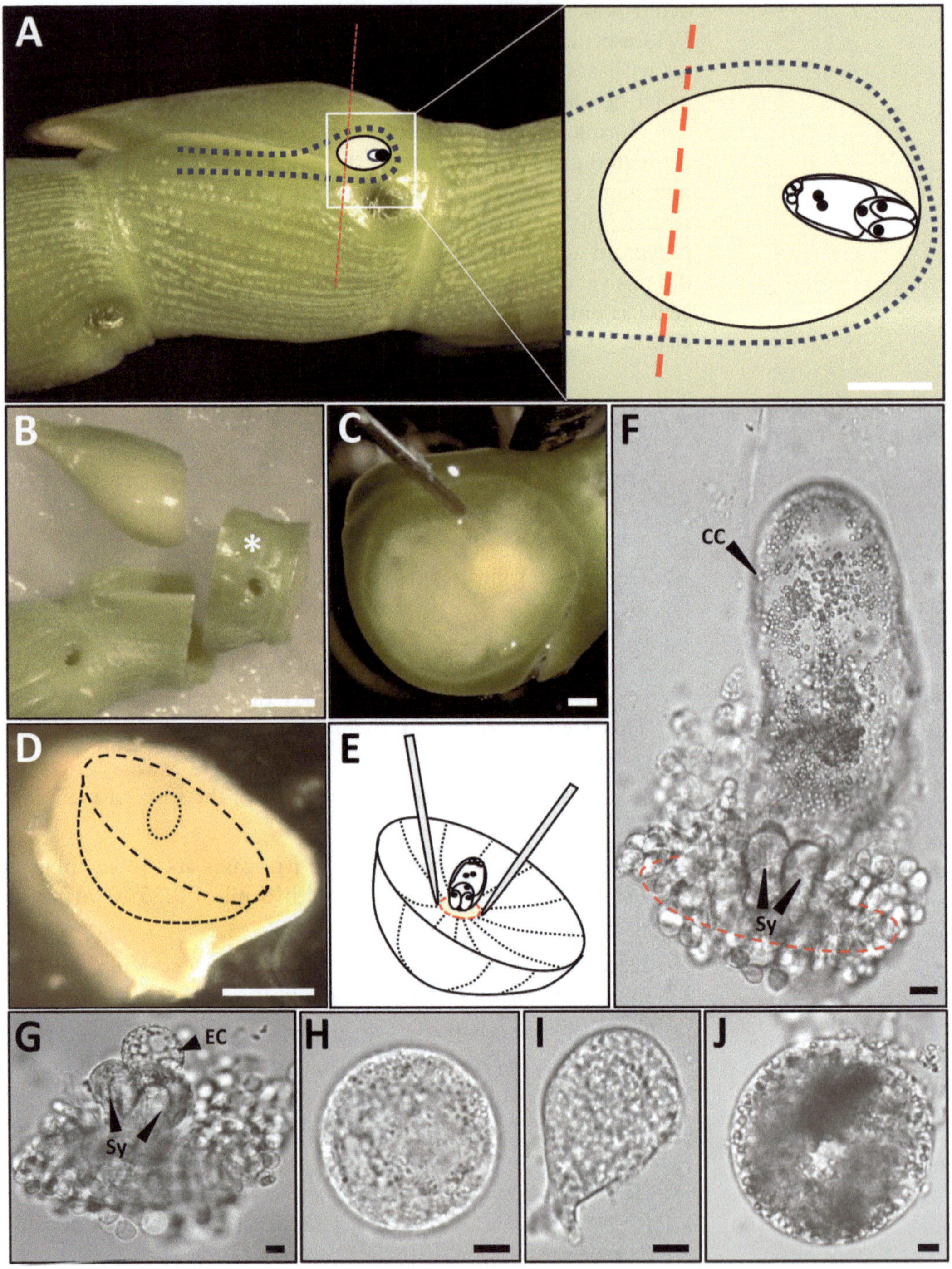

Fig. 3 Procedure to isolate parthenogenetic egg cells from Tripsacum. (**a**) Female spikelet from an apomictic polyploid line showing a fruitcase at the optimal stage (just before silk emergence) to perform egg cell isolation. A diagram of the ovary and the dissection angle is shown. (**b**) Fruitcase after dissection. Asterisks indicate the part that contains the embryo sac. (**c**) Removal of the lower portion of the embryo sac using a needle. (**d**) Lower portion of the dissected ovary. The dashed line indicates the ovule boundaries, while the dotted line indicates a darker area where the embryo sac is located. (**e**) Diagram of the lower portion of the cut ovule, showing the embryo sac in the middle, just on top of the micropylar region. The red dashed line

8. Repeat steps 6–7 until all the lower parts of the ovules from each fruitcase are released into the mannitol solution.
9. Move the released lower parts of the ovules carefully toward the border of the Petri dish with high-precision forceps and quickly transfer them into an Imaging Petri dish containing a drop of 20 μl of enzymatic solution. Make sure that samples are immersed inside the drop to ensure proper enzymatic digestion.
10. Place the Imaging Petri dish in a humid box and incubate samples for 45 min at room temperature.

3.3 Setting Up of Microscope Space

During the incubation time, prepare the area around the inverted microscope as follows:

1. Install the micromanipulator coupled to a glass capillary needle at the right place of the inverted microscope, considering you are right-handed; correspondingly, if you are left-handed (*see* **Note 5**). Moreover, install a holder to place the glass capillary and two sharp glass needles when not in use.
2. Load the double Imaging Petri dish stage on the inverted microscope to work with two Imaging Petri dishes at the same time.
3. In the left spot of the microscope stage, place an Imaging Petri dish containing 20 μl of fresh 0.55 M mannitol solution to wash and collect harvested cells (*see* **Note 6**).
4. Fill one sterile glass Petri dish with 100-μl ethanol absolute and place it close to the micromanipulator.
5. Fill another sterile glass Petri dish with 100 μl of 0.55 M mannitol solution and place it also close to the micromanipulator.
6. Place and empty a glass Petri dish close to the micromanipulator to discard liquids during the washings of the glass capillary.
7. Place close to you and at a safe place two sharp glass needles.
8. Take a 0.2-ml LoBind tube and place 3 μl of 0.55 M mannitol solution at its inner wall (*see* **Note 7**).

3.4 Manual Egg Cell Isolation

This procedure can be performed at room temperature.

Fig. 3 (continued) indicates the area to be kept after dissection with the help of glass needles. (**f**) Remaining part of the ovule showing the embryo sac on top of the micropylar nucellus region (red dashed line) after dissection. (**g**) Micropylar nucellus region showing the egg cell in focus adjacent to the two synergids on the back. (**h**) Isolated parthenogenetic egg cell. (**i**) Isolated synergid cell. (**j**) Isolated central cell. CC, central cell; EC, egg cell; Sy, synergids. Scale bars in A to D = 500 μm and in F to J = 10 μm

1. Take the Imaging Petri dish containing the digested lower part of the ovules and add 3 ml of fresh sterile 0.55 M mannitol solution.
2. Place the Imaging Petri dish under the stereomicroscope.
3. Hold a lower part of the ovule with high-precision forceps and use a hypodermic needle to carefully follow the dashed line as in Fig. 3d. The goal is to separate the ovule containing the embryo sac from the outer ovary tissues.
4. Repeat step 3 until all the lower part of the ovules are processed.
5. Discard the ovary tissues with high-precision tweezers, leaving only the ovule containing the embryo sac ready for isolation.
6. Place an Imaging Petri dish on the right spot of the double Imaging Petri dish holder of the inverted microscope, considering you are right-handed; correspondingly, if you are left-handed.
7. Focus on a dissected ovule containing the embryo sac (Fig. 3e) with the 10x objective.
8. Carefully dissect the surrounding area of the embryo sac as depicted in Fig. 3e by using two sharp glass needles (top in Fig. 2e). Make sure not to touch the embryo sac and leave just few nucellar ovule cells around the micropyle (red dashed line in Fig. 3e).
9. After removing almost all ovule cells, it is sometimes possible to see the whole embryo sac on top of the micropylar region (Fig. 3f), but often, only the synergids and the egg cell are visible while the central cell is disrupted (Fig. 3g).
10. Use one glass needle to hold the remaining ovule tissue, and the other glass needle to gently tap on the micropylar region to carefully detach the egg cell.

Please note that the morphology of grass embryo sac cells is considerably different, so they can be easily distinguished before collecting without usage of molecular markers: the egg cell of Tripsacum is about 50-μm diameter, shows many peripheral vacuoles, and a nice round nucleus (Fig. 3h). The synergids are approximately 45 μm in diameter, they contain many granules, and initially, a semicolon morphology before they round off (Fig. 3i), whereas the central cell (Fig. 3j) is almost 100 μm in diameter with many granules and contains two polar nuclei.

11. Focus on the egg cell to be collected and use a 20× objective to initiate the harvesting process.
12. Aspire 2–3 μl of sterile mannitol solution with a glass capillary coupled to the micromanipulator and immerse it in the

Imaging Petri dish. It is important to do this by looking through the eyepieces in order to better control the process.

13. Approach the glass capillary gently to carefully aspirate the egg cell. If the egg cell is surrounded by debris or other components, use the micromanipulator to clean the area by providing a smooth stream of liquid around the target cell (*see* **Note 8**).
14. Move the microscope stage with the Imaging Petri dish containing fresh mannitol solution in which the cells will be washed. Look through the eyepieces in order to check that the harvested cell is safely dispensed in fresh mannitol solution.
15. Carefully remove the glass capillary and aspirate 20 μl of ethanol absolute to wash away the remaining mannitol solution in the tip of the capillary.
16. Release all the ethanol in the discarding glass Petri dish and place the glass capillary back on its holder.
17. Repeat steps from 1 to 16 until the desired number of cells is collected.
18. After completing the cell harvest, aspirate egg cells carefully and dispense them in the 3-μl 0.55 M mannitol droplet placed inside the 0.2-μl LoBind tube (*see* **Note 9**).
19. Before freezing in liquid nitrogen, place the tip of the glass capillary back into the collecting Imaging Petri dish and check it through the eyepieces to confirm that no cells remain inside.

3.5 RNA Extraction, Library Preparation, and Sequencing

RNA extraction from few cells can be performed using the RNeasy Plus Micro Kit following the manufacturer's instructions with minor modifications. *See* Fig. 4 for the whole procedure.

1. First, after thawing Tripsacum egg and zygote cells, the entire volume for each sample is transferred into 350-μl buffer RLT Plus containing 1% beta-mercaptoethanol (included in the kit). Samples are homogenized by vortexing for 1 min.
2. Pass homogenized samples through gDNA Eliminator spin columns in order to remove genomic DNA contamination.
3. Add one volume of 70% ethanol to the flow-through and apply mixtures to RNeasy MinElute spin columns followed by several washing steps.
4. Finally, elute the total RNA in 12 μl of RNase-free water.
5. Extracted RNA should be assessed using, for example, the Agilent 2100 Bioanalyzer with the RNA 6000 Pico LabChip reagent set following the manufacturer's instructions.

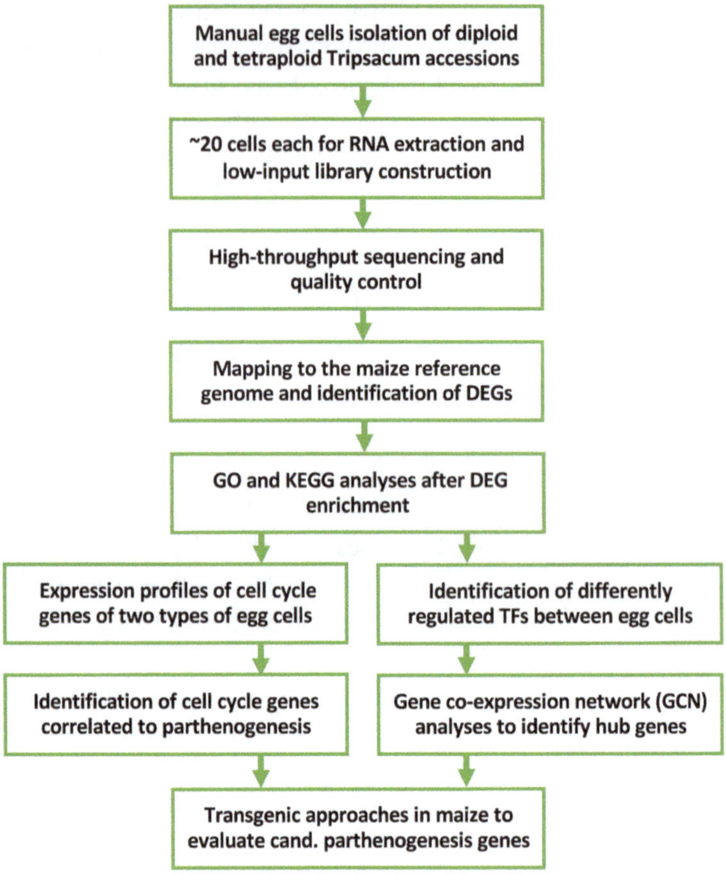

Fig. 4 Working scheme of the entire process to identify and validate candidate parthenogenetic genes from manual egg cell isolation using Tripsacum accessions via bioinformatic analyses and candidate gene validation in transgenic maize

RNA library preparation and sequencing should be conducted as described in the SMART-Seq mRNA LP User Manual, the Illumina NextSeq 2000 Sequencing System Guide, and the KAPA Library Quantification Illumina/ABI Prism Protocol.

6. According to the above-mentioned manufacturer's instructions, approximately 1.5 ng of total RNA for each sample is used to generate first-strand cDNA synthesis.
7. Amplify double-stranded cDNA using LD-PCR (12 cycles) and purify via magnetic bead clean-up.
8. Validate the quality and quantity of the obtained cDNA. Fragment the cDNA enzymatically and ligate stem-loop adapters. The libraries should then be PCR amplified (12–16 cycles) and indexed to generate Illumina-compatible libraries with unique dual indexes (UDIs).

9. After magnetic bead purification, quantify the libraries using the KAPA Library Quantification Kit.
10. Equimolar amounts of each library should then be sequenced on an Illumina NextSeq 2000 instrument following the standard guideline. Convert the resulting .cbcl files into .fastq files using the bcl2fastq software.

3.6 Mapping to a Reference Genome and Network Analyses

1. After obtaining raw RNA-seq data, assess their quality and status using FastQC based, for example, on a Linux platform.
2. Based on the FastQC's output information, use Cutadapt and Trimmomatic to remove adapter sequences and unpaired reads from the raw RNA-seq data, respectively. Generate clean reads for the following procedures (*see* **Note 10**).
3. Given the high evolutionary conservation between Tripsacum and maize, one can use Hisat2 to map the clean reads directly to the maize reference genome (Zm-B73-REFERENCE-NAM-5.0), utilizing the latest annotation (https://ensembl.gramene.org/Zea_mays/Info/Index). To account for the abundant SNPs present between Tripsacum and maize, the mismatch tolerance during mapping can be maximally relaxed.
4. Use, for example, SAMtools or another suitable tool to index and sort the resulting .sam or .bam files. For generating counting output files, .bam files are typically sorted by the gene model's name. Then use the HTSeq-count tool to count the reads mapped to the maize genome.
5. Additionally, .bam files are also sorted by genomic location, allowing gene expression and reads mapping to be visualized in Integrative Genomics Viewer (IGV).
6. The resulting count dataset, typically saved as a .csv file, could be imported into RStudio. Using the DESeq2 package, you should first conduct a principal component analysis (PCA) to evaluate similarities and differences among samples. Then, normalize the counts with the built-in size factor normalization method. Use the normalized counts to perform statistical analyses to identify differentially expressed genes (DEGs) and to generate a comprehensive list of gene expression levels, including Log2 Fold Change and adjusted P-values, calculated for various pairs of comparisons.
7. Based on gene IDs of identified DEGs, use online tools such as g:Profiler and KEGG Mapper Color to perform Gene Ontology (GO) and KEGG pathway enrichment analyses. Gene model tools in MaizeGDB (https://maizegdb.org/gene_center/gene) can be used as the translator between different types of gene IDs (*see* **Note 11**).

8. Parthenogenetic egg cells often exhibit activated cell cycle gene expression. To evaluate their cell cycle status, combine the gene expression list from step 6 with a cell cycle gene list derived from a previous study (https://www.pnas.org/doi/suppl/10.1073/pnas.2018788117/suppl_file/pnas.2018788117.sd01.txt) [9]. Categorize cell cycle genes enriched in the S-phase and G2/M-phase and plot their average normalized counts to compare expression between parthenogenetic and sexual egg cells.
9. Transcription factors and transcriptional reprogramming play critical roles in cell fate transitions. To analyze their involvement, integrate the gene expression list from step 6 with the maize transcription factor list available from the Grassius Maize TF database. Then, quantify how many transcription factors from each family are differentially expressed, and compare the normalized expression counts of key transcription factor DEGs between parthenogenetic and sexual eggs in a single plot.
10. Additionally, considering the significant regulation of numerous transcription factors between parthenogenetic and sexual eggs, a gene coexpression network can be constructed based on their expression counts. For this purpose, utilize a gene coexpression network script available on GitHub (https://github.com/Hokuto-GH/gene-coexpression-network-script). The data transformation required by this script is performed in RStudio, and the output data need to be imported into Cytoscape software to visualize the coexpression network. See Fig. 5 for an example.

3.7 From Candidate Genes to Transgenic Maize Lines

Once candidate parthenogenesis genes are identified, their functions in inducing embryogenesis can be validated in maize. Genes that are upregulated in parthenogenetic egg cells may function as activators of embryogenesis. To test this, candidate genes can be ectopically expressed using egg cell-specific promoters, such as the *EGG CELL1.2* (*EC1.2*) promoter from Arabidopsis [10] or the *EGG APPARATUS1* (*EA1*) promoter from maize [11] to mimic the behavior of parthenogenetic egg cells. For genes that are downregulated in parthenogenetic egg cells, CRISPR/Cas9 (Clustered Regularly Interspaced Short Palindromic Repeats/CRISPR-associated protein 9)-mediated knockouts can be used to determine if these genes act as repressors of embryogenesis initiation. If transgenic sexual egg cells can be successfully reprogrammed to initiate parthenogenesis, haploid egg cells should develop autonomously into a haploid embryo, confirming the role of the candidate genes in embryogenesis regulation.

1. Construct vectors for maize transformation using, for example, vectors described here [12, 13].

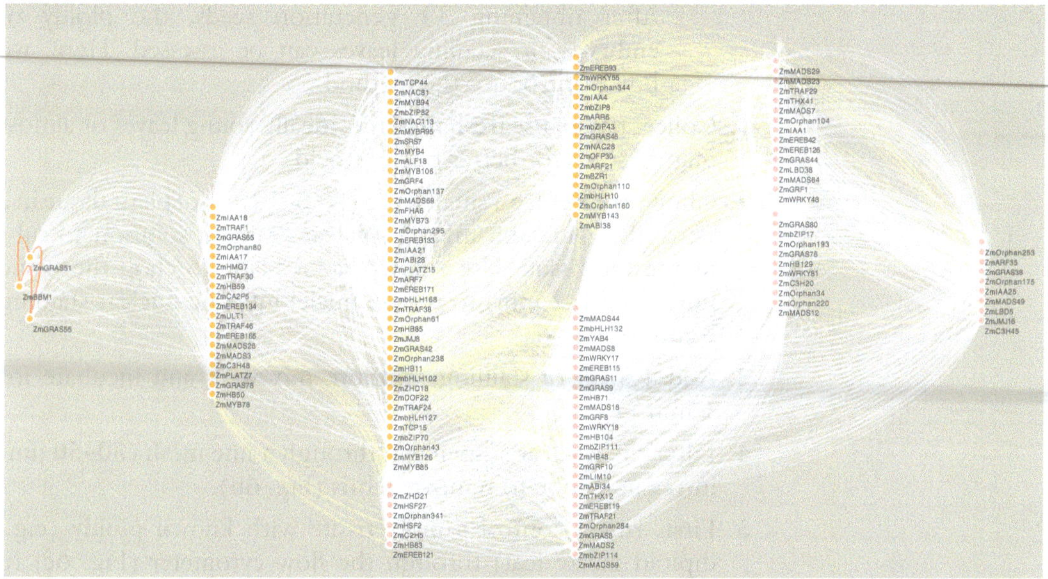

Fig. 5 Coexpression network of differentially regulated transcription factors (TFs) between parthenogenetic and sexual eggs of Tripsacum. The network features two main clusters, highlighted in orange and pink, with connections represented by different colors: white lines indicate links within each cluster, while yellow lines depict interactions between the clusters. Notably, the relationships among three major TFs, ZmGRAS51, ZmGRAS56, and ZmBBM1, are marked with red lines to emphasize their central role in the network

2. Stable transformation of maize can be done as described here [14].
3. Phenotyping for gene-induced parthenogenesis: the parthenogenesis phenotype becomes observable only after gametophyte development. Haploid embryos can be identified through various methods:
   (a) Chromosome counting: Staining and counting of chromosomes provide a direct approach; however, this method is impractical for large-scale plant analysis, as it is very labor-intensive.
   (b) Morphological traits: Characteristics such as dwarfism, smaller tassels and ears, as well as sterility are often associated with haploidy; however, these traits are only discernible in late vegetative stage of plants and can be influenced by environmental factors.
   (c) Molecular markers: Utilizing molecular markers can confirm haploidy, but requires prior knowledge of the parental genomes. For large-scale screening of haploid plants, anthocyanin markers like R1-nj and flow cytometry are widely employed [15].
   (d) Flow cytometry: This method is particularly advantageous due to its speed and accuracy in determining ploidy levels.

After obtaining T1 generation seeds, the ploidy of embryos or seedling leaves can be assessed. Here, we briefly introduce the method.

1. Collect tissue: use fresh material, ideally young leaves or almost mature seeds for better nuclear yield.
2. Chop tissue: place a small piece (~0.5–1 cm^2) of plant tissue in a Petri dish. Add 0.5 mL of Nuclear Staining solution to submerge the tissue. Use a sharp razor blade to chop the tissue finely (for 1–2 min) to release nuclei into the nuclear staining solution (Fig. 6a).
3. Add 1.5 ml of staining solution, mix well, and incubate for 1 min.
4. Filter the nuclear suspension through a fine mesh (30–50 μm) into a new tube to remove debris (Fig. 6b).
5. First, run a control sample tissue with known ploidy (e.g., diploid maize leaf) through the flow cytometer (Fig. 6c) to set baseline fluorescence peaks (*see* **Note 12**).
6. Load the transgenic sample into the flow cytometer. Compare sample fluorescence peaks to the diploid control; haploid plants will show half the fluorescence intensity of diploid plants (Fig. 6d).

4 Notes

1. Tripsacum is a short-day plant and grows best at temperatures between 20 °C and 30 °C; thus, other plant growth conditions work as well. We use the same long-day growth conditions as for maize to utilize the same greenhouse chambers for cultivation of both species and to accelerate growth.
2. This dissection angle proved to be the optimal one, as performing a lower cut damages the embryo sac, while a higher one will add unnecessary amount of tissue to complicate further dissection.
3. It is considerably easy to manipulate and apply the dissection angle in each fruitcase when they are still part of the spikelet, rather than cutting first all the fruitcases and then dissecting them separately.
4. It is important to keep the cuttings inside the mannitol solution and keep the Petri dish mostly closed as mannitol crystals appear very fast when in contact with air.
5. In case glass capillaries break, prepare a box with extra glass capillaries in proximity to the microscope so that they can be quickly replaced.

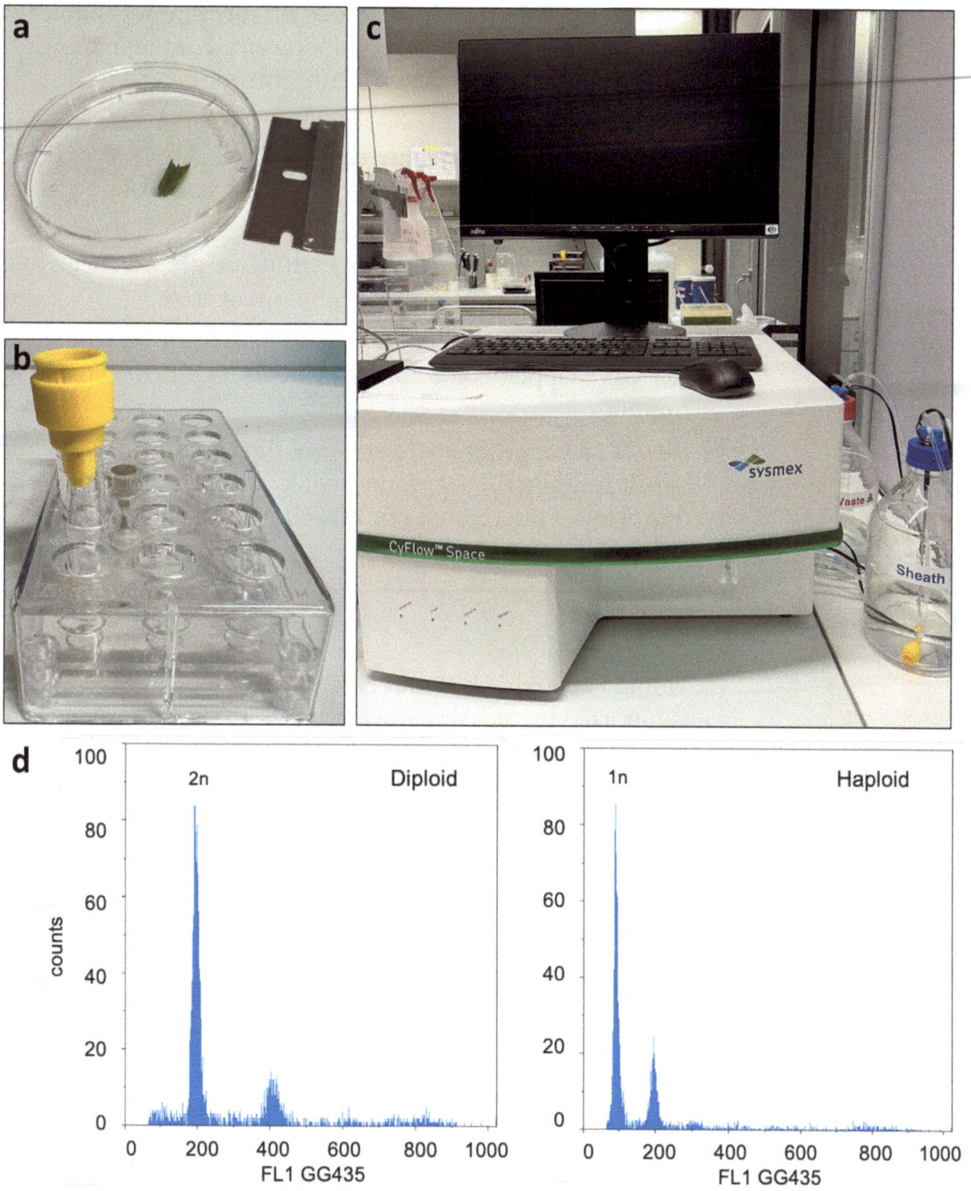

Fig. 6 Materials and equipment for flow cytometry analysis of plant ploidy. (**a**) A Petri dish and a single-edged blade used for chopping plant tissues. (**b**) Chopped samples are filtered through a CellTrics filter into sample tubes. (**c**) Filtered samples are ready for ploidy analysis using a flow cytometer. (**d**) DNA content analysis showing fluorescence peaks for diploidy (left) and haploidy (right)

6. In order to prevent fluctuations in the concentration of mannitol solutions containing harvested cells, it is recommended to place a wet paper on the inner part of the Imaging Petri dish lid.
7. Keep the 0.2-ml LoBind tube closed and in horizontal position to prevent evaporation and shifting of the mannitol drop to the bottom of the tube, respectively.

8. Avoid aspirating a big quantity of surrounding liquid when aspiring the cells to avoid unnecessary contamination of the washing solution in the collecting Imaging Petri dish.
9. Cells can be dispensed one by one or several at once, depending on if you plan to generate single-cell RNA-seq or bulk RNA-seq data. In both cases, try to keep the final volume as low as possible.
10. It is recommended to rerun FastQC on the cleaned reads to ensure high-quality data for subsequent steps.
11. g:profiler requires version 5 gene model for searching, while KEGG Mapper Color normally allows NCBI entry name as the input.
12. Setting an appropriate lower level threshold helps to eliminate small unwanted background while ensuring that signals from particles of interest are not missed.

Acknowledgments

We would like to thank Armin Hildebrand for plant care. This work was supported by the Alexander von Humboldt Foundation (to X. Z. and T.D.), the Fundação para a Ciência e a Tecnologia (FCT; grant 2021.00872.CEECIND to M.F.T.), and the HORIZON-MSCA-2021-PF program of the European Union (grant 101059247 to M.F.T).

Conflicts of Interest The authors have no conflicts of interest to declare that are relevant to the content of this chapter.

References

1. Hand ML, Koltunow AM (2014) The genetic control of apomixis: asexual seed formation. Genetics 197(2):441–450. https://doi.org/10.1534/genetics.114.163105
2. Goeckeritz CZ, Zheng X, Harkess A, Dresselhaus T (2024) Widespread application of apomixis in agriculture requires further study of natural apomicts. iScience 27(9):110720. https://doi.org/10.1016/j.isci.2024.110720
3. Cornaro L, Banfi C, Cucinotta M, Colombo L, van Dijk PJ (2023) Asexual reproduction through seeds: the complex case of diplosporous apomixis. J Exp Bot 74(8):2462–2478. https://doi.org/10.1093/jxb/erad054
4. Bantin J, Matzk F, Dresselhaus T (2001) Tripsacum dactyloides (Poaceae): a natural model system to study parthenogenesis. Sex Plant Reprod 14(4):219–226. https://doi.org/10.1007/s00497-001-0119-z
5. Chen J, Strieder N, Krohn NG, Cyprys P, Sprunck S, Engelmann JC, Dresselhaus T (2017) Zygotic genome activation occurs shortly after fertilization in maize. Plant Cell 29(9):2106–2125. https://doi.org/10.1105/tpc.17.00099
6. Voichek Y, Hurieva B, Michaud C, Schmucker A, Vergara Z, Desvoyes B, Gutierrez C, Nizhynska V, Jaegle B, Borg M, Berger F, Nordborg M, Ingouff M (2023) Cell cycle status of male and female gametes during Arabidopsis reproduction. Plant Physiol 194(1):412–421. https://doi.org/10.1093/plphys/kiad512

7. Yan L, Kenchanmane Raju SK, Lai X, Zhang Y, Dai X, Rodriguez O, Mahboub S, Roston RL, Schnable JC (2019) Parallels between natural selection in the cold-adapted crop-wild relative Tripsacum dactyloides and artificial selection in temperate adapted maize. Plant J 99(5): 965–977. https://doi.org/10.1111/tpj.14376
8. Gault CM, Kremling KA, Buckler ES (2018) Tripsacum De novo transcriptome assemblies reveal parallel gene evolution with maize after ancient polyploidy. Plant Genome 11(3). https://doi.org/10.3835/plantgenome2018.02.0012
9. Satterlee JW, Strable J, Scanlon MJ (2020) Plant stem-cell organization and differentiation at single-cell resolution. Proc Natl Acad Sci USA 117(52):33689–33699. https://doi.org/10.1073/pnas.2018788117
10. Sprunck S, Rademacher S, Vogler F, Gheyselinck J, Grossniklaus U, Dresselhaus T (2012) Egg cell-secreted EC1 triggers sperm cell activation during double fertilization. Science 338(6110):1093–1097. https://doi.org/10.1126/science.1223944
11. Marton ML, Cordts S, Broadhvest J, Dresselhaus T (2005) Micropylar pollen tube guidance by egg apparatus 1 of maize. Science 307(5709):573–576. https://doi.org/10.1126/science.1104954
12. Frame BR, Shou H, Chikwamba RK, Zhang Z, Xiang C, Fonger TM, Pegg SE, Li B, Nettleton DS, Pei D, Wang K (2002) Agrobacterium tumefaciens-mediated transformation of maize embryos using a standard binary vector system. Plant Physiol 129(1):13–22. https://doi.org/10.1104/pp.000653
13. Char SN, Neelakandan AK, Nahampun H, Frame B, Main M, Spalding MH, Becraft PW, Meyers BC, Walbot V, Wang K, Yang B (2017) An Agrobacterium-delivered CRISPR/Cas9 system for high-frequency targeted mutagenesis in maize. Plant Biotechnol J 15(2): 257–268. https://doi.org/10.1111/pbi.12611
14. Ishida Y, Hiei Y, Komari T (2007) Agrobacterium-mediated transformation of maize. Nat Protoc 2(7):1614–1621. https://doi.org/10.1038/nprot.2007.241
15. Gain N, Chhabra R, Chandra S, Zunjare RU, Dutta S, Chand G, Sarika K, Devi EL, Kumar A, Madhavan J, Muthusamy V, Hossain F (2023) Variation in anthocyanin pigmentation by R1-navajo gene, development and validation of breeder-friendly markers specific to C1-inhibitor locus for in-vivo haploid production in maize. Mol Biol Rep 50(3):2221–2229. https://doi.org/10.1007/s11033-022-08214-2

Chapter 9

Gene Discovery in Reproduction: RNA-seq Analysis of Isolated Pistils from Apomictic and Sexual *Eragrostis curvula*

Juan Pablo Selva, José Carballo, Diego Zappacosta, and Viviana Echenique

Abstract

This protocol describes a comprehensive approach to analyze the differential expression of genes in pistils at distinct developmental stages from sexual and apomictic genotypes of *Eragrostis curvula*. It integrates the development of reproductive calendars, methods for pistil isolation, and techniques for extracting high-quality RNA from limited tissue quantities. This protocol includes detailed steps for inflorescence collection, measurements and clarification of reproductive tissues, as well as the use of differential interference contrast (DIC) microscopy to ensure precise identification of developmental stages. Additionally, the methodology outlines the bioinformatics pipelines used to analyze gene expression patterns, providing insights into the molecular mechanisms underlying apomixis. This approach addresses key challenges, such as small tissue size and mixed developmental stages, offering a reliable framework for transcriptomic studies in *E. curvula*.

Key words Poaceae, Pistils, Reproductive mode, Gene expression

1 Introduction

Eragrostis curvula (Schrader) Nees, commonly known as weeping lovegrass, is a *Poaceae* family member native to Southern Africa. It has been introduced into several countries as forage crop due to its adaptability to marginal semiarid regions. Even though a few diploid and tetraploid individuals of this species reproduce sexually, most genotypes are facultative apomicts and polyploids [1, 2], while fully apomictic genotypes are rare and infrequent. The diploid genome size of *E. curvula* is around 1200 Mb, with a basic chromosome number of 10. The absence of meiosis and the occurrence of two rounds of mitosis during apomictic embryo sac development (Eragrostis type) make this model unique and suitable to be transferred to economically important crops, especially within the

Tatyana Radoeva and Peter J. van Dijk (eds.), *Apomixis: Methods and Protocols*, Methods in Molecular Biology, vol. 2987, https://doi.org/10.1007/978-1-0716-5001-1_9,

Poaceae family. In addition, the availability of sexual, facultative, and full apomictic genotypes makes this species an excellent model for the study of apomixis through comparative transcriptomic analysis.

Many studies have been performed in recent years to identify genes involved in apomixis control (reviewed by [3]). Traditionally, the study of these genes and the mechanisms in which they are involved were carried out using different strategies. One of these strategies is the functional genomics approach based on transcriptomics. In the case of *E. curvula*, as in other grass species, there are some limitations associated with the small size of the flowers and the specific requirements for RNA quality and quantity. Many studies have previously been conducted using complete spikelets [4–7]. However, the main constraints of this approach are the mix of developmental stages within the samples and the presence of undesirable tissues, such as vegetative or male tissues (e.g., anthers) that can mask the expression patterns of certain genes. In addition, variations in morphology and differences in the duration of embryo sac development across genotypes make the comparison between equivalent stages difficult. A possible solution for the listed challenges is the creation of a reproductive calendar combined with the development of techniques to isolate pistils and to subsequently obtain high quality RNA from small quantities of tissue. The latter would allow the capture of genes expressed at specific developmental stages. For that purpose, a series of techniques and protocols, involving clearing of reproductive tissues and the use of differential interference contrast (DIC) microscopy were developed. This simple and effective technique allows for the study of ovary and anther anatomy, as well as all stages of embryo sac and pollen grain development. It also enables the isolation of different developmental stages from both apomictic and sexual genotypes. Furthermore, it previously allowed the observation that meiosis in both micro- and megasporogenesis is generally synchronous. After evaluating various morphological measurements of the female gametophyte, pistil length was selected as the primary parameter to establish the reproductive timeline of *E. curvula*. Choosing this as the main parameter is essential since it maximizes the correlation with each female developmental stage [8], mainly due to its lower variability compared to other measures. In addition, the small size of the pistils samples requires a specialized RNA extraction protocol to achieve the quantity and quality required for sequencing.

This chapter introduces a protocol for pistils isolation at different developmental stages from sexual and apomictic genotypes of *E. curvula* coupled to extraction of small amount of RNA samples for sequencing and a methodology for the analysis of gene expression patterns.

2 Materials

2.1 Plant Materials

The following *Eragrostis curvula* genotypes were used for pistils isolation: Tanganyika (obligate apomictic), Don Walter (facultative apomictic), and Victoria (fully sexual).

2.2 Pistils Isolation for Morphological Observations

1. Farmer's solution: ethanol: glacial acetic acid (3:1).
2. 70% ethanol.
3. 85% ethanol.
4. 96% ethanol.
5. Eosin solution (0.1% eosin in 96% ethanol).
6. Absolute ethanol.
7. Methyl salicylate: ethanol (1:3).
8. Methyl salicylate: ethanol (1:1).
9. Methyl salicylate: ethanol (3:1).
10. Methyl salicylate.
11. Orbital shaker.
12. Stereomicroscope with digital camera equipment for pistils dissection, e.g., Leica S8APO or similar.
13. Leica software LAS EZ (version 3.4.0), or equivalent to the system used, to determine the sizes of different structures from images.
14. Microscope equipped with a Differential Interference Contrast (DIC) system, e.g., Leica DM2500 LED or similar and a digital camera, e.g., Leica MC170 HD or similar.
15. Microscope slides.
16. Scalpel #3 with blade #11, or an equivalent fine-scalpel and blade combination suitable for precise tissue dissections (*see* **Note 1**).
17. Very fine-tipped histological forceps (e.g., Dumont SS/45 and #5/45 forceps).
18. Syringes with fine hypodermic needles (e.g., insulin needles, 28G–32G) for precise handling.

2.3 Pistils Isolation for RNA Extraction

1. Very fine-tipped histological forceps (e.g., Dumont SS/45 and #5/45 forceps).
2. Syringes with fine hypodermic needles (e.g., insulin needles, 28G–32G) for precise handling.
3. Scalpel #3 with blade #11, or an equivalent fine-scalpel and blade combination suitable for precise tissue dissections.
4. Fine spatula.

5. 1.5 ml Eppendorf tubes.
6. Microscope slides and microscope cover slips (22 x 22).
7. Insulated polystyrene container.
8. Tube rack.
9. Sterile tissue paper.
10. 70% ethanol solution to frequently clean the forceps and scalpel.
11. Liquid nitrogen.
12. Stereomicroscope with an ocular reticle (10/0.1 mm) for direct measurements during observation and material dissection. The reticle saves time by eliminating the need to capture images for measurement.

2.4 RNA Extraction and Sequencing

1. NucleoSpin RNA XS RNA Isolation kit (Macherey Nagel) or similar to extract RNA for a low quantity of tissue.
2. Spectrophotometer capable of measuring small sample volumes (≤1 μL).
3. Bioanalyzer 2100 (Agilent) or any equivalent system for RNA quality assessment.

2.5 Transcriptome Analysis

1. 64-bit computer running Linux and at least 8 GB of RAM and 4 CPUs.
2. Raw data processing: Fastqc, Trimmomatic, cutadapt.
3. Mapping: Hisat2.
4. Transcript quantification: feature count.
5. Transcript expression: DEseq2.

3 Methods

To establish a relationship between *E. curvula* flower morphology and different female reproductive developmental stages, a series of measurements, such as pistil, ovary, and anther lengths, as well as ovary width, must be performed. Subsequently, it is essential to correlate these organ measurements with the corresponding ovule developmental stages observed under the microscope to accurately identify the correct stage to be analyzed. These measurements focus on pistils, as the components of apomixis are associated with the female gametophyte.

Once the correlation values are acceptable and the standard deviation is low, the reproductive developmental stages within the pistils can be inferred directly from pistil length. Finally, the number of pistils at the different developmental stages must be determined to ensure a sufficient amount of high-quality RNA for sequencing and gene expression analysis. For more detailed information on calendar construction, *see* Selva et al. [in progress]. The values provided here are specific to *E. curvula*, but the methodology can be applied to other species.

3.1 Pistil Collection to Establish the Reproductive Calendar and Verify the Correct Developmental Stage

The steps for fixation and methyl salicylate clearing follow the protocol reported by [9], with the modifications stated below. Observations should be conducted using a DIC microscope (*see* **Note 2**).

1. Collect the panicles when the apical spikelet is in anthesis. At this point, reproductive developmental stages ranging from archesporial cell to potentially fertilized egg and/or central cells can be observed in the different spikelets along the panicle [10]. Figure 1a shows an *E. curvula* panicle with the apical spikelet in anthesis.

Fig. 1 Panicle and spikelet morphology of *Eragrostis curvula.* (**a**) Panicle with spikelets in anthesis at the top. (**b**) Detail of a spikelet with five visible florets (b-1 to 5), showing anthesis on the first basal floret (b-1). (**c**) Measurement of floral parameters: al (anther length), pl (pistil length), ol (ovary length), and sl (style length)

2. Collect 20–30 spikelets from different parts of the panicle and place them in a 1.5-ml plastic tube containing Farmer's solution (*see* **Note 3**).
3. After 24 h of fixation, remove Farmer's solution and replace it with 70% ethanol.
4. Store samples at 4 °C (under these conditions, they can be stored for several months).
5. Replace 70% ethanol with 85% ethanol and shake for 1 h (*see* **Note 4**).
6. Replace 85% ethanol with 96% ethanol and shake for 1 h.
7. Replace 96% ethanol with eosin solution and shake for 30 min.
8. Replace eosin solution with 100% ethanol and shake for 1 h.
9. Replace 100% ethanol with methyl salicylate:100% ethanol (1: 3) and shake for 1 h.
10. Replace methyl salicylate:100% ethanol (1:3) with methyl salicylate: 100% ethanol (1:1) and shake for 1 h.
11. Replace methyl salicylate: 100% ethanol (1:1) with methyl salicylate:100% ethanol (3:1) and shake for 1 h.
12. Replace methyl salicylate:100% ethanol (3:1) with pure methyl salicylate and shake for 1 h. At this step, spikelets can be stored at 4 °C for several months.
13. Separate each flower from the spikelet by removing palea and lemma.
14. Mount the pistils along with one of the attached anthers onto a slide to observe both male and female reproductive structures (*see* **Note 5**).
15. Measure floral parameters as shown in Fig. 1c. The gynoecium length (gl) should be measured along the same axis, from the base of the carpel to the apex of the style. The ovary length (ol) should be measured from the base to the top of the carpel. The style length (sl) should be measured from its base (just above the ovary) to the tip. The anther length (al) should be measured from its base to the distal tip. It is recommended to make a general data matrix for all the measured parameters, including gynoecium, ovule, style, and anther lengths.
16. Observe under microscope the structures as shown in Fig. 2, where it is possible to distinguish all the different developmental stages.
17. Once all the measurements are taken and each developmental stage is assigned to each measurement, a box-and-whisker plot is created. If the stages are well separated by pistil length, the plot can be modified to include 75% of the data within the box. If this adjustment successfully separates the stages (as is shown

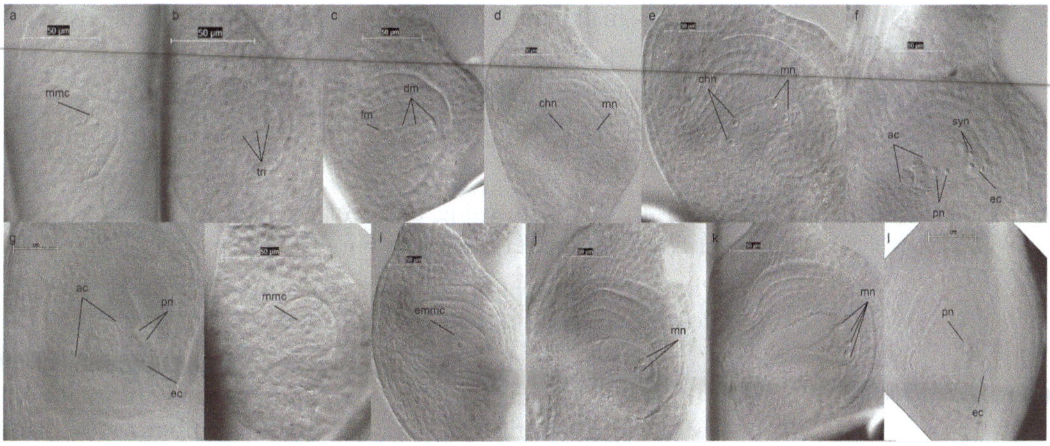

Fig. 2 Typical sexual (**a–g**) and apomictic (**h–l**) processes in *Eragrostis curvula* observed using DIC microscopy. (**a**) megaspore mother cell (mmc); (**b**) triad of megaspores (tri); (**c**) functional megaspore (fm) and degenerated megaspores (dm); (**d**) bi-nucleated embryo sac; (**e**) tetra-nucleated embryo sac; (**f**) octa-nucleated embryo sac; (**g**) mature embryo sac; (**h**) megaspore mother cell (mmc); (**i**) elongated megaspore mother cell (emmc); (**j**) bi-nucleated embryo sac; (**k**) tetra-nucleated embryo sac; (**l**) mature embryo sac. Chn: chalazal nucleus/nuclei, mn: micropillar nucleus/nuclei, ac: antipodal cell, ec: egg cell, pn: polar nucleus/nuclei

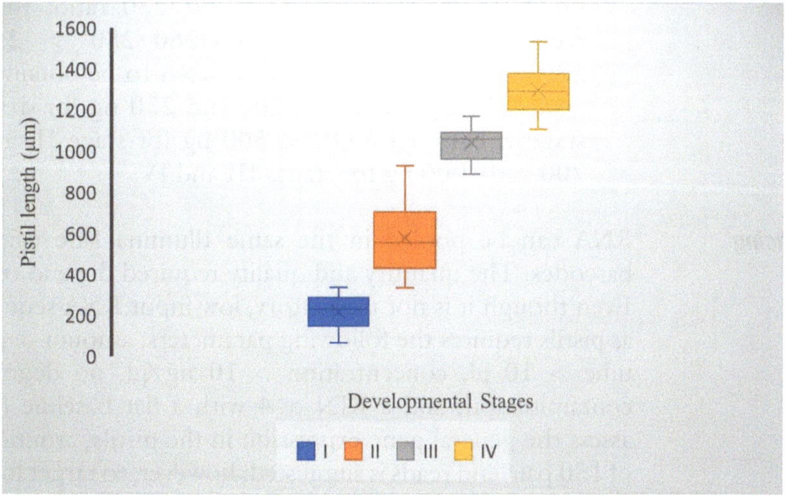

Fig. 3 Variation between pistil length (μm) and female developmental stage in *E. curvula* cv. Don Walter. The box represents 75% of data (interquartile range) for each of the four stages

in Fig. 3), the Q1 and Q3 values are then used as the lower and upper length limits for each stage.

3.2 Pistil Isolation

1. Place the flowering plant near the stereomicroscope where you will be working (*see* **Note 6**).
2. Carefully clean the working surfaces and instruments with 70% ethanol.

3. Take a small panicle branch containing 1–2 spikelets.
4. Under the stereomicroscope, remove the florets from the spikelet using forceps, scalpel, and a needle (*see* **Note 7**).
5. Quickly extract the entire pistil, taking care not to damage it (*see* **Note 8**).
6. Measure the pistil using the ocular reticle to estimate its size and developmental stage (*see* Fig. 3).
7. Place the pistil in a 1.5-ml microcentrifuge tube held in a rack, floating in liquid nitrogen (*see* **Note 9**).
8. Use an average of 200 pistils for stage I (archesporial cell), 100 pistils for stages II (meiosis) and III (immature embryo sac), and 50 pistils for stage IV (mature embryo sac). Each triplicate should consist of pistils collected from different plants or panicles.

3.3 RNA Extraction and RNA-Seq

1. Use an RNA extraction kit for small amounts of RNA, e.g., NucleoSpin RNA XS RNA isolation kit (Macherey Nagel).
2. Follow the manufacturer's protocol of the specific kit.
3. Quantify the amount of RNA obtained using a nanodrop, considering these quality parameters simultaneously: OD260/280 ratio and OD 260/230 ratio; the nanodrop recommended values are OD260/280 > 2, OD260/230 > 2. The total amount of RNA to be obtained with this protocol ranges between 200 and 250 ng for developmental stage I, between 300 and 500 ng for stage II, and between 700 and 1000 ng for stages III and IV.

3.4 Sequencing

RNA can be pooled in the same Illumina lane using different barcodes. The quantity and quality required depend on the input. Even though it is not mandatory, low input RNA sequencing such as pistils requires the following parameters: amount > 100 ng, volume > 10 μl, concentration > 10 ng/μl, no degradation and contamination, and a RIN > 4 with a flat baseline (Fig. 4). To assess the general gene expression in the pistils, around 50 million of 150 pair-end reads is suggested; however, to target low expressed transcripts, the number of reads should be doubled. These values could vary with the species being analyzed, taking into account the genome size, ploidy, and number of genes. To target larger regions, an insert size around 300 bp is desirable (Fig. 5).

3.5 Transcriptome Analysis

1. RNA-seq Quality Check

After sequencing, generate fastq files and perform quality check for the following parameters: Phred score distribution per base, per tile and per sequence, per base sequence content, and adaptor sequence content. The recommended thresholds for these

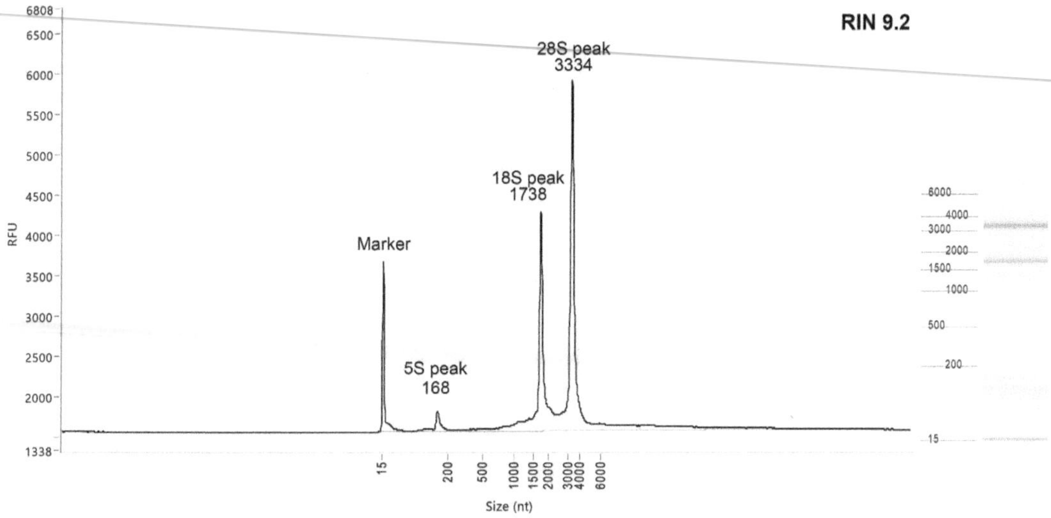

Fig. 4 Assessment of RNA integrity on an Agilent 2100 Bioanalyzer System

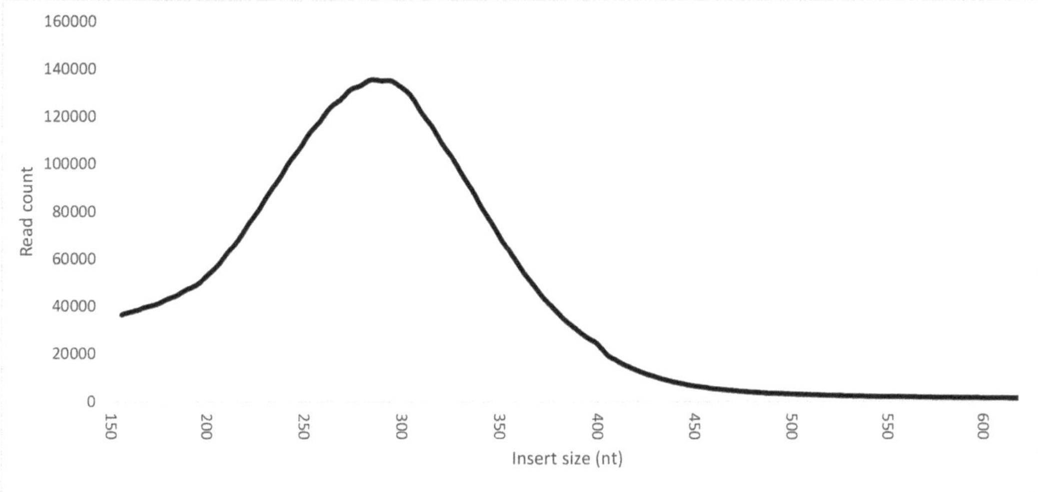

Fig. 5 Insert size distribution in the RNA-seq libraries

evaluations are as follows: per-base sequence quality >28, no deviation the quality per tile, increased distribution of sequences with a quality score > 30, and nonadaptor content in the reads. Perform these evaluations with FastQC software [11]. If a library does not meet any of these parameters, filter it using Trimmomatic or Cutadapt [12, 13].

(a) Running FastQC:

```
fastqc --threads number_of_threads  -o $OUTPUT_FILE
-f fastq $INPUT_FILE.fastq.gz
```

(b) Running trimmomatic:

```
java -jar trimmomatic.jar PE -threads <integer> $INPUT_-
PAIR1.fastq.gz $INPUT_PAIR2.fastq.gz $output_FILTERED_PAIR1.
fastq.gz $output_FILTERED_PAIR2.gz $output_UNPAIRED1.fastq.gz
$output_UNPAIRED2.fastq.gz ILLUMINACLIP:TruSeq3-PE.fa:2:30:7
SLIDINGWINDOW:4:20 MINLEN:100
```

(c) Running cutadapt:

```
Cutadapt -j number_of_threads -a $ADAPTER_SEQUENCE -o
$output__FILTERED_PAIR1.gz -p $output_FILTERED_PAIR2.gz $IN-
PUT_PAIR1.fastq.gz $INPUT_PAIR2.fastq.gz
```

2. Read Mapping

Reads must be mapped to the reference *E. curvula* genome assembly [14] with Hisat2 [15] and the bam files generated by this software are quality checked. The main metrics to evaluate here are the overall mapping rate and the reads aligned more than once. If the overall mapping rate is < ~75%, the remaining ~25% should be analyzed looking for contaminations. The number of allowed multiple-mapped reads depends on the ploidy level, i.e., in polyploid genomes, multiple-mapped reads are expected, while in diploids, no more than 5% is expected.

Running hisat2:

For each sample run:

```
hisat2 -x $reference.index -1 $output__FILTERED_PAIR1.gz
-2 $output__FILTERED_PAIR2.gz -p number_of_threads > $OUTPUT.
bam
```

3. Sample QC

After read mapping, the sampling and replicates must be evaluated. Firstly, transcript quantification must be carried out using featureCounts tool to count reads per transcript [16]. Then using DESeq2 rlog function, each sample is transformed and normalized with respect to library size [17]. Finally, to detect similarities between replicates and differences between tissues and stages, sample distance and variability are assessed using a heatmap and principal components analysis (PCA). The heatmap is generated with the Euclidean distance function in R, while the PCA is visualized using

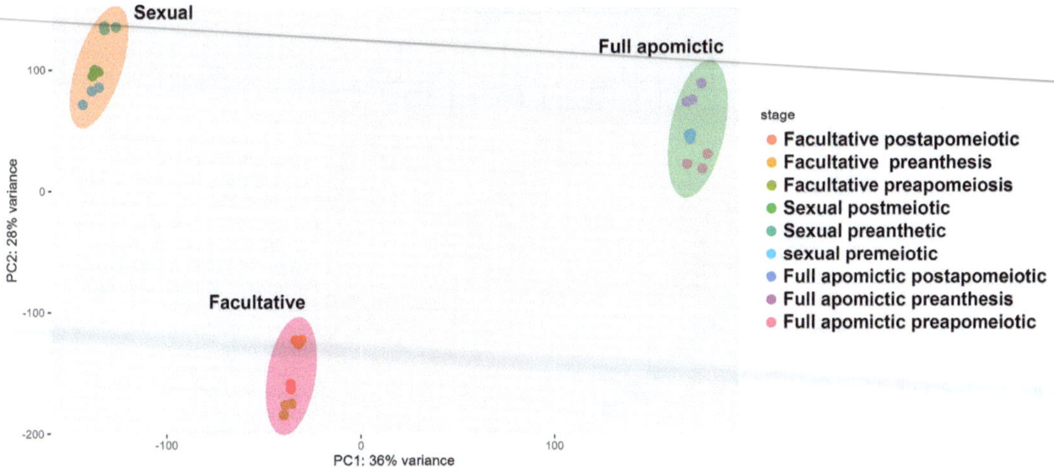

Fig. 6 Principal component analysis (PCA) of the different samples. Each point represents a sample, colored by stage. The variance explained by the first two principal components (PC1 and PC2) is shown on the axes, with PC1 capturing the highest variability

the DESeq2 *plotCounts* function [17, 18]. For data visualization, ggplot is used for the PCA (Fig. 6) and pheatmap for the heatmap (Fig. 7) [19]. In the heatmap, shorter distances are observed within replicates of the same stages and reproductive mode. Follow the steps below to perform the sample QC.

(a) Running featureCounts:

```
featureCounts --countReadPairs -p -M -T number_of_threads -t exon -g gene_id -o $OUTPUT.matrix.tsv -a $ANNOTATION_FILE.gff $INPUT_sample1.bam $INPUT_sample2.bam $INPUT_sample3.bam $INPUT_sampleN.bam
```

(b) Running DESeq2 and plotting in R:

```
count_table_genes <- read.table("$INPUT.matrix.tsv ", sep = "\t", header = T)
sample_table <- read.table("$INPUT.sample_table.tsv", sep = "\t", header = T)
coldata <- data.frame(apply(X = sample_table, MARGIN = 2, FUN = as.factor))
dds_genes <- DESeqDataSetFromMatrix(countData = count_table_genes, colData = coldata, design = ~stage)
dds_genes <- estimateSizeFactors(dds_genes)
dds_genes <- estimateDispersions(dds_genes)
rlog_dds_genes <- rlog(dds_genes)
```

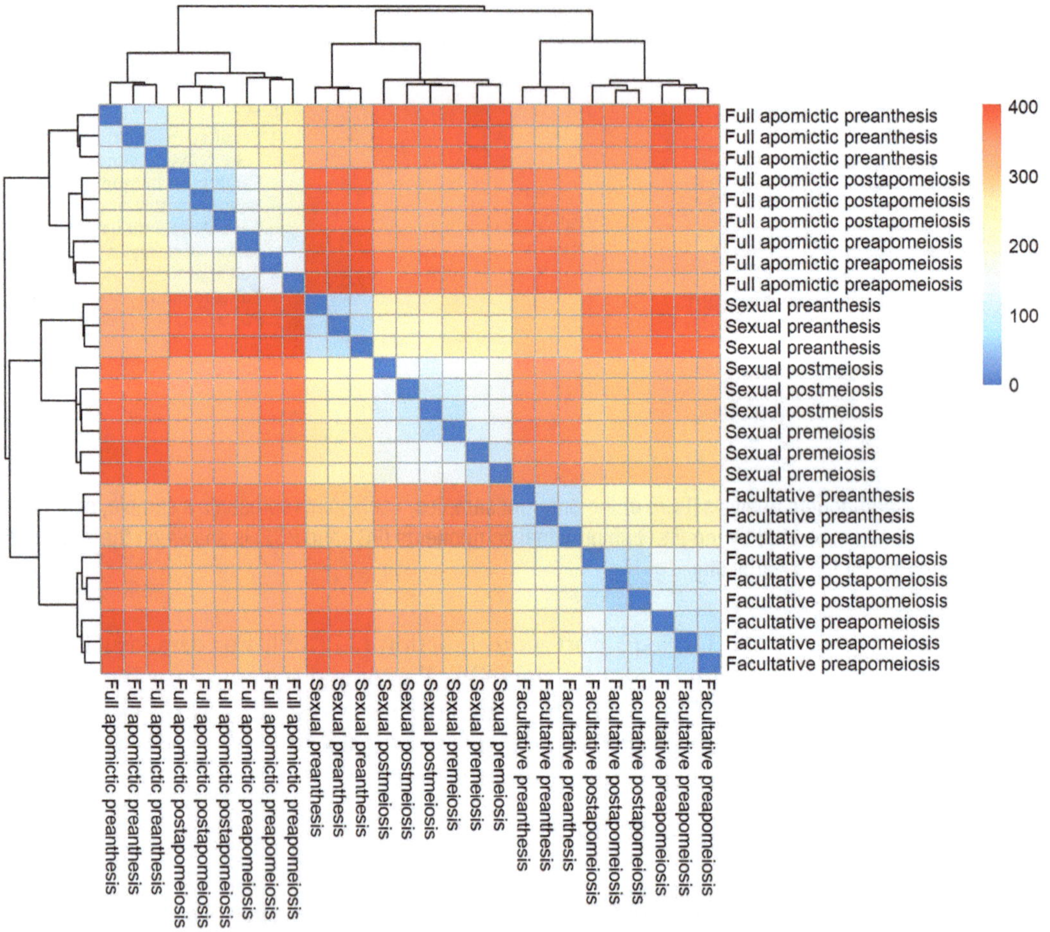

Fig. 7 Heatmap of the different samples showing the distance between replicates. The figure shows closer distance within replicates and the differences across different samples

```
dist <- as.matrix(dist( t( assay(rlog_dds_genes) ) ))
pheatmap(dist, scale = "none")
pca <- <-plotPCA(rlog_dds_genes, intgroup = "stage",
returnData=T)
percentVar <- round(100 * attr(pca, "percentVar"))
ggplot(pca, aes(PC1, PC2, color=stage)) + geom_point
(size=3) + xlab(paste0("PC1: ",percentVar[1],"% variance")) +
ylab(paste0("PC2: ",percentVar[2],"% variance"))
```

4. Differential Gene Expression Analysis

Differential gene expression analysis can be performed using DESeq2. The function "results" allows to do customized comparisons, set the p-value, and set the log2foldchange. Even though the p-value of 0.1 is widely used, here, it is advisable to raise it to 0.05 to increase confidence of the results. The genotypes and stages must be compared by pairs. Within each genotype, the different stages/tissues are compared to distinguish the genes that are specifically expressed during sexual, facultative, and full apomictic development. For instance, in this approach, premeiotic versus postmeiotic stages are compared to evaluate the expression pattern of genes involved in sexual embryo sac development, mainly meiotic genes. On the other hand, homologous stages between genotypes are compared to evaluate the changes on genes due to different reproductive modes. For example, a comparison could be obligate apomictic versus sexual and facultative versus sexual in premeiotic stage. Finally, differentially expressed genes in apomictic genotypes (full apomictic and facultative) are grouped together to analyze how they could be affecting the trait. Execute the following steps to perform differential gene expression analysis:

(a) Running differential gene expression in DESeq2 in R:

```
#continuing with the previous example
dds_genes <-DESeq(dds_genes)
#comparing stages different genotypes
   Full_apomicticVSsexual_premeiosis_genes <- results
(dds_genes, alpha=0.05, lfcThreshold=1, contrast = c
("stage","Full apomictic preapomeiotic", "Sexual premeiotic
"))
```

4 Notes

1. Use a scalpel with a sharp blade and small-tipped histological forceps (e.g., Dumont SS/45 #5/45 forceps). Insulin needles or similar histological tools can also facilitate the process.
2. The methyl salicylate-clearing technique was chosen due to its advantages over traditional histological methods (e.g., embedding, microtome sectioning, and staining). This approach significantly reduces the time required from sample collection to microscopic observation.
3. Always prepare fresh Farmer's solution when collecting plant material. Ensure the use of absolute ethanol (100%) instead of commercial ethanol (96%).
4. Samples were incubated on an orbital shaker at low speed to ensure continuous contact between the solution and the spikelets.
5. Pistils with two plumose stigmas must be positioned sagittally on the slide for optimal observation of ovule nuclei and integument size. Create a "channel" on the slide using two cover slips as supports, and place a third cover slip over the pistils for microscopic observation (Fig. 8).

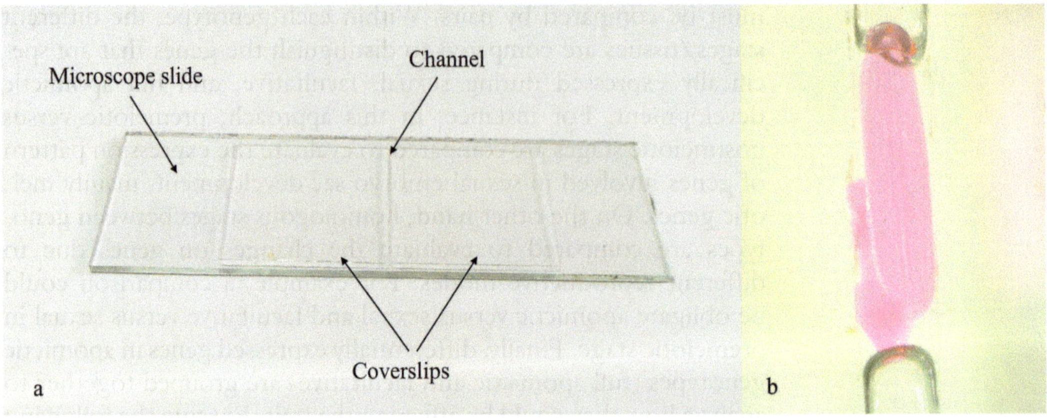

Fig. 8 Methodology for pistil and anther observation under the microscope. (**a**) Setup of a slide with cover slips to create a channel for proper positioning of pistils and anthers. (**b**) Detailed view showing the sagittal positioning of pistils within the channel for optimal microscopic observation

6. Keep the plant with the panicle close to the stereomicroscope to minimize delays in pistil extraction and prevent RNA degradation.
7. Spikelets should be detached from the rachis with a scalpel. Under the stereomicroscope, use forceps to grasp the spikelet from the top and a scalpel to cut the florets, extracting the pistil and anthers from each floret. Perform the dissection quickly to maintain RNA quality.
8. Handle pistils carefully during sampling to avoid damage and ensure accurate transcriptomic data. Conduct practice sessions prior to real sample collection.
9. Ensure that collected pistils remain in the tube when transferring, avoiding loss via the transportation/handling tools (e.g., needles or scalpels).

Acknowledgments

This protocol was developed in the frame of a project funded by the European Union's Horizon 2020 Research and Innovation Programme under the Marie Skłodowska-Curie Grant Agreement No 872417 (MAD) and NO 101007438 (POLYPLOID). Also, this was funded by Agencia Nacional de Promoción Científica y Tecnológica ANPCyT (PICT-2021-I-A-00576), Universidad Nacional del Sur (PGI 24/A261), Consejo Nacional de Investigaciones Científicas y Técnologicas CONICET (PIP 11220200101905CO), and CUIA/CONICET (24120190100015CO).

Conflicts of Interest The authors have no conflicts of interest to declare that are relevant to the content of this chapter.

References

1. Carballo J, Zappacosta D, Selva JP et al (2021) Eragrostis curvula, a model species for diplosporous apomixis. Plants 10:1818. https://doi.org/10.3390/plants10091818
2. Zappacosta D, Gallardo J, Carballo J et al (2019) A high-density linkage map of the forage grass eragrostis curvula and localization of the diplospory locus. Front Plant Sci 10:918. https://doi.org/10.3389/fpls.2019.00918
3. Schmidt A (2020) Controlling apomixis: shared features and distinct characteristics of gene regulation. Genes 11:329. https://doi.org/10.3390/genes11030329
4. Garbus I, Romero JR, Selva JP et al (2017) De novo transcriptome sequencing and assembly from apomictic and sexual Eragrostis curvula genotypes. PLoS One 12:e0185595. https://doi.org/10.1371/journal.pone.0185595
5. Garbus I, Selva JP, Pasten MC et al (2019) Characterization and discovery of miRNA and miRNA targets from apomictic and sexual genotypes of Eragrostis curvula. BMC Genomics 20:839. https://doi.org/10.1186/s12864-019-6169-0
6. Selva JP, Zappacosta D, Carballo J et al (2020) Genes modulating the increase in sexuality in the facultative Diplosporous grass Eragrostis curvula under water stress conditions. Genes 11:969. https://doi.org/10.3390/genes11090969
7. Pasten MC, Carballo J, Gallardo J et al (2022) A combined transcriptome – miRNAome approach revealed that a kinesin gene is differentially targeted by a novel miRNA in an apomictic genotype of Eragrostis curvula. Front Plant Sci 13:1012682. https://doi.org/10.3389/fpls.2022.1012682
8. Bisp P, Selva JP, Carballo J, et al (2022) A reproductive calendar of apomictic and sexual genotypes of Eragrostis curvula (Schrad.) ness. 26th international conference on sexual plant reproduction
9. Young BA, Sherwood RT, Bashaw EC (1979) Cleared-pistil and thick-sectioning techniques for detecting aposporous apomixis in grasses. Can J Bot 57:1668–1672. https://doi.org/10.1139/b79-204
10. Meier M, Zappacosta D, Selva JP et al (2011) Evaluation of different methods for assessing the reproductive mode of weeping lovegrass plants, Eragrostis curvula (Schrad.) Nees. Aust J Bot 59:253. https://doi.org/10.1071/BT10267
11. Andrews S (2010) Babraham bioinformatics – FastQC a quality control tool for high throughput sequence data. https://wwwbioinformaticsbabrahamacuk/projects/fastqc/Accessed 13 May 2025
12. Bolger AM, Lohse M, Usadel B (2014) Trimmomatic: a flexible trimmer for Illumina sequence data. Bioinformatics 30:2114–2120. https://doi.org/10.1093/bioinformatics/btu170
13. Martin M (2011) Cutadapt removes adapter sequences from high-throughput sequencing reads. EMBnet j 17:10. https://doi.org/10.14806/ej.17.1.200
14. Carballo J, Santos BACM, Zappacosta D et al (2019) A high-quality genome of Eragrostis curvula grass provides insights into Poaceae evolution and supports new strategies to enhance forage quality. Sci Rep 9:10250. https://doi.org/10.1038/s41598-019-46610-0
15. Kim D, Paggi JM, Park C et al (2019) Graph-based genome alignment and genotyping with HISAT2 and HISAT-genotype. Nat Biotechnol 37:907–915. https://doi.org/10.1038/s41587-019-0201-4
16. Liao Y, Smyth GK, Shi W (2019) The R package Rsubread is easier, faster, cheaper and better for alignment and quantification of RNA sequencing reads. Nucleic Acids Res 47:e47–e47. https://doi.org/10.1093/nar/gkz114
17. Love MI, Huber W, Anders S (2014) Moderated estimation of fold change and dispersion for RNA-seq data with DESeq2. Genome Biol 15:550. https://doi.org/10.1186/s13059-014-0550-8
18. R Development Core Team (2010) R a language and environment for statistical computing: reference index. R Foundation for Statistical Computing, Vienna
19. Kolde R, Kolde MR (2015) Pheatmap: pretty Heatmaps

Part IV

Genetic Analysis of Apomixis

Chapter 10

Genetic Mapping in Apomicts

Ross Bicknell, Andrew Catanach, and Chris Winefield

Abstract

Genetic mapping efforts in apomicts date back several decades, but significant progress only became possible with the application of molecular mapping tools. In this chapter, we present the approaches that have been taken to apply these techniques to apomictic species, the necessary experimental steps, caveats associated with interpreting the data, and potential new marker approaches. A brief overview of our understanding of the genetics of natural apomixis is also presented.

Key words Apomixis, Molecular mapping, Deletion mapping, Polyhaploid mapping

1 Introduction

Genetic mapping in sexual species is typically based on measurements of allelic segregation within populations of related individuals. Gene/trait associations are inferred through the co-segregation of genes and the traits they influence. Apomixis, however, leads to the generation of clonally identical progeny and therefore to the formation of highly uniform populations. In clonal populations, traits co-segregate with the entire genome, not with any subset of it, which confounds the use of standard approaches to inheritance.

In this chapter, we present an introduction to the approaches used to study inheritance in apomicts, with a specific focus on studies of the genetic architecture of apomixis itself. Both sporophytic and gametophytic apomicts have been studied and both will be discussed. However, most of the research effort, and therefore most of the data available, has focused on gametophytic systems, so these will be discussed in greater depth. As in sexual systems, inheritance studies in apomicts require two steps, first the generation and classification of meaningful variation, followed by a molecular analysis to map differences between the variant classes. The two steps will be presented separately as different approaches can be

Tatyana Radoeva and Peter J. van Dijk (eds.), *Apomixis: Methods and Protocols*, Methods in Molecular Biology, vol. 2987,
https://doi.org/10.1007/978-1-0716-5001-1_10,

successfully combined. Reference is also given to analogous efforts studying sex chromosomes in both plants and animals, which share similar experimental constraints.

2 Generating Variant Classes for Mapping in Apomicts

2.1 Segregation Analysis

In apomictic systems, as in sexual systems, inheritance is most frequently studied using segregation analysis. Intriguingly, the first to systematically study inheritance in an apomictic system was Gregor Mendel, who extensively recorded inheritance in the genus *Pilosella* (syn. *Hieracium*), in the years immediately following his studies in *Pisum* [1]. It is now understood that *Pisum* is sexual, while *Pilosella* is a gametophytic apomict, so it is unsurprising that Mendel was unable to reconcile the different results he obtained from the two systems [2, 3]. Curiously, the idea of two types of inheritance persisted even after the rediscovery of Mendel's work by de Vries [4], Correns [5], and von Tschermak [6].

As this demonstrates, interpreting inheritance in apomicts can be difficult. It is now understood that individual plants in the genus *Pilosella* (syn. *Hieracium*) may be either entirely sexual or facultatively apomictic. In a facultative apomict, some of the egg cells progress directly into embryogenesis without fertilization ("parthenogenesis"), while others require fertilization before further development will occur ("syngamy'). In *P. piloselloides* (syn. *H. piloselloides*) and *P. aurantiaca* (syn. *H. aurantiacum*), the levels of sexual seed formed following pollination were 2.0 and 2.4%, respectively [7]. Facultative apomixis is the most common form of apomixis among both sporophytic and gametophytic apomicts. Therefore, if inheritance is studied using progeny gathered from an apomictic parent, the products of fertilization must first be separated from the products of parthenogenesis before the data can be analyzed for trait/marker associations.

Apomixis is a developmental pathway characteristic of the plant gynoecium. The processes of male meiosis and pollen formation in apomictic species most commonly appear to be similar to those of sexual relatives [8]. Crosses using an apomictic pollen parent and a sexual recipient, therefore, often result in segregating progeny, which can be used to study inheritance, although the unidirectional nature of these crosses prevents the consideration of parent-of-origin effects. The choice of sexual recipient is a critical parameter. This plant needs to be sufficiently similar to the apomictic pollen donor to ensure fertility in the progeny, as seed-set in the progeny is likely to be a measured parameter. However, many apomicts have diverged from their sexual relatives to the point where they are no longer regarded as conspecific, so these crosses are typically interspecific in nature. The need to use a sexual plant of sufficient relatedness to an apomict also means that care must be taken that the chosen plant does not have any latent tendencies toward apomixis. Very often, they do [9].

Additionally, almost all gametophytic apomicts are polyploid and this must be considered in the choice of parents used. Ideally, the selected apomictic parent will have an even ploidy status (4x or 6x) as these predominantly produce a single class of euploid pollen. Similarly, the ideal sexual recipient would have a similar ploidy to ensure uniformity and fertility in the progeny. Even when the parents are of a similar ploidy, the progeny of polyploid plants often vary significantly in chromosome number, so they must be tested by flow cytometry to confirm the mechanism of their formation [10]. Finally, apomixis is a complex trait involving several developmental processes. Progeny should, therefore, be assessed for paternity, and then for the components of the mechanism rather than for apomixis as an intact trait. Correctly, this would involve an embryological examination and germination test, together with tests for ploidy and marker segregation [11–13].

2.2 Deletion Mapping

Deletion mutation has a long history of use in biology. Following the pioneering discoveries of early researchers [14, 15], it was generally accepted that exposure to radiation, in particular X-rays, ultraviolet light and radium, influenced the expression of heritable traits in both plants and animals [16]. Most of these early studies, however, lacked objective, quantifiable measures of genetic change due to mutation. Muller [17, 18], working on the fruit fly *Drosophila melanogaster*, proposed the use of specific heterozygous stocks carrying known, mapped, recessive markers, to accurately quantify the impact of a range of mutagens. Using this approach, he confirmed and quantified the mutagenic influence of X-rays on several fly genes [19]. Stadler [20, 21] reported a similar influence of X-ray exposure on maize and barley, also using characterized heterozygous stocks to quantify the effect.

Within an individual cell, exposure to ionizing radiation influences the DNA via two types of reactions: directly through DNA cleavage and indirectly through the production of free radicals, principally hydroxy radicals, which cause oxidative damage to the DNA molecule [22]. Repair of the damaged DNA, particularly of double-stranded breaks, can result in permanent changes in code ranging from single base changes to the loss of entire chromosomes [23, 24].

In animal systems, deletion mapping has been extensively used in cancer research. Many forms of cancer are associated with somatic mutation, allowing candidate loci to be identified by comparing genomic sequence from a tumor with that of healthy somatic tissue from the same individual [25, 26]. In plants, deletion mutagenesis has been widely applied to induce variation for plant breeding purposes [27]. In cereals, most notably bread wheat, it has also been used to develop genetic stocks for the placement of genes on the homologous chromosomes of this allo-hexaploid species [28]. For apomixis research, deletion mapping has the advantages that it can be applied to native apomictic genotypes, it does not require a sexual parent, it is phenotype-based (permitting forward genetic screening), it is not influenced by suppressed recombination, and it can be applied to species without prior need of a genetic map.

The application of deletion mutation to apomixis research requires five prerequisites: evidence that target components of apomixis (e.g., apomeiosis, parthenogenesis and/or autonomous endosperm) are controlled by dominant determinants, the availability of genetically characterized stocks where the target dominant allelic determinant(s) are simplex, a method for vegetative increase of the mutants to permit the isolation of nonchimeric plants and the perpetuation of sterile mutants, a mapping tool that can be applied to identify common genetic lesions, and finally, a reference genome. In practical terms, this means that deletion mapping is typically performed on plants that have already been studied using other mapping approaches, most commonly, segregation analysis. In the past, the main mapping approach used was AFLP [29], but this is likely to be replaced by whole-genome approaches for future studies (*see* below and [24]). Two apomictic genera have been studied using deletion mapping: *Pilosella piloselloides* (syn. *Hieracium caespitosum*) [30] and *Taraxacum officinale* [31]. The two are closely related members of the Asteraceae, both expressing a form of gametophytic apomixis. In both studies, mutants were generated through the gamma-irradiation of seed, using a ^{60}Co source, and AFLP was used to analyze the mutants for regions of common genome loss (*see* Chapter 14 for further experimental details).

Gamma irradiation results in cell-specific mutation. Within an irradiated tissue, such as a meristem, each irradiated cell has an independent mutation history, so a treated meristem gives rise to a mosaic of different mutant sectors. As this complicates both phenotypic and molecular screening, a method for purifying mutant genotypes is required. In sexual species, this is typically achieved by outcrossing. With asexual plants, however, outcrossing is either not possible or undesirable, as it results in the loss of the uniform (clonal) genetic background of the asexually derived starter stock. In both the *Pilosella* and *Taraxacum* studies, tissue culture was used to regenerate plants from mutant sectors of the M0 plants. Regions of common genome loss/rearrangement were then identified by aligning mapping data for similar mutants, and candidate genes inferred. The power of this approach is critically dependent on the number of mutants used. Deletion mutation often results in very large chromosomal rearrangements (which provide less resolution) and it also results in a large number of unlinked lesions. In practical terms, therefore, ideally, at least 30 mutants with a common phenotype are required.

2.3 Polyhaploid Mapping

Gametophytic apomixis is the combined outcome of a series of component processes, notably, the avoidance of meiosis ("apomeiosis"), the spontaneous division of the unfertilized egg cell ("parthenogenesis"), and a mechanism that enables the formation of a compatible endosperm tissue to ensure the survival and development of the embryo [32]. These components appear to be under separate genetic control in the systems studied (*see* below). Commonly, they are also not completely penetrant in apomictic genotypes. As a result, the progeny of a gametophytic apomict may include a small number of seedlings that arose following meiosis and/or following fertilization, as well as the majority that arose strictly by maternal replication [32]. Polyhaploids are a progeny type that result from the parthenogenic development of reduced egg cells [8]. Accordingly, they are classified as n + 0 progeny (*see* Chapter 1 for terminology). As polyhaploids are the product of parthenogenesis, they are entirely of maternal origin, with each a representing a subset of the maternal genotype. As they arise following meiotic reduction, polyhaploids have half the maternal chromosome number. Being the product of meiosis, each is genetically unique, both in terms of the inheritance of maternal homologues but also through the influence of crossover. This means that polyhaploids represent a segregating progeny class that can be used for molecular mapping in apomictic systems. The main difficulties experienced with using polyhaploids for molecular mapping are that they are weaker than maternal progeny (so are often lost) [8] and they are typically rare, often representing less than 2% of the total progeny formed [7].

Polyhaploid-based mapping has only been reported in one apomictic system, *Pilosella piloselloides*, where it has been separately used to map the *LOA* locus controlling apomeiosis [33] and the *LOP* locus controlling parthenogenesis [34]. As *LOP* function is essential for polyhaploid formation, polyhaploid mapping was used to map both the physical and functional limits of this locus in *P. piloselloides* [34].

Polyhaploid mapping is feasible in *Pilosella* because methods have been devised to isolate large numbers of this rare progeny class with minimum selection bias. Two approaches are reported. In the first, a target apomictic genotype was transformed with a cassette containing *npt*II, a positive selectable marker (kanamycin resistance) and *cod*A, a negative selectable marker (5-flurocytosine sensitivity) [35]. Approximately 98% of the progeny of the transformed apomict were maternal [7]. As they carried both markers, they were susceptible to 5-flurocytosine and easily identified when seed was germinated on a medium containing the selection agent. Due to the action of meiosis, half of the polyhaploid progeny also inherited the transgene cassette, and half did not. Germination on selection medium, therefore, rescued approximately 1% of the total seeds sown. Although this is a low yield, the selection system is simple, and the numbers of seed produced by *Pilosella* are so large that several thousand could be isolated easily. The second technique used a mutant of *P. piloselloides* (syn. *H. caespitosum*) with lost *LOA* function, identified in a deletion mutation screen [30]. In this mutant, all egg cells arose following meiosis. As 50% of them inherited the *LOP* determinant that promotes parthenogenesis, the polyhaploidy rate in the progeny approached 50%, making it very easy to produce a mapping population. Care needs to be taken when using these populations, however, as the original plant carries an induced mutant load that might influence inheritance in the polyhaploid populations derived from it.

3 Molecular Analyses of Variant Classes

3.1 Non-reference-Based Approaches: AFLP

Early attempts at identifying genetic determinants controlling apomixis pre-dated the era of ready access to genome sequencing and assembly, so the molecular tools applied were used without a reference genome for the plant under study. Even when plant genome assemblies initially became available, they often proved to be of limited value, as they focused on model species, such as *Arabidopsis*, or on important crop species. None of these were apomicts and most not even closely related to an apomictic species. Two notable exceptions were *Boechera* (an apomictic relative of *Arabidopsis*) [36] and members of the Cichorioideae, a tribe of the family *Asteraceae*. Even in these cases, however, the use of related sequence from a sexual species has proved challenging. Schranz et al. [37] reported that the alignment of *Arabidopsis* and *Boechera* genomic sequence revealed that the highly reduced state of the *Arabidopsis* genome, and apparent chromosome reshuffling compared to relatives, reduced the power of comparative genomic approaches for mapping in *Boechera*. Improved utility of *Arabidopsis* sequence was achieved by considering it in terms of 24 conserved chromosomal regions. Similarly, the Cichorioideae includes lettuce (*Lactuca sativa*) as well as the apomictic genera *Taraxacum*, *Pilosella*, *Hieracium*, *Chondrilla*, *Ixeris*, and *Crepis*. A genome assembly for lettuce was published in 2017 [38], but it was available as a draft chromosome-level assembly in 2013, so the lettuce genome assembly helped to identify loci associated with parthenogenesis in *Taraxacum* and *Pilosella* [31]. As in *Boechera*, however, while local alignment to the lettuce genome proved helpful for designing markers, significant chromosomal rearrangement was also apparent, so prior knowledge of locus positions in the apomicts was necessary.

Amplified fragment length polymorphism (AFLP) [39], typically combined with a form of bulked segregant analysis [40], was a particularly useful reference-free marker class. In addition to enabling the identification of markers linked to the trait without prior genomic knowledge [29], AFLP is highly reproducible, scalable, and it can be automated. Any AFLP markers found in association with a trait could be isolated as bands from a gel, sequenced by the Sanger method, and used as initial probes for bacterial artificial chromosome (BAC) walking. A BAC walk aimed to generate a contiguous series of overlapping BAC clones across a putative genomic region (often several megabases), which could then be sequenced. While powerful, given the technology available at the time, BAC libraries were expensive to produce and screening them was time-consuming, whether it be by probing blots of BAC clones or by PCR of multidimensional pools of BAC plasmid DNA. Although considerable progress was made using these techniques, it was the advent of DNAseq and RNAseq using Illumina technology, coupled with the availability of relevant genome assemblies, which finally enabled the full elucidation of apomixis at the molecular level in apomixis model systems.

3.2 k-mer Analysis

Apomixis determinants have dominant allelic action and often reside in nonrecombinant genomic regions that have undergone augmentation and allele sequence divergence when compared to syntenic haplotypes [31, 41]. These are features also shared by sex-determining loci in both plants and animals, which have received more research attention. One model of the transition from hermaphroditism to dioecy in plants involves two determinants, one to suppress female floral development and the other to promote male floral development, in a context where male development does not proceed in female plants [42]. This is analogous to gametophytic apomixis, with determinants for apomeiosis and parthenogenesis, although clonal reproduction in apomicts allows the two genes to be unlinked, whereas recombination in dioecious species between the two components quickly results in sterility, hence they remain tightly linked in a single nonrecombinant region of the sex chromosome.

Notwithstanding the additional challenges presented due to polyploidy, the identification of sex loci and their determinants share many of the challenges faced when identifying apomixis loci, and k-mer based approaches have proven to be particularly successful in sex locus research. Two notable examples are the sex loci of persimmon and kiwifruit. In persimmon, Illumina genomic and cDNA sequence data were generated from pools of male and female F1 plant samples, and k-mers unique to males (hemizygous regions within the Y chromosome) were catalogued and used for sequence assembly of male-specific genomic contigs. RNAseq reads were also aligned to the genomic contigs assembled from male-specific genomic k-mers, along with differential expression, enabling a narrowing of gene candidates along with an elegant verification of results obtained from each approach. The gene determinant was found to be a male-specific small RNA, carrying multiple stop-codons, but highly conserved in the genus, which represses a transcription factor that normally causes feminization through androecia sterilization.

The kiwifruit sex locus was similarly identified via male-specific k-mers. While genome assemblies of female kiwifruit cultivars were available, the first male genome assembly was yet to be published. The approach was therefore essentially reference-free, although alignments of contigs constructed from assembled male-specific k-mers of putative male-specific regions could be made displaying Y and X chromosome-specific polymorphisms. However, of most interest were male-specific contigs that showed no alignment, and these proved to be the source of the sex-determining genes of the Y chromosome.

Brukhin et al. [36] used k-mers to identify differences between apomictic and sexual accessions of *Boechera*, noting that the technique highlighted regions of heterozygosity unique to the apomicts. Although this research lead is promising, we are unaware of any further reports of k-mers being used to identify apomixis loci and therefore recommend that this approach be considered for other apomictic species. We also expect that deletion mutation would also be well suited for k-mer mapping, as deletion mutants are derived from clones, with samples sharing most of their DNA sequence.

3.3 Reference-Based Approaches: Comparative Genomics

The advent of next-generation sequencing (NGS) dramatically accelerated the pace and reduced the cost of producing genome assemblies. Apomicts, however, remained poorly represented in the growing collection of available genome assemblies due to their polyploid and highly heterozygotic nature. Loci associated with apomixis are also typically rich in transposable elements and other repeated sequences, complicating genome assembly and the phasing of alleles. Illumina data do, however, enable the use of reference-free k-mer approaches to compare apomicts with sexual or mutant counterparts as described above.

Sequencing technology, however, is progressing at an unprecedented rate, as are advances in the computational tools for analyzing sequence data. It is now possible for single labs to generate high-quality, annotated, whole-genome sequences. Long-read sequencing technologies also assist with allele phasing in polyploid backgrounds, overcoming many of the limitations that have delayed the elucidation of apomictic genomes [43]. Reference-based approaches identify features that are specific to apomictic samples by mapping differential sample data against an apomictic reference genome. Short-read Illumina sequence data may be mapped, followed by conventional downstream variant calling and read-depth analysis. The detection of extended regions containing SNPs that are heterozygous in apomictic samples, while homozygous in sexual samples, accompanied by a test such as

Fisher's Exact Test, provides evidence of apomixis-associated regions. Hemizygous regions, present in apomicts and absent in sexuals, are also likely to be of great interest. These regions can be identified using read-depth analysis, with a tool such as SAMtools [44]. Similar techniques have been used to identify sex chromosomes, and sex-determining genes in animals, such as in the fish species *Pseudocaranx georgianus* [45].

3.4 Transcriptomics

Transcriptomics in apomixis is being addressed in Part III in this book; however, in the context of mapping, transcript data provide a valuable resource for marker design and has been used to direct candidate approaches in gene function studies. In *Pilosella piloselloides (*syn. *Hieracium praealtum*), the output of an RNAseq experiment of [46] was used to identify alternative gene and allele sequences, which later informed mapping studies defining the site, size, and allelic phasing of *LOP*, the locus controlling parthenogenesis in this species [34]. Similarly, the reference transcriptome of *Paspalum notatum* [47] informs the identification and mapping of apomixis-associated genes in this model apomict [48]. In *Citrus*, Shimada et al. [49] constructed a custom microarray based on *Citrus* mRNA sequence, with particular emphasis on genes within the citrus polyembryonic locus. Transcript probes covering a developmental series throughout fruit development were then used to identify differences in expression between a monoembryonic and a polyembryonic variety. In this way, a candidate locus was identified that appears to be critical for apomixis in this system.

4 The Genetic Control of Apomixis

As Part V of this book covers genetic control in more detail for individual systems, this topic is covered here in an overview. Gametophytic apomixis is reported to be relatively simply acquired in most systems, through the inheritance of either one or a small number of factors, most commonly as dominant determinants. In the Asteraceae, the component processes of apomeiosis and parthenogenesis are separately inherited in the diplosporous apomicts *Taraxacum* [50, 51] and *Erigeron* [52–54] and in the aposporous apomict *Pilosella* (syn. *Hieracium*) [55, 56, 30]. Conversely, apospory is reported to be inherited as a single dominant factor in the dicotyledonous genus *Ranunculus* [12]. Among the aposporic grasses, simple dominant inheritance is reported in *Panicum* [57–60], *Pennisetum* [61], *Brachiaria* [62], and *Paspalum* [63, 64] and for the diplosporous grasses *Eragrostis curvula* [65] and *Tripsacum dactyloides* [66]. In *Poa pratensis*, recombination has been reported between apospory and parthenogenesis [67, 68]. There is evidence of segregation distortion in some of these systems, with dominant factor(s) associated with apomixis appearing to confer gamete lethality [12, 69–71, 34]. Autonomous endospermy is only known in a small number of apomictic systems, most notably in the Asteraceae, where it appears to be separately inherited to apomeiosis and parthenogenesis. van Dijk et al. [72] reported that autonomous endospermy was unlinked to either diplospory or parthenogenesis in *Taraxacum*. In *Pilosella piloselloides* (syn. *H. caespitosum*), Catanach et al. [30] reported that autonomous endospermy was lost along with parthenogenesis in a collection of deletion mutants, indicating that the two are either closely linked or pleiotropic. Ogawa et al. [73], however, reported identifying recombinants between autonomous endospermy and parthenogenesis in the related species *Pilosella officinarium* (syn. *Hieracium pilosella*), indicating that the linkage can be broken.

Less information is available regarding the genetics of sporophytic apomixis. Garcia et al. [74] reported the detection of six quantitative trait loci (QTLs) contributing to the total variation associated with the expression of apomixis in a cross between *Poncirus trifoliata* and *Citrus volkameriana*. As mentioned above, Shimada et al. [49] identified a gene, *CitRKD1*, at the polyembryonic locus of *Citrus* that appears to be responsible for polyembryony in this system. Intriguingly, the allele governing polyembrony differs from the corresponding monoembryonic allele by the presence of a MITE insertion upstream of the promoter. It is interesting to note that MITE promoter insertions are also known to promote parthenogenesis in the daisy genera *Taraxacum*, *Pilosella* and *Hieracium* [31, 34].

There is a highly significant association between the expression of gametophytic apomixis and the occurrence of polyploidy. Of the hundreds of taxa known to include gametophytic apomicts [75], naturally occurring diploid apomicts have only been clearly identified in *Boechera* [76]. Müntzing [77] reported diploid apomicts in the genus *Potentilla*, but this result is contested [78]. Even in *Boechera*, most apomictic field collections are reported to be polyploids [79]. The reason for the close association between gametophytic apomixis and polyploidy is unknown. Among sporophytic apomicts, diploidy is common. Similarly, other forms of clonal increase, such as stolon proliferation, are not always associated with polyploidy, so it appears unlikely that polyploidy arises as a consequence of cloning. Also, diploid gametophytic apomicts have been identified in experimental populations of some gametophytic species [80–82], indicating that polyploidy is not a prerequisite for the expression of gametophytic apomixis. These individuals, however, are typically reported to be pollen sterile, so although they are capable of clonal replication, they cannot pass the trait on via their pollen. Consequently, it has been suggested that while polyploidy is not necessary for the expression of apomixes, it may be required for the transmission of the trait via reduced gametes. This would, over evolutionary time, lead to its near exclusive expression in polyploid races [12]. Nogler [83] further proposed that a genetic element involved in the expression of gametophytic apomixis may act as a recessive gamete lethal factor. The action of such an element would be exposed in haploid gametes, but shielded in heterozygotic diploid gametes, thus promoting the formation of polyploid apomictic races. Evidence of this type of transmission bias has been reported in *Tripsacum* [69], *Pennisetum* [70, 71], and *Pilosella* [34]. Using this observation, along with other experimental evidence, Bicknell et al. [34] presented a model for the evolution of gametophytic apomixis in *Pilosella* that details how the trait may have evolved independently so many times in long-lived, polyploid, flowering plant species.

5 Conclusions

Molecular mapping is possible in apomictic species, but it presents unique challenges. The seedlings of apomicts fall into different groups, depending on the nature of their genesis. Therefore, when considering allelic segregation patterns within progeny populations, the seedlings first need to be pooled by type, before the data can be analyzed. Pollen transmission of apomixis determinants is possible in many systems, but care must be taken both in the selection of the sexual recipient and in the analysis of the progeny. In particular, the impact of polyploidy needs to be considered as apomixis does not appear to transmit in haploid gametes, either via the egg or via the pollen. A special case of segregation analysis is polyhaploid mapping, which avoids the need for a sexual recipient, but, for experimental reasons, this approach is currently limited to *Pilosella*.

Deletion mutation provides an alternative approach to molecular mapping in apomicts. Mutation has the advantage that it avoids meiosis, so is well-suited to asexual systems, but it does typically require an experimental step to reduce the influence of genetic chimerism in the treated tissues.

The methods used to analyze variation in mapping populations are similar to those used in segregating systems with the caveat that apomicts often have large, polyploid genomes with significant transposon load. This complicates the analysis of segregation/mutation. It is notable that a complete, phased, genome of an apomict has yet to be published. However, we expect that advances in sequencing and genome assembly will overcome this issue soon.

Acknowledgments

The authors would like to thank the former New Zealand Foundation for Science and Technology (now the New Zealand Ministry of Business Innovation and Employment) for their financial support (grant number C10X0404).

Conflicts of Interest The authors have no conflicts of interest to declare that are relevant to the content of this chapter.

References

1. Mendel G (1869) Ueber einige aus künstlicher Befruchtung gewonnene *Hieracium*-Bastarde. Verhandlungen des naturforchenden Vereins in Brünn 8:26–31
2. Bicknell R, Catanach A, Hand M, Koltunow A (2016) Seeds of doubt: Mendel's choice of *Hieracium* to study inheritance, a case of right plant, wrong trait. Theor Appl Genet 129(12):2253–2266
3. Nogler GA (2006) The lesser-known Mendel: his experiments on *Hieracium*. Genetics 172(1):1–6
4. de Vries H (1900) The law of segregation of hybrids. In: Stern C, Sherwood ER (eds) The origin of genetics. A Mendel source book. W. H. Freeman and Company, San Francisco, pp 107–118
5. Correns C (1900) Gregor Mendel's Regel über das Verhalten der Nachkommenschaft der Rassenbastarde. Berichte der Deutschen Botanischen Gesellschaft 18:158–168
6. von Tschermak E (1900) Über künstliche Kreuzung bei *Pisum sativum*. Berichte der Deutschen Botanischen Gesellschaft 18:232–239
7. Bicknell RA, Lambie SC, Butler RC (2003) Quantification of progeny classes in two facultatively apomictic accessions of *Hieracium*. Hereditas 138:11–20
8. Asker SE, Jerling L (1992) Apomixis in plants. CRC Press, Boca Raton
9. Savidan Y (2000) Apomixis: genetics and breeding. In: Janick J (ed) Plant Breeding Review, vol 18. John Wiley & Sons, Inc, New York, pp 13–86
10. Matzk F, Meister A, Schubert I (2000) An efficient screen for reproductive pathways using mature seeds of monocots and dicots. Plant J 21(1):97–108
11. Nogler GA (1984) Gametophytic apomixis. In: Johri BM (ed) Embryology of angiosperms. Springer-Verlag, Berlin, pp 475–518
12. Nogler GA (1984) Genetics of apospory in apomictic *Ranunculus auricomus*. V. Conclusion. Bot Helv 94(2):411–422
13. Grimanelli D, Leblanc O, Perotti E, Grossniklaus U (2001) Developmental genetics of gametophytic apomixis. Trends Genet 17(10):597–604
14. Guyénot E (1914) Action des rayons ultraviolets sur *Drosophila ampelophila*, Low. Bull Scient Prance Belg 48:160–169
15. Mohr OL (1918) Mikroskopische Untersuchungen zu Experimenten über den Einfluss der Radiumstrahlen und der Kältewirkung auf die Chromatinreifung und das Heterochromosom bei *Decticus verruccivorus* (♂). Arch Mikrosk Anat 92(1):300–368
16. Reichert NA (2021) History of plant genetic mutations±human influences. In Vitro Cell Dev Biol Plant 57(4):554–564
17. Muller HJ (1928) The measurement of gene mutation rate in *Drosophila*, its high variability, and its dependence upon temperature. Genetics 13(4):279
18. Muller HJ (1928) The production of mutations by X-rays. Proc Natl Acad Sci 14(9):714–726
19. Muller HJ (1927) Artificial transmutation of the gene. Science 66(1699):84–87
20. Stadler LJ (1928) Genetic effects of X-rays in maize. Proc Natl Acad Sci 14(1):69–75
21. Stadler L (1928) Mutations in barley induced by X-rays and radium. Science 68(1756):186–187
22. Dowlath MJH, Karuppannan SK, Sinha P, Dowlath NS, Arunachalam KD, Ravindran B, Chang SW, Nguyen-Tri P, Nguyen DD (2021) Effects of radiation and role of plants in radioprotection: a critical review. Sci Total Environ 779:146431
23. Norbury CJ, Hickson ID (2001) Cellular responses to DNA damage. Annu Rev Pharmacol Toxicol 41(1):367–401
24. Meier B, Volkova NV, Gerstung M, Gartner A (2020) Analysis of mutational signatures in *C. elegans*: implications for cancer genome analysis. DNA Repair 95:102957
25. Stratton MR, Campbell PJ, Futreal PA (2009) The cancer genome. Nature 458(7239):719–724
26. Hanahan D, Weinberg RA (2000) The hallmarks of cancer. Cell 100(1):57–70
27. Yali W, Mitiku T (2022) Mutation breeding and its importance in modern plant breeding. J Plant Sci 10(2):64–70

28. Endo TR, Gill BS (1996) The deletion stocks of common wheat. J Hered 87(4):295–307
29. Lin JJ, Kuo J (1995) AFLP: a novel PCR-based assay for plant and bacterial DNA fingerprinting. Focus 17(2):66–70
30. Catanach AS, Erasmuson SK, Podivinski E, Jordan BR, Bicknell RA (2006) Deletion mapping of genetic elements associated with apomixis in *Hieracium*. Proc Natl Acad Sci USA 103(49):18650–18655
31. Underwood C, Vijverberg K, Rigola D, Okamoto S, Oplaat C, Camp R, Radoeva T, Schauer S, Fierens J, Jansen K, Mansveld S, Busscher M, Xiong W, Datema E, Nijbroek K, Blom E-J, Bicknell E, Catanach A, Erasmuson S, Winefield C, van Tunen A, Prins M, Schranz E, van Dijk P (2022) A *PARTHENOGENESIS* allele from apomictic dandelion can induce egg cell division without fertilization in lettuce. Nat Genet 54:84–93
32. Crane CF (2001) Classification of apomictic mechanisms. In: Savidan Y, Carman JG, Dresselhaus T (eds) Flowering of apomixis: from mechanisms to genetic engineering. CIMMYT, IRD, European Commission DG VI, Mexico, pp 24–34
33. Kotani Y, Henderson ST, Suzuki G, Johnson SD, Okada T, Siddons H, Mukai Y, Koltunow AM (2014) The *LOSS OF APOMEIOSIS* (*LOA*) locus in *Hieracium praealtum* can function independently of the associated large-scale repetitive chromosomal structure. New Phytol 201(3):973–981
34. Bicknell R, Gaillard M, Catanach A, McGee R, Erasmuson S, Fulton B, Winefield C (2023) Genetic mapping of the *LOSS OF PARTHENOGENESIS* locus in *Pilosella piloselloides* and the evolution of apomixis in the Lactuceae. Front Plant Sci 14. https://doi.org/10.3389/fpls.2023.1239191
35. Stougaard J (1993) Sustrate-dependant negative selection in plants using a bacterial cytosine deaminase gene. Plant J 3(5):755–761
36. Brukhin V, Osadtchiy JV, Florez-Rueda AM, Smetanin D, Bakin E, Nobre MS, Grossniklaus U (2019) The Boechera genus as a resource for apomixis research. Front Plant Sci 10:392
37. Schranz ME, Windsor AJ, Song B-h, Lawton-Rauh A, Mitchell-Olds T (2007) Comparative genetic mapping in Boechera stricta, a close relative of Arabidopsis. Plant Physiol 144(1): 286–298
38. Reyes-Chin-Wo S, Wang Z, Yang X, Kozik A, Arikit S, Song C, Xia L, Froenicke L, Lavelle DO, Truco M-J, Xia R, Zhu S, Xu C, Xu H, Xu X, Cox K, Korf I, Meyers BC, Michelmore RW (2017) Genome assembly with in vitro proximity ligation data and whole-genome triplication in lettuce. Nat Commun 8(1):14953
39. Vos P, Hogers R, Bleeker M, Reijans M, Vandelee T, Hornes M, Frijters A, Pot J, Peleman J, Kuiper M, Zabeau M (1995) Aflp – a new technique for DNA-fingerprinting. Nucleic Acids Res 23(21):4407–4414
40. Michelmore RW, Paran I, Kesseli RV (1991) Identification of markers linked to disease resistance genes by bulked segregant analysis: a rapid method to detect markers in specific genomic regions using segregating populations. Proc Natl Acad Sci USA 88:9828–9832
41. Akiyama Y, Conner JA, Goel S, Morishige DT, Mullet JE, Hanna WW, Ozias-Akins P (2004) High-resolution physical mapping in *Pennisetum squamulatum* reveals extensive chromosomal heteromorphism of the genomic region associated with apomixis. Plant Physiol 134: 1733–1741
42. Charlesworth B, Charlesworth D (1978) A model for the evolution of dioecy and gynodioecy. Am Nat 112(988):975–997
43. Sun H, Jiao W-B, Krause K, Campoy JA, Goel M, Folz-Donahue K, Kukat C, Huettel B, Schneeberger K (2022) Chromosome-scale and haplotype-resolved genome assembly of a tetraploid potato cultivar. Nat Genet 54(3):342–348
44. Li H, Handsaker B, Wysoker A, Fennell T, Ruan J, Homer N, Marth G, Abecasis G, Durbin R, Subgroup GPDP (2009) The sequence alignment/map format and SAMtools. Bioinformatics 25(16):2078–2079
45. Catanach A, Ruigrok M, Bowatte D, Davy M, Storey R, Valenza-Troubat N, López-Girona E, Hilario E, Wylie MJ, Chagné D (2021) The genome of New Zealand trevally (Carangidae: Pseudocaranx georgianus) uncovers a XY sex determination locus. BMC Genomics 22:1–16
46. Bräuning S, Catanach A, Lord JM, Bicknell R, Macknight RC (2018) Comparative transcriptome analysis of the wild-type model apomict *Hieracium praealtum* and its loss of parthenogenesis (*lop*) mutant. BMC Plant Biol 18(1): 206
47. Ortiz JPA, Revale S, Siena LA, Podio M, Delgado L, Stein J, Leblanc O, Pessino SC (2017) A reference floral transcriptome of sexual and apomictic *Paspalum notatum*. BMC Genomics 18(1):318
48. de Oliveira FA, Vigna BBZ, da Silva CC, Fávero AP, de Matta FP, Azevedo ALS, de Souza AP (2020) Coexpression and transcriptome analyses identify active Apomixis-related

genes in *Paspalum notatum* leaves. BMC Genomics 21(1):78
49. Shimada T, Endo T, Fujii H, Nakano M, Sugiyama A, Daido G, Ohta S, Yoshioka T, Omura M (2018) MITE insertion-dependent expression of CitRKD1 with a RWP-RK domain regulates somatic embryogenesis in citrus nucellar tissues. BMC Plant Biol 18:1–19
50. Van Dijk PJ, Tas ICQ, Falque M, Bakx-Schotman T (1999) Crosses between sexual and apomictic dandelions (*Taraxacum*). II. The breakdown of apomixis. Heredity 83 (Part 6):715–721
51. van Dijk PJ, Bakx-Schotman JMT (2004) Formation of unreduced megaspores (diplospory) in apomictic dandelions (*Taraxacum officinale*, s.l) is controlled by a sex-specific dominant locus. Genetics 166(1):483–492
52. Noyes RD (2004) Inheritance of apomeiosis (diplospory) in fleabanes (*Erigeron*, Asteraceae). Heredity 94(2):193–198
53. Noyes RD, Rieseberg LH (2000) Two independent loci control agamospermy (apomixis) in the triploid flowering plant *Erigeron annuus*. Genetics 155(1):379–390
54. Noyes RD (2000) Diplospory and parthenogenesis in sexual x agamospermous (apomictic) *Erigeron* (Asteraceae) hybrids. Int J Plant Sci 161(1):1–12
55. Koltunow AM, Johnson SD, Rodrigues J, Okada T, Hu Y, Tsuchiya T, Wilson S, Fletcher P, Ito K, Suzuki G (2011) Sexual reproduction is the default mode in apomictic *Hieracium* subgenus *Pilosella*, in which two dominant loci function to enable apomixis. Plant J 66(5):890–902
56. Hand ML, Koltunow AMG (2014) The genetic control of apomixis: asexual seed formation. Genetics 197(2):441–450
57. Savidan Y (1980) Chromosomal and embryological analyses in sexual x apomictic hybrids of *Panicum maximum* Jacq. TAG Theor Appl Genet 58(2):153–156
58. Ebina M, Nakagawa H, Yamamoto T, Araya H, Si T, Takahara M, Nakajima K (2005) Co-segregation of AFLP and RAPD markers to apospory in Guineagrass (*Panicum maximum* Jacq.). Grassl Sci 51(1):71–78
59. Yamada-Akiyama H, Akiyama Y, Ebina M, Xu Q, Tsuruta S-i, Yazaki J, Kishimoto N, Kikuchi S, Takahara M, Takamizo T (2009) Analysis of expressed sequence tags in apomictic guineagrass (*Panicum maximum*). J Plant Physiol 166(7):750–761
60. Kaushal P, Dwivedi KK, Radhakrishna A, Saxena S, Paul S, Srivastava M, Baig M, Roy A, Malaviya D (2018) Ploidy dependent expression of apomixis and its components in guinea grass (*Panicum maximum* Jacq.). Euphytica 214:1–22
61. Sherwood RT, Berg CC, Young BA (1994) Inheritance of apospory in buffelgrass. Crop Sci 34:1490–1494
62. Valle CB, Glienke C, Leguizamon GOC (1994) Inheritance of apomixis in *Brachiaria*, a tropical forage grass. Apomixis Newsl 7:42–43
63. Martinez EJ, Acuna CA, Hojsgaard DH, Tcach MA, Quarin CL (2007) Segregation for sexual seed production in *Paspalum* as directed by male gametes of apomictic triploid plants. Ann Bot 100(6):1239–1247
64. Martinez EJ, Urbani MH, Quarin CL, Ortiz JPA (2001) Inheritance of apospory in bahiagrass, Paspalum *notatum*. Hereditas 135(1): 19–25
65. Voigt PW, Burson BL (1981) Breeding of apomictic *Eragrostis curvula*. In: Smith JA, Hayes VW (eds) 14th International Grasslands Congress, Lexington Ky. Westview Press, Boulder, pp 160–163
66. Leblanc O, Grimanelli D, Gonzalezdeleon D, Savidan Y (1995) Detection of the apomictic mode of reproduction in maize-*Tripsacum* hybrids using maize Rflp markers. Theor Appl Genet 90(7–8):1198–1203
67. Matzk F, Prodanovic S, Bäumlein H, Schubert I (2005) The inheritance of apomixis in *Poa pratensis* confirms a five locus model with differences in gene expressivity and penetrance. Plant Cell 17(1):13–24
68. Albertini E, Porceddu A, Ferranti F, Reale L, Barcaccia G, Romano B, Falcinelli M (2001) Apospory and parthenogenesis may be uncoupled in *Poa pratensis*: a cytological investigation. Sex Plant Reprod 14(4):213–217
69. Grimanelli D, Leblanc O, Espinosa E, Perotti E, De Leon DG, Savidan Y (1998) Non-Mendelian transmission of apomixis in maize-*Tripsacum* hybrids caused by a transmission ratio distortion. Heredity 80:40–47
70. Roche D, Chen ZB, Hanna WW, Ozias-Akins P (2001) Non-Mendelian transmission of an apospory-specific genomic region in a reciprocal cross between sexual pearl millet (*Pennisetum glaucum*) and an apomictic F1 (*P. glaucum* x *P. squamulatum*). Sex Plant Reprod 13(4):217–223
71. Jessup RW, Burson BL, Burow GB, Wang YW, Chang C, Li Z, Paterson AH, Hussey MA (2002) Disomic inheritance, suppressed recombination, and allelic interactions govern apospory in buffelgrass as revealed by genome mapping. Crop Sci 42(5):1688–1694

72. Van Dijk PJ, Op den Camp R, Schauer SE (2020) Genetic dissection of apomixis in dandelions identifies a dominant parthenogenesis locus and highlights the complexity of autonomous endosperm formation. Genes 11(9):961
73. Ogawa D, Johnson SD, Henderson ST, Koltunow AMG (2013) Genetic separation of autonomous endosperm formation (*AutE*) from the two other components of apomixis in *Hieracium*. Plant Reprod 26(2):113–123
74. Garcia R, Asins MJ, Forner J, Carbonell EA (1999) Genetic analysis of apomixis in *Citrus* and *Poncirus* by molecular markers. Theor Appl Genet 99(3–4):511–518
75. Carman JG (1997) Asynchronous expression of duplicate genes in angiosperms may cause apomixis, bispory, tetraspory, and polyembryony. Biol J Linn Soc 61(1):51–94
76. Kantama L, Sharbel TF, Schranz ME, Mitchell-Olds T, de Vries S, de Jong H (2007) Diploid apomicts of the *Boechera holboellii* complex display large-scale chromosome substitutions and aberrant chromosomes. Proc Natl Acad Sci USA 104:14026–14031
77. Müntzing A (1928) Pseudogamie in der Gattung *Potentilla*. Hereditas 11:267–283
78. Holm S, Ghatnekar L (1996) Sexuality and no Apomixis found in crossing experiments with diploid *Potentilla argentea*. Hereditas 125:77–82
79. Schranz ME, Dobe C, Koch MA, Mitchell-Olds T (2005) Sexual reproduction, hybridization, apomixis, and polyploidization in the genus *Boechera* (Brassicaceae). Amer J Bot 92:1797–1810
80. Nogler GA (1982) How to obtain diploid apomictic *Ranunculus auricomus* plants not found in the wild state. Bot Helv 92:13–22
81. Bicknell RA (1997) Isolation of a diploid, apomictic plant of *Hieracium aurantiacum*. Sex Plant Reprod 10:168–172
82. Kojima A, Nagato Y (1997) Discovery of highly apomictic and highly amphimictic dihaploids in *Allium tuberosum*. Sex Plant Reprod 10(1):8–12
83. Nogler GA (1986) Genetics of apomictic reproduction. G Bot Ital 120(2):49–52

Chapter 11

Chromosome Preparation and Fluorescent In Situ Hybridization

Terezie Mandáková

Abstract

Chromosome preparation and fluorescent in situ hybridization (FISH) are indispensable methodologies in cytogenetics, serving as a vital link between molecular biology and genome visualization. In the era of next-generation sequencing and near-complete genome assemblies, these techniques remain crucial for elucidating chromosomal architecture, dynamics, and evolutionary relationships. This chapter presents robust protocols for preparing mitotic and meiotic chromosomes from root tips and floral tissues, optimized for diverse plant species. Detailed procedures for enzymatic digestion, chromosome squashing and spreading, and FISH are described, ensuring reproducible and high-quality results. Furthermore, protocols for probe labeling and fluorescence detection highlight the adaptability of FISH for various cytogenetic applications, including chromosome painting or genomic in situ hybridization. By mastering these approaches, researchers can explore chromosomal organization with unprecedented resolution. These step-by-step protocols are designed to be accessible to both beginners and experienced researchers, ensuring wide applicability across plant species and cytogenetic contexts.

Key words Cytogenetics, Fluorescent in situ hybridization, Mitotic and meiotic chromosomes, Plant genomes

1 Introduction

In the era of advanced genome sequencing and telomere-to-telomere assemblies, cytogenetics remains essential for connecting sequence data with structural and functional genome analysis. While sequencing technologies have revolutionized our understanding of the linear arrangement of DNA, they are limited in resolving the spatial organization of chromosomes, detecting structural variations across large genomic regions, and capturing chromosomal behavior in situ. Cytogenetics bridges these gaps by enabling direct visualization of chromosomal structures and their dynamics within the nucleus, offering critical insights into genome stability, gene regulation, and chromosomal evolution. Techniques such as chromosome preparation and fluorescent in situ

Tatyana Radoeva and Peter J. van Dijk (eds.), *Apomixis: Methods and Protocols*, Methods in Molecular Biology, vol. 2987,
https://doi.org/10.1007/978-1-0716-5001-1_11,

hybridization (FISH) are instrumental in advancing our understanding of complex chromosomal phenomena, including the precise positioning of specific loci, mapping of repetitive DNA sequences, and the identification of structural variations such as inversions, translocations, and aneuploidies. FISH also allows for the detailed investigation of meiotic chromosome behavior during pairing and recombination, the determination of parental subgenomes in hybrids and allopolyploids, and the elucidation of evolutionary changes in karyotype structure. Furthermore, this versatile methodology has proven invaluable for studying chromosomal interactions within the three-dimensional nuclear architecture, detecting chromosomal abnormalities in plants, and characterizing polyploid genomes with intricate chromosomal configurations. These cytogenetic approaches are crucial for validating genome assemblies, integrating genetic and physical maps, and unraveling the functional interplay between genome architecture and its biological roles (*see* Fig. 1) [1–13].

This chapter provides a comprehensive overview of two primary methods for chromosome preparation: squashing and spreading. Chromosome squashing of root tips involves compressing digested meristematic tissues under a coverslip to produce dense chromosome spreads ideal for basic analyses, such as chromosome counting or mitotic stage studies. Chromosome spreading of anthers and pistils, in contrast, employs circular dispersion of chromosomes and nuclei on a slide, ensuring a uniform and cytoplasm-free distribution. This technique is particularly suited for high-resolution FISH applications, where clarity and reduced background are paramount [14]. Both methods share foundational steps, including fixation with Carnoy's fixative and enzymatic digestion, but differ in their end goals and suitability for specific applications.

The chapter also details the principles and execution of FISH, a versatile technique for mapping specific DNA sequences on chromosomes. FISH employs DNA probes labeled with either fluorochromes (e.g., Cy3, Alexa Fluor) or haptens (e.g., biotin, digoxigenin). Fluorochrome-labeled probes are directly visualized, while hapten-labeled probes require secondary detection using fluorescent antibodies [15]. Chromosome pretreatment steps, such as RNase and pepsin digestion, are critical for minimizing background noise and enhancing probe accessibility. Following hybridization, stringent washing steps using formamide-based buffer remove unbound probes, ensuring specificity and reducing nonspecific signals [16].

These protocols were initially developed for crucifer species (Brassicaceae) [1–5, 7, 10–14] but are broadly applicable across plant taxa [6, 8, 17–24]. By following these protocols, researchers

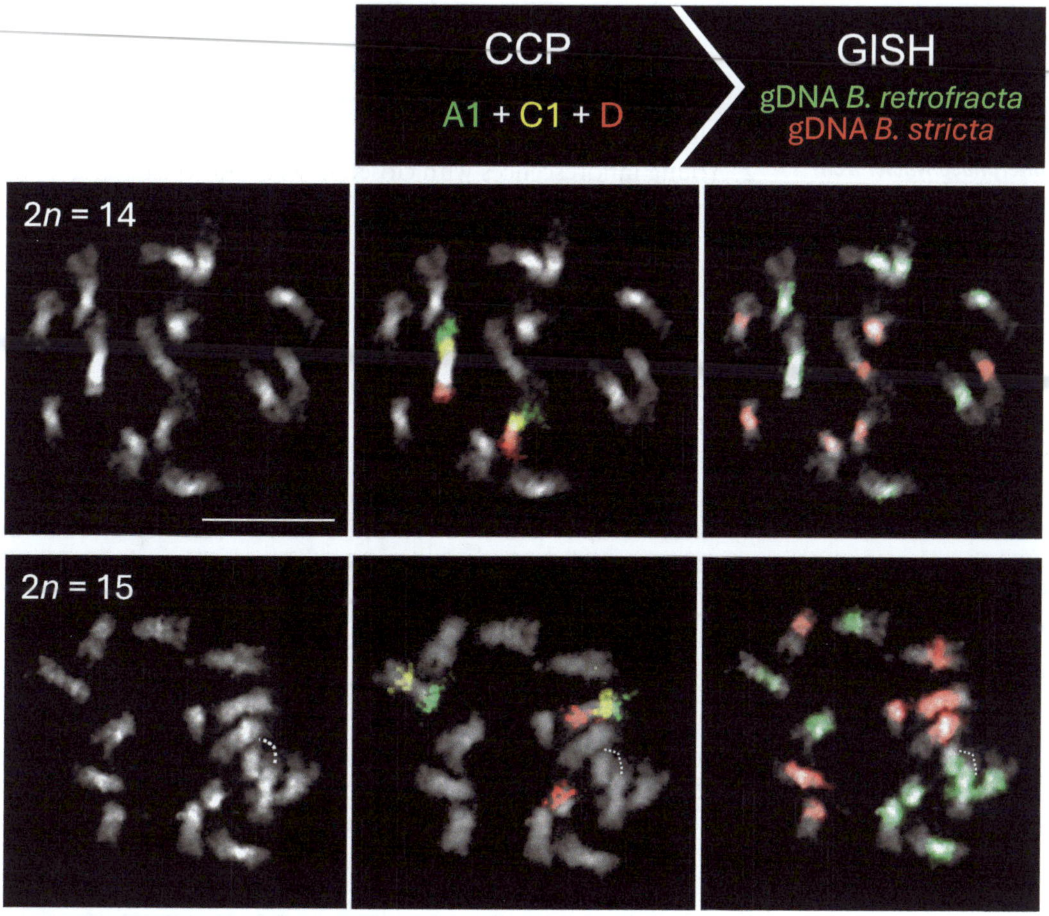

Fig. 1 Comparative chromosome painting (CCP) and genomic in situ hybridization (GISH) performed on mitotic chromosomes of two diploid apomictic accessions of *Boechera retrofracta* × *stricta* with differing chromosome numbers ($2n = 14$ and $2n = 15$). Using CCP, homeologs of *Boechera* chromosome 1 were identified with differentially labeled BAC contigs from *Arabidopsis thaliana*, representing genomic blocks A1, C1 and D [1, 2]. GISH was employed to distinguish the two subgenomes within the hybrid chromosome complements, using differentially labeled total genomic DNA from the parental species, *B. retrofracta* and *B. stricta*. Chromosomes were counterstained with 4′,6-diamidino-2-phenylindole (DAPI). Scale bar, 10 μm

can achieve high-quality chromosome spreads and robust FISH results, facilitating studies in genome organization, chromosomal evolution, and species diversity. The integration of these methods into genomic research underscores their enduring relevance and adaptability, ensuring their continued importance in cytogenetics and beyond.

2 Materials

2.1 Preparation of Chromosomes and Interphase Nuclei from Root Tip Meristems and Generative Organ Tissues

2.1.1 Plant Materials

1. Root tips from 2- or 3-day-old seedlings or potted plants (*see* Fig. 2a).
2. Young inflorescences with closed flower buds containing white anthers (*see* Fig. 2b).

2.1.2 Solutions and Buffers

1. Carnoy's fixative: 3:1 (v/v) ethanol: glacial acetic acid. Freshly prepared.
2. 1× citrate buffer prepared from 10× stock. 10× stock: mix 40 mL of 100 mM citric acid and 60 mL of 100 mM trisodium citrate. Adjust pH to 4.8 using 1 M HCl. Dilute to 1× with deionized H_2O. Store 1× citrate buffer in aliquots at −20 °C until use.
3. 60% (v/v) glacial acetic acid in distilled water.
4. 4% formaldehyde diluted from 37% formaldehyde solution. Freshly prepared.
5. Pectolytic enzyme mixture: 0.3% (w/v) pectolyase, 0.3% (w/v) cellulase, 0.3% (w/v) cytohelicase. Prepare from frozen 1% (w/v) mixture stock in 1× citrate buffer (store up to 1 year at −20 °C) by diluting with 1× citrate buffer.

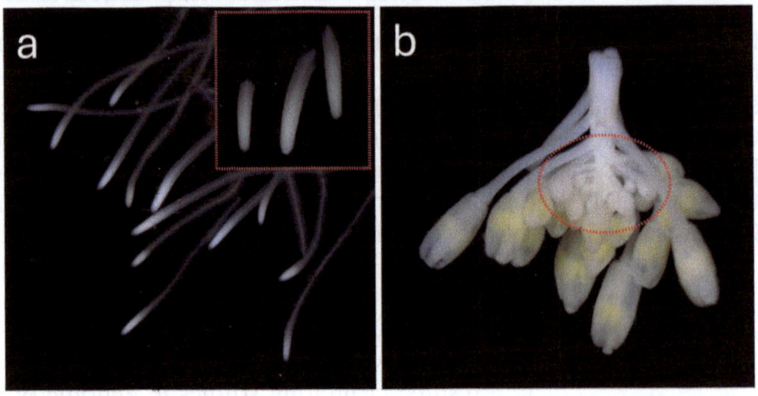

Fig. 2 Plant materials used for chromosome preparation. (**a**) Harvested, pretreated, and fixed root tips with visible root-tip meristems, shown as terminal white areas and highlighted in the inset, *see* Subheading 3.1). (**b**) Harvested and fixed inflorescence (*see* Subheading 3.2). Yellow flower buds, which contain anthers with pollen, should be discarded prior to enzymatic digestion. Only the inner portion containing white flower buds, marked by the red ellipse, is suitable for further processing

2.1.3 Equipment and Supplies

1. Liquid nitrogen or block of dry ice.
2. Filter paper.
3. Petri dishes.
4. Fine-tipped forceps and long forceps.
5. Glass vials or microcentrifuge tubes.
6. Black staining blocks with glass cover or small glass Petri dishes.
7. Orbital shaker.
8. Humid box (e.g., a slide storage box filled with wet paper tissue).
9. Dissecting needles.
10. SuperFrost microscope slides.
11. Glass coverslips (24 × 24 mm).
12. Alcohol burner.
13. Tube rack or similar holder.
14. Dust-free storage box for slides.
15. Coplin jars.
16. 50 °C heating block.
17. Hair dryer.
18. Stereomicroscope.
19. Phase-contrast light microscope.

2.2 Fluorescent In Situ Hybridization

2.2.1 Chromosome Preparations

1. High-quality chromosome preparations on slides. Refer to Subheadings 3.1 and 3.2 for preparing chromosomes from root tip meristems and generative organ tissues. High-quality chromosome preparations are essential for achieving optimal results. Ideally, chromosomes and nuclei should be evenly spread in a single layer, with minimal or no physical contact between them to ensure clarity and accuracy during analysis.

2.2.2 Solutions and Buffers

1. 2× saline sodium citrate (SSC) prepared from 20× SSC. 20× SSC: 3 M sodium chloride, 300 mM trisodium citrate, pH 7.0. Store up to several months at 4 °C.
2. RNase solution: 100 μg/ml ribonuclease A. Prepare from a 1 mg/ml stock solution in distilled water. Ensure that the RNase solution is DNase-free. If not, inactivate DNase by boiling the RNase stock solution in water for 15 minutes. Store in aliquots at −20 °C.
3. Pepsin solution: 0.1 mg/ml. Dilute from a 100 mg/ml pepsin stock prepared in 10 mM HCl. Prepare a stock solution using porcine gastric mucosa in 10 mM HCl. Store aliquots for up to several months at −20 °C.

4. Hybridization buffer: 50% formamide, 10% dextran sulfate in 2× SSC. Store up to several months at −20 °C.
5. 4% formaldehyde in 2× SSC: Freshly prepared before use.
6. 20% deionized formamide in 2× SSC: Adjust to pH 7.0 as needed.
7. 4 T buffer: 4× SSC (pH 7.0) with 0.05% Tween-20. Store up to several months at 4 °C.
8. Blocking solution: 5% bovine serum albumin (BSA), 0.2% Tween-20 in 4× SSC. Store up to several months at −20 °C.
9. TNB buffer: 100 mM Tris-Cl (pH 7.5), 150 mM NaCl, 0.5% blocking reagent. Store up to several months at −20 °C.
10. TNT buffer: 100 mM Tris-Cl (pH 7.5), 150 mM NaCl, 0.05% Tween-20. Store up to several months at 4 °C.

2.2.3 DNA Labeling and Hybridization

1. DNA for hybridization probes (*see* **Note 1**).
2. 10× NT buffer: 500 mM Tris-Cl (pH 7.5), 50 mM $MgCl_2$, 0.05% bovine serum albumin. Store up to several months at −20 °C.
3. Nucleotide mixture: 2 mM dATP, dCTP, dGTP, and 400 mM dTTP.
4. 1 mM X-dUTP: X = biotin, digoxigenin, Cy3, or other hapten/fluorochrome.
5. 0.1 M β-Mercaptoethanol.
6. DNase I: Dilute 1:250 from a 1 mg/mL stock in 0.15 M NaCl (50% glycerol).
7. DNA Polymerase I: 10 U/μL.

2.2.4 Antibodies

1. Avidin-Texas Red.
2. Biotinylated goat anti-avidin.
3. Mouse anti-digoxigenin.
4. Goat anti-mouse Alexa Fluor 488.

2.2.5 Additional Reagents and Consumables

1. 70%, 80%, and 96% ethanol.
2. 96% ice-cold ethanol.
3. Vectashield antifade with 2 μg/mL DAPI (4,6-diamidino-2-phenylindole). Store at 4 °C.
4. Rubber cement.
5. 1% agarose gel.
6. 100-bp ladder DNA marker.
7. 0.5 M EDTA (pH 8.0).
8. 3 M sodium acetate (pH 5.2).

2.2.6 Laboratory Equipment

1. Coplin (holding up to 9 slides with 50 ml of solution) or Hellendahl (holding up to 15 slides with 80 ml of solution) jars for holding slides vertically.
2. Orbital shaker.
3. Humid box: For example, a slide storage box filled with wet paper towels.
4. Coverslips of sizes 24 × 24 mm, 24 × 32 mm, and 24 × 50 mm.
5. 37 °C incubator.
6. 37 °C and 42 °C water bath.
7. Plastic slide rack.
8. 0.5- and 2-ml microcentrifuge tubes.
9. Vortex mixer.
10. Programmable temperature-controlled heating block or thermal cycler: Set at 15 °C and 60 °C.
11. Electrophoresis system.
12. Refrigerated centrifuge.
13. Desiccator or vacuum centrifuge.
14. 37 °C thermomixer.
15. 80 °C heating block with precise temperature control.
16. Forceps.
17. Paper or plastic slide case.
18. Epifluorescence microscope equipped with optical filters for DAPI and other fluorochromes, a digital CCD camera, and image acquisition software.

3 Methods

3.1 Squash Preparation of Chromosomes and Interphase Nuclei from Root Tip Meristems

Root tips are the preferred plant tissue for preparing mitotic chromosomes due to their high density of actively dividing cells. These meristematic tissues can be easily harvested from germinating seeds or actively growing plants. The preparation protocol involves three main steps: collection and pretreatment of root tips; enzymatic digestion to remove cell walls and soften tissues; and chromosome squash preparation for microscopic observation.

3.1.1 Collection and Pretreatment of Root Tips

1. Germinate seeds on filter paper soaked in distilled water within a Petri dish at 20–22 °C. Alternatively, collect root tips from actively growing, healthy plants cultivated in pots. For potted plants, apply fresh soil around the root ball to stimulate the growth of new roots within a few days. Older roots may be trimmed to encourage the formation of young, actively growing roots.

2. Using fine-tipped forceps, harvest 1-cm-long root tips from seedlings. For potted plants, gently loosen the root ball and identify young, white root tips (approximately 1-cm long; *see* Fig. 2a). Carefully cut these tips from the exposed areas of the root ball using fine-tipped forceps.
3. Place the harvested root tips immediately into ice-cold distilled water held in glass vials or microcentrifuge tubes. Submerge these vials in a box containing ice-cold water and ice flakes or cubes (*see* **Note 2**). Store the vials in a refrigerator for approximately 24 hours. Adjust the pretreatment duration as needed to obtain an adequate number of dividing cells.
4. Remove the root tips from the vials or tubes using fine-tipped forceps, briefly blot them dry on filter paper, and transfer them into freshly prepared Carnoy's fixative. Use glass vials or microcentrifuge tubes for this step.
5. Keep the root tips in the fixative at 4 °C overnight. For long-term storage, root tips can be stored in Carnoy's fixative at − 20 °C until use.

*3.1.2 Enzymatic Digestion of Root Tips (See **Note** 3)*

1. Wash the selected root tips twice with distilled water in a black staining block, each wash lasting 5 minutes. Ensure that the volume of water is sufficient to fully submerge the root tips.
2. Use a glass Pasteur pipette to remove the distilled water. Add 1× citrate buffer to the staining block and wash the root tips twice for 5 minutes on an orbital shaker.
3. Remove the citrate buffer using a Pasteur pipette. Add 0.3% pectolytic enzyme mixture to the staining block and incubate the root tips in a humid box at 37 °C for 90 minutes or longer (*see* **Note 4**).
4. Replace the enzyme mixture with 1× citrate buffer using a Pasteur pipette. If further processing is delayed for up to 1 hour, keep the staining block containing the digested root tips refrigerated. For longer delays, overnight storage is recommended to soften the material further. Be cautious, as prolonged storage (beyond overnight) may lead to overdigestion of the root tips.

3.1.3 Chromosome Preparation

1. Place one or two root tips on a clean microscope slide. Using a stereomicroscope, carefully cut a 1- to 2-mm piece of meristematic tissue from the terminal white-dense area of the root tip. Discard the remaining portion of the root tip.
2. Use a Pasteur pipette to remove the citrate buffer surrounding the meristematic tissue. Add 10–30 μl of 60% acetic acid, depending on the size and number of root tips, and let the tissue incubate for 1–2 minutes.

3. Using dissecting needles, gently disintegrate the meristematic tissue to form a uniform suspension. Discard any debris, such as rhizodermis or root cap.
4. Cover the tissue suspension with a coverslip. Tap the coverslip gently and vertically with a needle to spread the cells evenly.
5. Briefly move the slide above the flame of an alcohol burner three times to soften the material but avoid boiling the suspension. Optionally, use forceps to hold the slide for precise control over heating.
6. Place the slide between two sheets of filter paper on a flat surface. Firmly press the coverslip with your thumb to flatten the cells into a monolayer. Apply gentle pressure without horizontal movement to avoid smearing. Ensure that the volume of acetic acid is sufficient to cover the coverslip area. Too much liquid will cause overflow, while too little will result in uneven dispersion of chromosomes and nuclei.
7. Use a phase-contrast microscope to check the quality of the chromosome squash and estimate the frequency of mitotic divisions. Adjust preparation steps if needed to ensure high-quality results.
8. Dip the slide into liquid nitrogen for a few seconds using long forceps. Quickly flick off the coverslip with a razor blade. Ensure that the entire coverslip is submerged during freezing. Alternatively, place the slide on a block of dry ice for a few minutes before removing the coverslip.
9. Immerse the slide in freshly prepared Carnoy's fixative using a Coplin jar and incubate for 10 minutes.
10. Air dry the slide in a tube rack or similar holder.
11. Store the dried slides in a dust-free box at 4 °C. Although fixed slides can be stored for several weeks or months, freshly prepared or only a few-days-old slides yield the best results.

3.2 Surface Spreading Preparation of Chromosomes and Nuclei from Generative Organ Tissues

When actively growing roots are unavailable or preserving a valuable seedling is critical, floral tissues provide a practical alternative for preparing mitotic chromosome spreads. Mitotic chromosomes can be routinely obtained from pistils and anther tapetal cells, while pre-meiotic nuclei and chromosomes across all meiotic stages can be prepared from young anthers using this surface spreading protocol. This technique is particularly well-suited for visualizing pachytene chromosomes, which are 10- to 40-fold longer than mitotic chromosomes, offering greater detail for cytogenetic analysis. This streamlined protocol consists of three main steps: collection of floral material, enzymatic digestion, and chromosome spreading, ensuring efficient and high-quality preparation of meiotic and mitotic chromosomes.

3.2.1 Collection of Floral Material

1. Using fine-tipped forceps, collect whole inflorescences containing predominantly closed flower buds with white anthers (*see* Fig. 2b). Place them into 15- or 50-ml plastic tubes or similar vials filled with freshly prepared Carnoy's fixative (3:1 ethanol: glacial acetic acid) (*see* **Note 5**).
2. Fix the material at room temperature or at 4 °C overnight (up to 24 hours). Change the fixative several times during this period until the flower buds turn white or yellow-white and the fixative remains clear.
3. After fixation, replace the fixative with 70% ethanol. Store the material at −20 °C until use. Although fixed material remains usable for several months or even years, the best-quality preparations are typically achieved with freshly fixed inflorescences.

3.2.2 Enzymatic Digestion (See **Note 3**)

1. Place the selected inflorescences into a black staining block and wash them twice with distilled water, each wash lasting 5 minutes.
2. Under a stereomicroscope, carefully remove and discard any unwanted parts, such as flower buds with yellow anthers containing pollen.
3. Ensure that the volume of solutions used in subsequent steps is proportional to the amount of material being processed. Inflorescences should remain fully submerged throughout.
4. Remove the distilled water using a glass Pasteur pipette. Add 1× citrate buffer to the staining block and wash the inflorescences twice for 5 minutes on an orbital shaker.
5. Remove the citrate buffer with a Pasteur pipette and add a 0.3% pectolytic enzyme mixture to the staining block. Ensure that the inflorescences are fully submerged.
6. Place the staining block inside a humid box and incubate in an incubator at 37 °C for at least 3 hours (*see* **Note 6**).
7. Replace the enzyme mixture with 1× citrate buffer using a Pasteur pipette. If the material cannot be processed immediately (within 1 hour), store the staining block containing digested inflorescences in a refrigerator. Overnight storage can further soften the tissue. However, avoid storing the material for longer than overnight, as this may lead to overdigestion and compromise sample quality.

3.2.3 Chromosome Surface Spreading

1. Using a Pasteur pipette, select a flower bud from the enzymatically digested inflorescence and place it on a clean microscope slide. Although flowers consist of various floral organs (sepals, petals, stamens, and carpels), their small size can make separating individual parts challenging, especially for beginners. In such cases, entire flower buds can be used to prepare mitotic

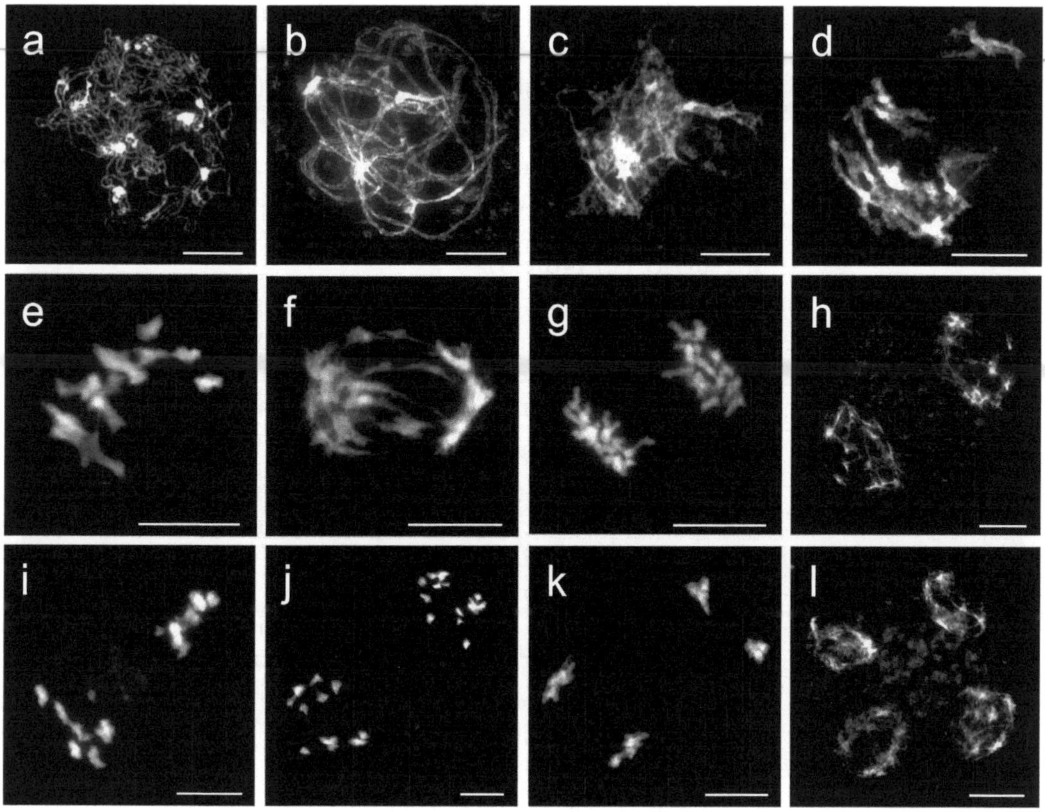

Fig. 3 Pollen meiosis of diploid apomict *Boechera retrofracta* × *stricta* ($2n = 15$). (**a**) Leptotene, (**b**) Pachytene, (**c**) Diplotene, (**d**) Diakinesis, (**e**) Metaphase I, (**f**) Anaphase I, (**g**) Telophase I, (**h**) Prophase II, (**i**) Metaphase II, (**j**) Anaphase II, (**k**) Telophase II, and (**l**) Four meiotic products forming pollen nuclei. Chromosomes were counterstained with DAPI. Scale bars: 10 μm

and meiotic chromosome spreads (*see* Fig. 3). Alternatively, petals can be separated to prepare chromosomes in various stages of meiosis or pre-meiotic nuclei. Additionally, the tapetal cells of petals contain mitotic chromosomes and interphase nuclei, making petals an excellent choice for preparing all these stages (*see* **Note 7** for developmental stages). Alternatively, the pistil can be separated from the rest of the flower and used specifically for preparing mitotic chromosomes and interphase nuclei.

2. If necessary, remove excess fluid using a Pasteur pipette.
3. Under a stereomicroscope, tap the entire flower bud or separated pistil with dissecting needles until a fine cell suspension is formed. If the tissue is resistant to disintegration, remove the citrate buffer, add the 0.3% pectolytic enzyme mixture, and extend the enzymatic digestion. For anthers or individual microsporangia, dissect them transversely with a dissecting

needle to create smaller sections, allowing multiple slides to be prepared from a single flower bud or anther.

4. Add 20 μL of 60% acetic acid to the cell suspension on the microscope slide.
5. Place the slide on a heating block set to 50 °C. Gently spread the suspension in a circular motion with a dissecting needle for 30 seconds, ensuring that the needle does not touch the slide surface (*see* **Note 8**).
6. Fix the chromosomes by pipetting 100 μL of Carnoy's fixative around the suspension in a stepwise manner. Apply the final drop to the center of the suspension, and then tilt the slide to discard excess fluid.
7. Dry the slide using a hair dryer.
8. Use a phase-contrast light microscope to assess the quality of the chromosome spread. Ensure that chromosomes and nuclei are not obscured by cytoplasm (*see* Fig. 4). Slides with insufficient chromosome spreads or excessive cytoplasmic debris can be discarded. If cytoplasmic coverage is excessive, consider a pepsin treatment as described in Subheading 3.3).
9. Post-fix the slide in 4% formaldehyde (prepared in distilled water) in a Coplin jar for 10 minutes. Perform this step in a ventilated fume hood.
10. Air-dry the slide in a tube rack or similar holder.
11. Store the dried slides in a dust-free box at 4 °C (*see* **Note 9**).

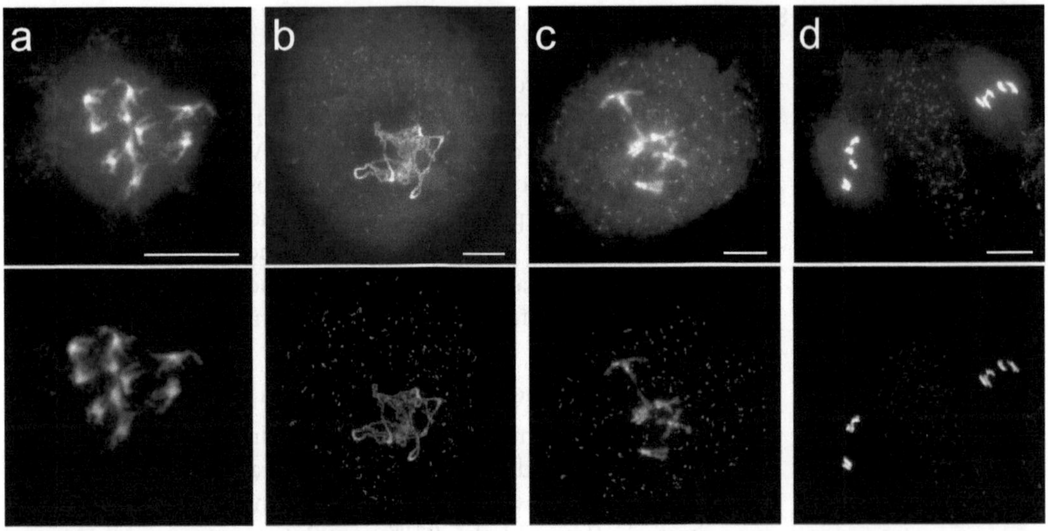

Fig. 4 Chromosome preparations before (upper image) and after (lower image of each pair) RNase and pepsin treatments. (**a**) Mitosis, (**b**) Pachytene, (**c**) Diakinesis, (**d**) Metaphase II. Chromosomes were counterstained with DAPI. Scale bar: 10 μm

3.3 Chromosome Pretreatment

Effective slide treatment using RNase and pepsin is critical for ensuring successful hybridization. Following the treatment, slides must be free of cytoplasmic residues (as shown in Fig. 4). Residual cytoplasm surrounding chromosomes and nuclei can impede the access of DNA probes to their chromosomal targets. This not only reduces the intensity of hybridization signals but also increases background noise, potentially compromising the accuracy and quality of the results.

1. Wash slides in 2× SSC in a Coplin or Hellendahl jar twice, for 5 minutes each time (*see* **Note 10**), shaking on an orbital shaker. Ensure complete coverage of the slides with the solution.
2. Pipette a 100-μL drop of RNase solution onto a 24 × 50 mm coverslip. Pick up the coverslip using a slide by capillary force to transfer the RNase drop.
3. Place the slide in a humid box and incubate at 37 °C for 1 hour.
4. Tilt the slide to allow the coverslip to fall off and wash the slide as described in **step 1**.
5. Treat the slide with pepsin solution for 5 minutes in a jar placed in a 37 °C water bath to remove cytoplasm.
6. Wash the slide as described in step 1.
7. Dehydrate the slide in an ethanol series (70%, 80%, and 96% ethanol, 3 minutes each).
8. Allow the slides to air-dry in a vertical position in a plastic rack.
9. Apply 15 μL of Vectashield with DAPI to the slide.
10. Cover the slide with a 24 × 24 mm coverslip.
11. Examine the slide using an epifluorescence microscope to ensure that chromosomes are undamaged and free of cytoplasm. If cytoplasm persists (*see* Fig. 4), remove the coverslip under a flow of tap water and repeat **steps 5–7**.
12. Remove the coverslip under a flow of tap water and wash the slide as described in **step 1**.
13. Post-fix the slide for 10 minutes in a jar containing 4% formaldehyde in 2× SSC in a ventilated fume hood (*see* **Note 11**).
14. Wash the slide as described in **step 1**.
15. Dehydrate the slide again in an ethanol series (70%, 80%, and 96% ethanol, 3 minutes each).
16. Allow the slides to air-dry in a vertical position in a plastic rack.

3.4 Probe Labeling by Nick Translation

Nick translation is a molecular technique used to label DNA by incorporating modified nucleotides into its structure. The process involves the sequential action of DNase I, which introduces single-strand nicks, and DNA polymerase I, which uses its 5′ to 3′

exonuclease activity to remove nucleotides downstream of the nick and its polymerase activity to replace them with labeled nucleotides such as biotin-dUTP or fluorophore-conjugated dUTPs. This technique produces labeled DNA probes for in situ hybridization and other applications where detection of specific DNA sequences is required. The labeled probes are highly sensitive and versatile, enabling fluorescence-based detection in various genomic studies [15].

1. In a 0.5-mL microcentrifuge tube, combine the following: 1 μg of DNA sample (*see* **Notes 1** and **12**) in 32 μL distilled water (or 29 μL if using custom-made x-dUTPs); 5 μL of 10× NT buffer; 5 μL of nucleotide mixture; 1 μL of 1 mM commercial x-dUTP or 4 μL of 1 mM custom-made x-dUTP; 5 μL of 0.1 M β-mercaptoethanol; 1 μL DNase I; and 1 μL DNA polymerase I. Ensure that the total reaction volume is 50 μL by adjusting the volume of distilled water based on the type of x-dUTP used (*see* **Note 13**).
2. Gently vortex the tube to mix and briefly spin to collect contents.
3. Incubate the reaction mixture in a thermal cycler at 15 °C for 90 minutes.
4. After incubation, transfer the tube to ice.
5. Load 5 μL of the reaction mixture onto a 1% agarose gel alongside a 100-bp DNA ladder marker. If the smear of labeled DNA fragments is between 200 and 500 bp (*see* Fig. 5 and **Note 14**), stop the nick translation reaction by adding 1 μL of

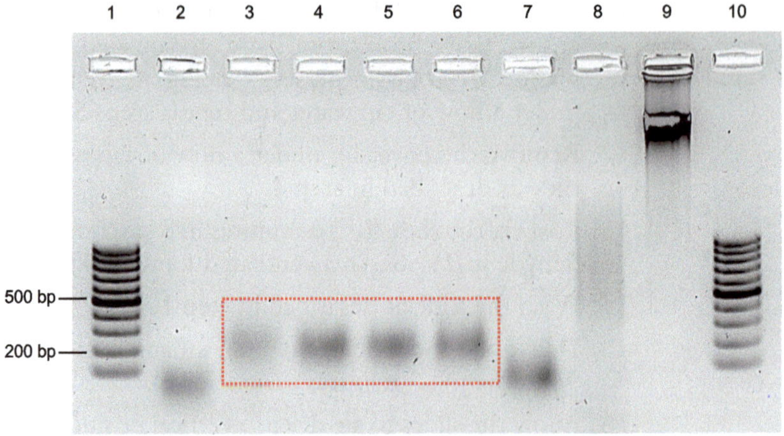

Fig. 5 Agarose gel analysis of probe fragment sizes following nick translation. Lanes 1 and 10: 100-bp DNA ladder. Lanes 3–6: optimal probe fragment sizes ranging between 200 and 500 bp. Lanes 2 and 7: excessively sheared DNA fragments (<200 bp), caused by <1 μg of degraded BAC DNA and/or >1 μL of DNase I in the reaction. Lanes 8 and 9: DNA fragments that are too long, resulting from >1 μg of BAC DNA and/or <1 μL of DNase I in the nick translation reaction. *See* Subheading 3.4 and **Note 14**

0.5 M EDTA and incubating at 60 °C for 10 minutes. If fragments are larger than 500 bp, extend the incubation at 15 °C for an additional 30 minutes and repeat gel analysis to reassess fragment size. For fragments still exceeding 500 bp, add an additional 1 μL of DNase I and repeat the process. Conversely, if the DNA fragments are too short, reduce the amount of DNase I in a subsequent experiment. Optimize DNase concentration and incubation time empirically for each probe.

6. Store the labeled probe at −20 °C until use.

3.5 Preparation of Probe

1. Combine individually labeled DNA samples by pipetting 5 μL (≈100 ng DNA) of each into a 2-mL microcentrifuge tube.
2. Add 0.1× volume of 3 M sodium acetate and 2.5× volume of ice-cold 96% ethanol to the tube. Vortex briefly and keep the mixture on ice or at −20 °C for at least 30 minutes.
3. Centrifuge at 13,000× g for 30 minutes at 4 °C.
4. Carefully discard the supernatant.
5. Add 500 μL of 70% ethanol to the pellet and centrifuge again for 5 minutes.
6. Repeat the step of discarding the supernatant.
7. Dry the DNA pellet using a desiccator or vacuum centrifuge. The labeled DNA pellet typically appears as a 1–2 mm oval smudge.
8. Resuspend the pellet in 20 μL of hybridization buffer.
9. Incubate the tube in a 37 °C thermomixer at 400 rpm. Simple probes usually dissolve within minutes to hours, while complex probes with dozens or hundreds of DNA samples may require overnight incubation. Ensure that the DNA is fully dissolved to avoid high background noise during hybridization.

3.6 Fluorescence In Situ Hybridization

1. Add 20 μL of the probe solution to a prepared chromosome slide. Pipette carefully to avoid introducing air bubbles. Cover with a 24 × 24 mm coverslip and seal the edges with rubber cement (*see* **Note 15**).
2. Place the slide on an 80 °C heating block for 2 minutes to denature both the probe and chromosomal DNA.
3. Transfer the slide to a humid box and incubate overnight at 37 °C (*see* **Note 16**).

3.7 Post Hybridization Washing and Detection of Fluorescence

This protocol exemplifies signal detection and amplification for hapten-labeled probes (e.g., biotin- and digoxigenin-dUTP) visualized by indirect immunofluorescence via Texas Red and Alexa Fluor 488, respectively. It can also accommodate fluorochrome-labeled probes (e.g., Cy3- or DEAC-dUTP), which require only post-hybridization washing prior to microscopic evaluation. Perform all washing steps in Coplin or Hellendahl jars placed in a 42 °C water bath shaker. Carry out incubation steps on microscope slides covered with 24 × 50 mm coverslips (*see* **Note 17**), placed in a humid box at 37 °C. Avoid slide drying throughout the procedure to minimize nonspecific background signals.

1. Use forceps to carefully remove the rubber cement frame and allow the coverslip to fall off.
2. Wash slides for 2 minutes in 2× SSC.
3. Remove unbound probes by washing slides in 20% formamide stringent wash buffer three times for 5 minutes each.
4. Wash slides for 2 minutes in 2× SSC.
5. If only fluorochrome-labeled probes are used, proceed to **step 14**.
6. Wash slides for 5 minutes in 4 T buffer.
7. Incubate slides for 30 minutes with 100 μL of blocking solution applied on a 24 × 50 mm coverslip (*see* **Note 18**).
8. Pick up the coverslip by capillary force and place slides in a humid box.
9. Wash slides twice for 5 minutes each in 4 T buffer.
10. Incubate slides for 30 minutes with 100 μL of avidin–Texas Red (1:1000) in TNB buffer.
11. Wash slides twice for 5 minutes each in TNT buffer.
12. Incubate slides for 30 minutes with 100 μL of a mixture of biotinylated goat anti-avidin (1:200) and mouse anti-digoxigenin (1:250) in TNB buffer.
13. Repeat the washing step in TNT buffer as described in **step 9**.
14. Incubate slides for 30 minutes with 100 μL of a mixture of avidin–Texas Red (1:1000) and goat anti-mouse Alexa Fluor 488 (1:200) in TNB buffer.
15. Wash slides twice for 5 minutes each in TNT buffer.
16. Dehydrate slides in an ethanol series (70%, 80%, and 96%), 3 minutes each step, and allow them to air-dry in a plastic slide rack.
17. Apply 15 μL of Vectashield with DAPI to each slide and cover with a 24 × 32 mm coverslip.
18. Observe and photograph slides using a fluorescence microscope equipped with appropriate filters, a CCD camera, and image acquisition software.
19. Store slides in a paper or plastic slide case at 4 °C (*see* Fig. 1 and **Note 19**).

4 Notes

1. DNA for hybridization probes should be selected based on the design and objectives of the experiment. Depending on the experimental goals, various types of DNA probes can be prepared and used. For example, repetitive DNA sequences such as satellite repeats or ribosomal DNA are commonly employed for mapping heterochromatic regions. Bacterial artificial chromosome (BAC) clones can be used for chromosome painting or for identifying specific chromosomal regions (*see* Fig. 1). Synthetic oligonucleotides are ideal for targeting short, specific DNA motifs, while whole-genome DNA can be used for genomic in situ hybridization (GISH) to identify parental subgenomes in hybrids and allopolyploids (*see* Fig. 1). For broader applications, labeled plasmid inserts can also serve as probes for detecting unique sequences or genes. The flexibility in probe design ensures adaptability across diverse cytogenetic studies, from simple sequence localization to complex comparative genome analyses.
2. Preparation of ice-cold water: Fill glass vials or microcentrifuge tubes halfway with distilled water and freeze. Before use, add fresh distilled water to partially melt the ice and create ice-cold water. The ice-cold water pretreatment effectively arrests dividing cells at metaphase. This method is inexpensive, nontoxic, and reliable compared to spindle-inhibiting drugs, such as colchicine, α-bromonaphthalene, or 8-hydroxyquinoline, which are also commonly used for chromosome preparations.
3. A critical factor in achieving high-quality chromosome preparations is the enzymatic digestion of plant cell walls. The precise use of cellulase, pectolyase, and cytohelicase ensures efficient cell wall degradation while maintaining chromosomal integrity. In root tip meristems, the combination of cellulase and pectolyase effectively breaks down cellulose and pectin components of the cell wall. For pollen mother cells, however, the presence of thicker callose walls necessitates the addition of cytohelicase to the enzyme mix. This tailored enzymatic treatment is carefully optimized to produce well-dispersed chromosomes that are free from cytoplasmic interference, ensuring their suitability for high-resolution hybridization and downstream cytogenetic applications. The balance between effective digestion and chromosomal preservation is essential to maximize both the clarity and utility of the preparations (*see* Fig. 4).
4. After 90 minutes of digestion, remove one or two root tips using a Pasteur pipette and place them on a clean microscope slide. Under a stereomicroscope, use dissecting needles to gently disintegrate the tissue. A fine cell suspension should

form. If the tissue remains hard or difficult to disintegrate, extend the digestion time and reassess. Root tips stored in fixative for extended periods (months to years) or harvested from potted plants often require longer digestion times than those harvested from seedlings. If processing only a single root tip, apply the standard 90-minute digestion time without performing intermediate checks. When handling multiple root tip samples, checking digestion progress for one representative sample is usually sufficient. The 0.3% enzyme mixture can be reused multiple times. After each use, store the mixture at −20 °C. Note that digestion times may need to be incrementally increased with each reuse to maintain effectiveness.

5. Since typically only one flower bud per inflorescence contains pachytene chromosomes or other meiotic stage, it is advisable to harvest and process multiple inflorescences simultaneously to increase the chances of obtaining the desired stage.
6. Digestion times can vary significantly depending on the type of inflorescence, plant species, or how long the material has been stored in fixative. For older material or species with tougher tissues, longer digestion times may be necessary. After a standard 3-hour digestion, reassess progress by disintegrating the tissue and checking for the formation of a fine cell suspension. If the tissue remains resistant or contains large amounts of cytoplasm, extend the digestion time by 30 minutes or more, then check again. Repeat as needed until the desired consistency is achieved. The 0.3% enzyme mixture used for inflorescences can also be reused multiple times. After each use, store the mixture at −20 °C and adjust digestion times accordingly for subsequent uses to account for reduced enzymatic activity.
7. Meiotic chromosomes are commonly prepared from pollen mother cells within young anthers. Identifying the correct meiotic stage, typically between pachytene and metaphase I, requires practice and experience. For those less familiar with interpreting meiotic stages, it is advisable to prepare a series of slides from individual flower buds or anthers. Start with the oldest and largest buds, working progressively down to the youngest and smallest ones to capture a range of developmental stages. In *Boechera* (*see* Fig. 3), yellow flower buds contain anthers with pollen. The largest white flower buds may contain pollen or pollen tetrads. The second-largest white flower buds usually have pollen mother cells completing meiosis II. Prophase I and metaphase I chromosomes can be found in the third or fourth white flower buds. Mitotic chromosome spreads and interphase nuclei from anther tapetal cells can be prepared from flower buds of all sizes.

8. If needed, increase the volume of acetic acid to 40–60 μl and extend the spreading time to 2 minutes for larger buds. For very small buds, the spreading duration can be reduced. During this process, acetic acid dissolves acid-soluble proteins and cytoplasmic components, clearing the preparation. Insufficient acetic acid volume or spreading time may result in chromosomes and nuclei surrounded by cytoplasm.
9. Although slides can be stored for several weeks or months, freshly prepared or recently made slides yield the best results. Label slides on the frosted area with a pencil (not a marker pen) for identification.
10. Washing steps are typically recommended to last 5 minutes each but can be extended to 10 minutes or more if necessary. Conduct all washing steps at room temperature unless otherwise specified. Ensure that slides are completely submerged during washing steps to prevent dirt accumulation or uneven washing. When changing solutions, carefully pour off the old solution and slowly replace it with fresh solution. Alternatively, use forceps to transfer slides to a separate jar containing the new solution. Avoid allowing the slides to dry out at any point during solution changes or transitions between steps. Handle slides gently to avoid scratching chromosomes or nuclei. Ensure that slides do not touch each other during washing steps in Coplin or Hellendahl jars to prevent damage. Always maintain consistent hydration of the slides throughout the process to preserve sample integrity.
11. Post-fixation stabilizes chromosome morphology and prevents material loss during subsequent steps.
12. DNA can be isolated using the standard alkaline lysis method [25, 26] or with commercial DNA isolation kits. A milky appearance of the isolated DNA (indicating protein or polysaccharide contamination) does not affect hybridization efficiency. Avoid eluting BAC DNA in EDTA-containing buffers (e.g., TE buffer), as EDTA inhibits enzyme activity.
13. While the selection of haptens for indirect immunofluorescence detection is relatively limited—primarily biotin- and digoxigenin-dUTPs, with dinitrophenyl (DNP)-dUTP being rarely used—there are numerous fluorescent dyes available for direct DNA labeling. The choice of fluorochromes should be guided by their emission spectra and compatibility with the optical filter sets, ensuring minimal spectral crosstalk. Compared to fluorochrome-labeled DNA probes, hapten-labeled painting probes with immunofluorescence detection typically yield lower signal-to-noise ratios for chromosome-specific fluorescence. This reduction in signal quality is often due to nonspecific interactions between antibodies and unbound

probe or unincorporated nucleotides. Commercially labeled nucleotides are more efficiently incorporated into nascent DNA than custom-made x-dUTPs. When using custom-made x-dUTPs, add 4 μl to the reaction instead of 1 μl. Custom x-dUTPs can be synthesized by coupling allylamine-dUTP to succinimidyl-ester derivatives of haptens or fluorochromes [29].

14. The nick translation protocol described here is optimized to achieve high incorporation of labels into DNA within 90 minutes, producing probe fragments of approximately 200–300 bp (*see* Fig. 4). Probe fragment size is a critical factor—fragments that are too long may fail to hybridize effectively to chromosomes, while excessively short probes can lead to nonspecific hybridization and elevated background noise.
15. Rubber cement prevents drying out and dilution of the probe due to condensation during long hybridization periods.
16. For homologous (intraspecies) sequences, overnight incubation is typically sufficient. For homeologous hybridizations (interspecific), extend the incubation to 48–72 hours to ensure optimal hybridization.
17. To avoid drying out during post-hybridization incubations on the slide, 100-μl volumes are applied to fit the area of large coverslips (24 × 50–mm).
18. Blocking prevents nonspecific binding of detection reagents.
19. Over time (weeks, months, or years), slides may lose utility as fluorescent signals fade or bacterial growth occurs under the coverslip. Regular monitoring and proper storage can extend slide longevity. Several factors can significantly influence the hybridization efficiency of DNA probes. For experimenters encountering suboptimal results, particularly those less experienced, consulting specialized publications and reputable websites can provide valuable insights into the effects of various parameters and how to adjust them effectively. The denaturation temperature (80 °C in this protocol) may need to be adjusted depending on the outcome of the hybridization process. Lowering the temperature might be necessary if chromosome damage is observed, while increasing it could help if the hybridization of a probe appears compromised. Similarly, the concentration of formamide used for stringent post-hybridization washing (20% in this protocol) and the temperature of all post-hybridization washes (42 °C in this protocol) can be altered to achieve more or less stringent hybridization conditions, optimizing the overall process.

Acknowledgments

This study was funded by the Czech Science Foundation (24-11371S).

References

1. Mandáková T, de Jong H, Schranz ME, Lysak MA (2015) Karyotype evolution in apomictic Boechera: the origin of the aberrant chromosomes. Plant J 82:785–793
2. Mandáková T, Price B, Ashby K, Hloušková P, Windham MD, Mitchell-Olds T, Carman J, Lysak MA (2020) The origin of the Boechereae (Brassicaceae) was preceded by descending dysploidy, whereas intra-tribal cladogenesis was marked by overall genome stasis. Front Plant Sci 11:514
3. Shan W, Kubová M, Mandáková T, Lysak MA (2021) Nuclear organization in crucifer genomes: nucleolus-associated telomere clustering is not a universal interphase configuration in Brassicaceae. Plant J 108:528–540
4. Naish M, Alonge M, Wlodzimierz P, Tock AJ, Abramson BW, Schmücker A, Mandáková T, Jamge B, Lambing C, Kuo P, Yelina N, Hartwick N, Colt K, Smith L, Ton J, Kakutani T, Martienssen RA, Schneeberger K, Lysak MA, Berger F, Bousios A, Michael TP, Schatz MC, Henderson IR (2021) The genetic and epigenetic landscape of the Arabidopsis centromeres. Science 374:eabi7489
5. Bayat S, Lysak MA, Mandáková T (2021) Pericentric inversions as the main genome-diversifying driver within a cruciferous tribe. Plant J 108:1768–1785
6. Zuo S, Mandáková T, Kubová M, Lysak MA (2022) Genomes, repeatomes and interphase chromosome organization in the meadowfoam family (Limnanthaceae, Brassicales). Plant J 110:1462–1475
7. Mau M, Mandáková T, Ma X, Ebersbach J, Zou L, Lysak MA, Sharbel TF (2022) Convergent evolution of an apomixis-specific allele class of the tapetum-active UPGRADE2 gene in supernumerary chromatin of apomicts. Front Plant Sci 13:890038
8. Bartha L, Mandáková T, Kovařík A, Bulzu P-A, Rodde N, Mahelka V, Lysák MA, Fustier M-A, Šafář J, Cápal P, Keresztes L, Banciu HL (2022) Intact rDNA arrays of Potentilla-origin detected in Erythronium nucleus suggest recent eudicot-to-monocot horizontal transfer. New Phytol 235:1246–1259
9. Mandáková T, Lysak MA (2022) The identification of the missing maternal genome of the allohexaploid camelina (Camelina sativa). Plant J 112:622–629
10. Wlodzimierz P, Rabanal FA, Burns R, Naish M, Primetis E, Scott A, Mandáková T, Gorringe N, Tock AJ, Holland D, Fritschi K, Habring A, Lanz C, Patel C, Schlegel T, Collenberg M, Mielke M, Roux F, Shirsekar G, Alonso-Blanco C, Lysak MA, Novikova P, Bousios A, Weigel D, Henderson IR (2023) Rapid cycles of satellite homogenization and retrotransposon invasion drive Arabidopsis pancentromere evolution. Nature 618:557–565
11. Huang Y, Guo G, Zhang K, Mandáková T, Cheng F, Lysak MA (2023) The meso-octoploid Heliophila variabilis genome sheds a new light on the impact of polyploidization and diploidization on the diversity of the cape flora. Plant J 116:446–466
12. Sakamoto T, Ikeamtsu S, Nakayama H, Mandáková T, Gohari G, Sakamoto T, Li G, Hou H, Matsunaga S, Lysak MA, Kimura S (2024) Chromosome-level genome assembly of amphibious plant Rorippa aquatica revealed its allotetraploid origin and mechanisms of phenotypic plasticity on leaf shape upon submergence. Commun Biol 7:431
13. Mandáková T, Krumpolcová A, Matyášek R, Volkov R, Lysak MA, Kovařík A (2024) Uniparental silencing of 5S rRNA genes in plant allopolyploids—insights from Cardamine (Brassicaceae). Plant J 119:1313–1326
14. Lysak MA, Mandáková T (2013) Analysis of plant meiotic chromosomes by chromosome painting. Methods Mol Biol 990:13–24
15. Rigby PW, Dieckmann M, Rhodes C, Berg P (1977) Labeling deoxyribonucleic acid to high specific activity in vitro by nick translation with DNA polymerase I. J Mol Biol 113(2): 237–251
16. Meinkoth J, Wahl G (1984) Hybridization of nucleic acid immobilized on solid support. Anal Biochem 138:267–284
17. Ambrozova K, Mandáková T, Bureš P, Neumann P, Leitch IJ, Macas J, Lysak MA (2011) Molecular insight into the giant genomes of Fritillaria lilies. Ann Bot 107:255–268
18. Chumová Z, Záveská E, Mandáková T, Krak K, Trávníček P (2017) The Mediterranean: the

cradle of Anthoxanthum (Poaceae) diploid diversity. Ann Bot 120:285–302

19. Filiault D, Ballerini ES, Mandáková T, Aköz G, Derieg N, Schmutz J, Jenkins J, Grimwood J, Shu S, Hayes RD, Hellsten U, Barry K, Yan J, Mihaltcheva S, Karafiátová M, Nizhynska V, Lysak MA, Hodges SA, Nordborg M (2018) The Aquilegia genome: adaptive radiation and an extraordinarily polymorphic chromosome with a unique history. elife 7:e36426
20. Peška V, Mandáková T, Ihradská V, Fajkus J (2019) Comparative dissection of three giant genomes: Allium cepa, Allium sativum, and Allium ursinum. Int J Mol Sci 20:733
21. Peška V, Mátl M, Mandákova T, Vitales D, Fajkus P, Fajkus J, Garcia S (2020) Human-like telomeres in the marine plant Zostera marina. J Exp Bot 71:5786–5793
22. Chumová Z, Záveská E, Hloušková P, Ponert J, Schmidt P-A, Čertner M, Mandáková T, Trávníček P (2021) Repeat proliferation and partial endoreplication jointly drive the genome size evolution of orchids. Plant J 107:511–524
23. Eriksson MC, Mandakova T, McCann J, Temsch EM, Chase MW, Hedrén M, Weiss-Schneeweiss H, Paun O (2022) Repeat dynamics across timescales: a perspective from sibling allotetraploid marsh orchids (Dactylorhiza majalis s.l.). Mol Ecol Evol 39:msac167
24. Slovák M, Melichárková A, Gbúrová Štubňová E, Kučera J, Mandáková T, Smyčka J, Lavergne S, Giuseppe Passalacqua N, Vďačný P, Paun O (2022) Pervasive introgression during the explosive diversification of the European mountain genus Soldanella L. (Primulaceae). Syst Biol 72:491–504
25. Heilig J, Elbing KL, Brent R (2001) Large-scale preparation of plasmid DNA. Curr Protoc Mol Biol 41:1.7.1–1.7.16
26. Henegariu O, Bray-Ward P, Ward DC (2000) Custom fluorescent nucleotide synthesis as an alternative method for nucleic acid labeling. Nat Biotechnol 18:345–348

Chapter 12

Targeted Mutagenesis in Natural Apomicts

Tatyana Radoeva, Diana Rigola, Rik H. M. Op den Camp, Peter J. van Dijk, and Charles J. Underwood

Abstract

Apomixis—clonal reproduction through seeds—is an alternative reproductive strategy that takes place in less than 0.1% of plant species and has evolved independently in diverse plant lineages. To date, the genetic basis of apomixis has been unraveled genetically in only a few genera. The identification of causal apomixis genes is technically challenging, as apomictic species are typically polyploid and the genetic loci associated with apomixis are often in low-recombination regions limiting conventional fine-mapping. In triploid apomictic dandelion (*Taraxacum officinale*), after conventional genetic mapping, deletion mapping, and complete apomixis loci haplotype assembly, we made use of targeted mutagenesis using CRISPR/Cas9 technology to identify the *Taraxacum officinale PARTHENOGENESIS* (*ToPAR*) gene that is responsible for embryogenesis in the absence of fertilization. Here, we report the methods used to clone the *ToPAR* gene by targeted mutagenesis and we expect that the general principles could be applied in other systems to identify novel apomixis genes.

Key words Apomixis, Taraxacum, Parthenogenesis, CRISPR/Cas9, Genome editing, Targeted mutagenesis

1 Introduction

Taraxacum officinale, the common dandelion, is a member of the *Asteraceae* family and occurs as both sexual diploid ($2n = 2x = 16$) and apomictic polyploid (mostly triploid, $2n = 3x = 24$) cytotypes [1]. Next to its well-known distinct yellow flowers that turn into round, white, puffy seed heads, *Taraxacum* is also one of the best-known and most widely spread natural apomicts. Consequently, the triploid dandelion has been developed and used as a model species for the genetic and molecular analysis of gametophytic apomixis [1–4]. Even though studying apomixis-related loci is challenging due to low recombination rates and the polyploid nature of apomictic species [5], significant progress has been made in recent years. This progress includes the identification and functional

Tatyana Radoeva and Peter J. van Dijk (eds.), *Apomixis: Methods and Protocols*, Methods in Molecular Biology, vol. 2987, https://doi.org/10.1007/978-1-0716-5001-1_12,

characterization of novel apomixis-related genes in the apomicts *Pennisetum*, *Taraxacum*, and *Pilosella* [4, 6].

When a locus associated with apomixis (or a component of it) is identified through genetic mapping, the causal gene responsible for apomixis still needs to be pinpointed. In practice, due to low recombination rates in apomixis loci in both natural and experimental populations, it is difficult to narrow down the mapped region to a causal gene. By sequencing the apomixis locus—using bacterial artificial chromosome (BAC) sequencing or whole-genome sequencing (WGS)—and applying bioinformatics, putative functions can be assigned to the genes within the region.

In some cases, a clear candidate gene is present, such as the *BABYBOOM-like* (*BBML*) gene for parthenogenesis in *Pennisetum*, whose *Brassica* homolog is a known somatic embryogenesis inducer [7]. More often, however, the causal apomixis gene is less obvious. In such cases, an unbiased approach to identifying the functional apomixis gene within the locus is to mutate *all* or a selection of (based on predicted gene functions) the genes in the region. This can be done through targeted mutagenesis using clustered regularly interspaced short palindromic repeats/CRISPR-associated protein 9 (CRISPR/Cas) technology, which enables specific gene knockouts. Over the past decade, this technology has become the most powerful method for targeted DNA manipulation [8] and is now routinely used in both animal and plant research.

The CRISPR/Cas9 system makes double-stranded DNA breaks (DSBs), guided to a precise location in the host genome by a single-guide RNA (20-bp DNA target sequence immediately upstream of a 5′-NGG protospacer adjacent motif [PAM]) [8]. This triggers a nonhomologous end joining (NHEJ) repair mechanism, which introduces short insertions or deletions at the DSB site, resulting in efficient targeted mutations that can knock out gene functions [9]. Such a disruption of a functional apomixis-related gene is expected to lead to loss of the apomictic phenotype. Screening by amplification followed by (Sanger) sequencing of the target locus or DNA sequencing of loss-of-apomixis (LoA) plants will then lead to the identification of the disrupted apomixis gene. Genetic studies have shown that apomixis is often controlled by a dominant allele; therefore, knockout mutations will need to be allele-specific.

In addition, CRISPR/Cas9 can be used to introduce larger deletions, covering multiple genes simultaneously. Thus, in a situation where there are many candidate genes to be tested, a two-stage approach can be employed: (i) deletion of smaller regions consisting of a few putative candidate genes, followed by (ii) targeting the individual genes from the region found to be associated with LoA.

In this chapter, we describe a procedure for targeted mutagenesis in dandelions used to ultimately identify the *PAR* gene as essential for apomixis [4]. The overall procedure involves selection and design of guide RNAs, vector construction and cloning, genotyping and selection of the transgenic plants, and finally, a detailed phenotyping procedure for loss of parthenogenesis.

2 Materials

2.1 Plant Materials

Taraxacum officinale, A68 apomictic line ($2n = 3x = 24$) and FCH72 sexual line ($2n = 2x = 16$) are vegetatively propagated in vitro and grown in the greenhouse at 22 °C and 16-h day:8-h night photoperiod.

2.2 Gene Editing Construct Design and Cloning

1. Plasmid vectors—standard shuttle vectors (e.g., pUC19 or derivative) carrying (the synthesized) guide RNA fragments and binary vector carrying PcUBI::AtCas9 with cloning sites (e.g., GW or others) to insert the guide RNAs.
2. Standard reagents for molecular cloning, e.g., recombination-assisted Gateway Cloning [10, 11], such as LR Recombinant Clonase kit (*see* **Note 1**).
3. *Escherichia coli* TOP10 (or alternative strain) electrocompetent cells (*see* **Note 2**) and all standard reagents required for *E. coli* transformation and selection of positive colonies.
4. *Agrobacterium tumefaciens* EHA105 electrocompetent cells and all standard reagents required for *A. tumefaciens* transformation and selection of positive colonies.

2.3 Genotyping of Transgenic Plants and Confirmation of Successful Edits

1. Putative transgenic (A68) *Taraxacum* plants.
2. 96-well plate or 1.5 microcentrifuge tubes for collection of leaf tissue material.
3. Standard reagents for cetyltrimethyl ammonium bromide (CTAB)-based DNA extraction from young leaf material [12, 13].
4. Cas9-specific and gene-specific (for amplification of guide RNA target regions) primers.
5. DNA polymerase (e.g., Phusion Flash High-Fidelity PCR Master Mix; or any other commercially available Taq DNA polymerase).
6. Standard equipment required for PCR experiments.
7. Standard equipment and reagents for gel electrophoresis.
8. Reagents for (genome) sequencing analysis (*see* **Note 3**).

2.4 Flow Cytometry Analysis

1. Selected transgenic *Taraxacum* plants.
2. Otto I extraction buffer: 0.1 citric acid monohydrate, 0.5% (v/v) Tween 20. Filter sterilize and store at 4 °C.
3. Otto II staining buffer: 0.4 M HPO_4 (as a $Na_2HPO_4.2H_2O$ or $Na_2HPO_4.12H_2O$), filter sterilize, aliquot in 50 mL portions and store at room temperature. Add 4 μg/mL DAPI before use.
4. 200 μg/mL DAPI stock solution; dissolved in Milli-Q water, filter sterilized, aliquoted in 1 ml portions, and stored at −20 ° C.
5. Micronic 96-well plates or *alternatively* 6-cm Petri dishes.
6. 4-mm steel beads or *alternatively* sharp razors for chopping.
7. High-throughput tissue homogenizer (e.g., Geno/Grinder).
8. Deep well plates or *alternatively* sample tubes for flow cytometry (e.g., Sysmex).
9. 20- or 40-micron MultiScreen-MESH 96-Well Filter Plate or *alternatively* 30-μm nonsterile CellTrics® filters.
10. (Multichannel) pipette.
11. Wide pipette tips.
12. Flow cytometer (e.g., Quantum P or alternative).

2.5 Phenotyping of Gene-Edited Lines for Loss of Parthenogenesis

1. 9-cm petri dishes.
2. 8.5-cm Whatman filter paper disks.
3. Autoclaved tap or distilled water.

3 Methods

3.1 Selection and Design of Guide RNAs

1. Select candidate target genes based on, e.g., putative functional annotation. In our case, 13 genes from the *PAR* locus of the apomictic dandelion A68 line were selected for targeting (Fig. 1a).
2. Design apomictic allele-specific guide RNAs to knock-out the selected candidate genes using a webtool such as CRISPR-P (http://crispr.hzau.edu.cn/CRISPR2/) [14] or CHOP-CHOP (https://chopchop.cbu.uib.no/). The efficient use of such a webtool is only possible when a draft or a complete genome sequence of the target species is available online. Alternatively, guide RNAs can be designed manually by searching for a 20-bp sequence directly adjacent to a PAM site (5′-$(N)_{20}$-NGG-3′). If possible, design at least two-guide RNAs per target gene.

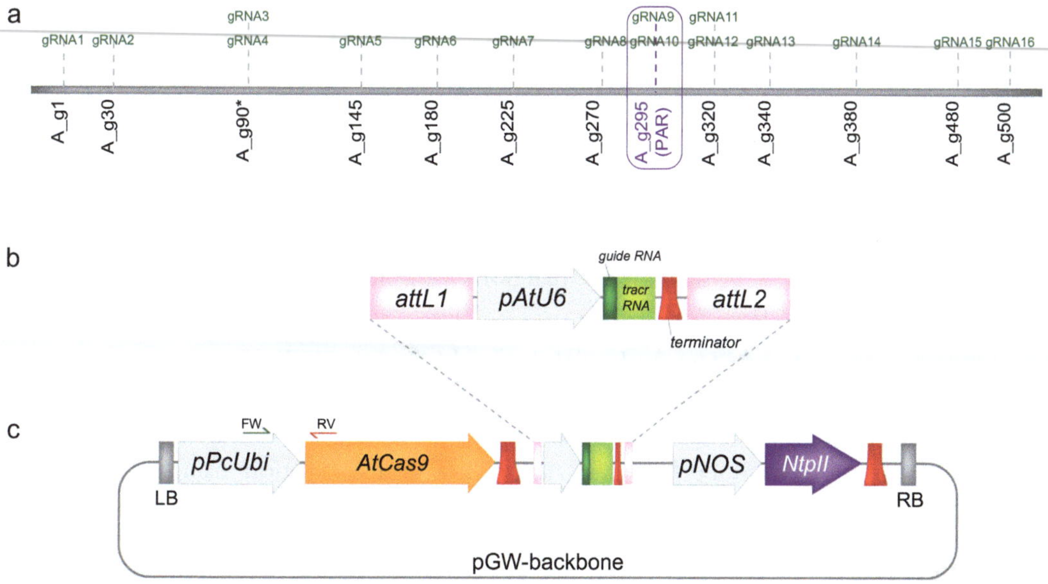

Fig. 1 Gene editing construct design and cloning. Schematic depiction of (**a**) the *PAR* locus with the 13 selected candidate target genes and the designed guide RNAs (the *PAR* gene is shown in a purple box), (**b**) a guide RNA construct for recombination-assisted Gateway Cloning, and (**c**) the pGW-pPcUBI::AtCas9 destination vector. The drawings are not made to scale

3. If the guide RNAs are designed manually, it is important to check for off-targets (*see* **Note 4**) by performing a BLAST search using the designed guide RNAs (including the PAM) against the target species genome.

3.2 Gene Editing Construct Design and Cloning

The design of the guide RNA constructs will depend on the chosen cloning method. Here, a Gateway Cloning-based workflow is described.

1. Assemble the guide RNA constructs in silico. A guide RNA construct for recombination-assisted Gateway Cloning consists of the following elements (*see* Fig. 1b):
   (a) attL1- and attL2-sites.
   (b) AtU6 promoter.
   (c) Guide RNA, consisting of the designed 20-bp unique sequence (excluding the PAM) and a tracrRNA (a scaffold RNA common to all guides).
   (d) a short terminator sequence.
2. Order the in silico-designed constructs for synthesis (*see* **Note 5**).
3. Assemble the guide RNA constructs into the pGW-pPcUBI::AtCas9 destination vector by using the Gateway LR Clonase kit. Set up a separate reaction for each guide RNA as follows:

pGW-pPcUBI::AtCas9 (destination vector)	X μL (150 ng)
guideRNA synthesized construct (entry clone)	X μL (150 ng)
LR Clonase	2 μL
TE buffer, pH 8	Up to 10 μL

Incubate the reactions for 2 hr. at room temperature (*see* **Note 6**).

4. Use 2 μL of the assembled reaction for *E. coli* transformation. Perform single colony identification, followed by plasmid isolation and sequencing (Sanger or whole-plasmid sequencing).
5. Perform *A. tumefaciens* transformation using 1 μL of the sequenced plasmid, followed by a single colony identification.

3.3 Agrobacterium-Mediated Genetic Transformation of Taraxacum Officinale

Agrobacterium-mediated transformation of common dandelion (*Taraxacum officinale*) is a lengthy, time-intensive process that takes about 14 months to complete and obtain transgenic plants. The transformation procedure involves many steps that cannot be described concisely in this chapter and thus deserves its own protocol. Therefore, we refer the reader to a detailed protocol describing genetic transformation technologies for the common dandelion by Dinkeloo et al. [15].

3.4 Genotyping of Transgenic Plants

1. Carefully cut ~1 cm leaf material from the recovered putative transgenic lines, transfer to a 96-well plate or a 1.5-mL microcentrifuge tube (depending on the number of samples), freeze (-dry) for DNA extraction.
2. Perform DNA extraction by following a CTAB-based DNA extraction protocol as described by [13].
3. Measure DNA concentration and dilute samples to reach a uniform concentration (e.g., 10 ng/μL).
4. Perform PCR to check for presence of Cas9 using Cas9-specific primers, e.g., PcUBI-Cas9-FW and Cas9-RV (*see* Fig. 1c; or Cas9-FW and terminator-RV could be another option). Prepare the PCR reaction as follows:

Reaction component	Volume [μL]	Thermocycling conditions
Phusion flash high-Fidelity PCR master mix	12.5	Step 1: 98 °C, 2 min
10 μM forward (FW) primer	1	Step 2: 98 °C, 10 sec
10 μM reverse (RV) primer	1	Step 3: 56 °C, 20 sec

(continued)

Reaction component	Volume [μL]	Thermocycling conditions
DMSO	0.75	Step 4: 72 °C, 15 sec/ 1 kb
DNA	2.5	Back to step 2, x 35
Milli-Q	7.25	Step 5: 72 °C, 5 min
Total	25	Step 6: 12 °C ∞

5. After completion, run the PCR products on a 1% agarose gel and identify Cas9-positive and negative lines.
6. Discard all Cas9-negative lines.

3.5 Ploidy Analysis by Flow Cytometry

Genetic transformation and tissue culture techniques, such as organogenesis and somatic embryogenesis, often lead to changes in ploidy levels. This can impact the genetic uniformity of the regenerated plants and makes ploidy analysis after genetic transformation a crucial and common practice. Flow cytometry is widely used in various plants and tissues for ploidy determination as it is a simple, fast, and cost-effective method. Here, a detailed protocol for flow cytometry-based ploidy analysis is described.

1. Carefully collect a small piece of leaf material (~ 1cm^2) from the selected transgenic lines (*see* **Note 7**).
2. Add a 4-mm steel bead to each well of a 96-well plate.
3. Add 500 μL of OTTO I buffer to each well.
4. Homogenize samples on a high-throughput tissue homogenizer for 0:15 seconds at 1500 RPM (*see* **Note 8**).
5. Take a deep well plate and place a 20- or 40-μm MultiScreen-MESH 96-Well Filter Plate on top of it (*see* **Note 9**).
6. Use a multichannel pipette with wide tips, collect sample homogenate, and filter it through the filter plate to remove tissue debris.
7. Take a new micronic plate and add 150 μL of the filtered sample.
8. Add 300 μL of OTTO II buffer containing 4 μg/mL DAPI.
9. Homogenize by carefully shaking the plate.
10. Incubate for 30 minutes on ice in the dark.
11. Measure the samples on a flow cytometer (e.g., Quantum P). Operate the flow cytometer at the slow or medium flow rate setting.
12. Plot FL1 (log) on the X-axis and particle count on the Y-axis to visualize the results. Sample quality should remain good for at least 2 hours.

13. Transfer plants with unaltered ploidy (*see* **Note 10**) to soil and grow them in a greenhouse or a growth chamber for further phenotypic analysis.

3.6 Phenotyping for Loss of Parthenogenesis

1. When the gene-edited plants begin to flower, start screening for the loss-of-parthenogenesis (LOP) phenotype [4]. LOP seed heads contain empty seeds and thus lack a dark center (Fig. 2b, d), normally formed by full, viable seeds, as seen in the apomictic line A68 (Fig. 2a, c).
2. Screen and collect the seeds of five independent seed heads per plant (*see* **Note 11**).
3. Perform a germination test with seeds from at least three independent seed heads per phenotyped LOP plant/line (use all obtained lines) and plants with unaltered phenotype (*see* **Note 12**). The procedure is as follows:
   (a) Line Petri dishes with filter paper.
   (b) Wet the filter paper well with autoclaved tap or distilled water.
   (c) Spread the seeds onto the filter paper and close the Petri dish. Sealing the dish with parafilm is not necessary.

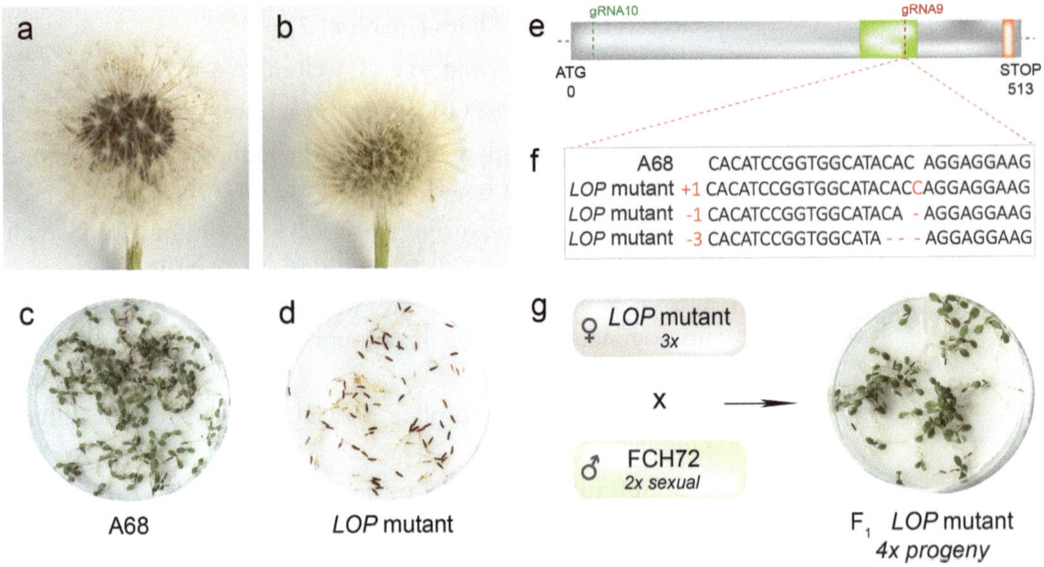

Fig. 2 Identification of the *PAR* gene by targeted mutagenesis. A seed head from (**a**) wild-type apomictic dandelion A68 and (**b**) CRISPR-Cas9 *LOP* mutant. Germination of (**c**) A68 and (**d**) CRISPR–Cas9 *LOP* mutant seed heads. (**e**) Schematic representation of the *PAR* gene, with guide RNA targets indicated; the green and orange boxes mark the zinc finger and the EAR domains, respectively. (**f**) Example genotypes at gRNA9, target of 4 CRISPR-Cas9 *LOP* mutants. (**g**) The crossing scheme for confirmation of reversion to sexuality of the CRISPR-Cas9 LOP mutants. Germination of the resulting F_1, LOP mutant × FCH72 tetraploid (4x) progeny is shown

(d) Check the plates daily to ensure that the filter paper stays moist. Add more water if necessary. Seeds should begin germinating within a week to ten days.

4. Proceed with sequencing of the confirmed LOP plants to determine specific targeted mutations and to identify the causal gene(s). Include also some of the plants with unaltered phenotypes. Go to Subheading 3.7 for the procedure.
5. Pollinate LOP plants, with confirmed edits, using pollen from a diploid sexual line (e.g., FCH72 [4]; Fig. 2g). Subsequently, assess again if the obtained seeds are viable by performing a germination test as described in step 3 (*see* **Note 13**). If that is the case, the LOP mutants have lost their parthenogenetic potential and reverted to sexuality. Therefore, they require pollination for seed production.
6. Measure the ploidy of the produced offspring (follow Subheading 3.5; *see* **Note 13**).

3.7 Confirmation of Successful Edits

1. Use the diluted genomic DNA extracted from the recovered transgenic lines as described in Subheading 3.4.
2. Perform PCR to amplify the guide RNA target site, *see* **step 4** in Subheading 3.4 for PCR reaction assembly and conditions.
3. Prepare the amplicons for sequencing according to the chosen sequencing platform (e.g., Illumina MiSeq; *see* **Note 3**).
4. After sequencing, align the obtained sequencing data to reference gene sequences and determine the size and type of the Cas9-induced mutations (Fig. 2e, f).

4 Notes

1. Any other preferred alternative cloning method can be used, e.g., Golden Gate, Gibson assembly, or traditional cloning. Adapt the construct design accordingly.
2. We use electrocompetent cells for better efficiency; however, chemically competent cells can be used as well.
3. We used Illumina MiSeq platform, but any other sequencing platform to confirm targeted mutations can be used.
4. In the case of apomicts, most of which are polyploids [16], performing a BLAST search is essential to check for allele-specific guides (e.g., apomictic versus sexual alleles).
5. Order the constructs to be synthesized and delivered either in a basic shuttle vector or as a linear fragment. Both can be directly used as an entry clone for the GW LR reaction.

6. We often leave our Gateway LR reactions to incubate overnight, but 2-hr incubation should be sufficient for successful reaction.
7. We advise taking along a leaf tissue from a plant with known ploidy (e.g., diploid sexual dandelion FCH72 [4]) to serve as a control.
8. *Alternatively*, if only a few samples are to be analyzed, i.e., less than 10, place the excised leaf tissue in a small Petri dish, add 500 μl Otto I buffer, and chop the tissue using a sharp razor blade.
9. You may need to apply some pressure to press the filter plate into the deep well plate.
10. *Taraxacum officinale* apomictic line A68 is triploid.
11. This is needed to confirm the phenotype and to exclude chimerism.
12. Seeds collected from the LOP plants should not be capable of germination due to the lack of a (parthenogenetic) embryo, whereas those from plants with unaltered phenotypes should germinate normally (*see* Fig. 2 a–d).
13. Pollination of a triploid LOP mutant by a sexual diploid would be expected to produce viable tetraploid seeds, and offspring, because the diplospory gene is present and the egg cells are therefore unreduced, and triploid [4].

Acknowledgments

The pUBI::Cas9 construct was a kind gift of H. Puchta.
Keygene N.V. owns patents and/or patent applications relating to the work described here.

References

1. Van Dijk PJ, Op Den Camp R, Schauer SE (2020) Genetic dissection of Apomixis in dandelions identifies a dominant parthenogenesis locus and highlights the complexity of autonomous endosperm formation. Genes 11:961. https://doi.org/10.3390/genes11090961
2. van Dijk PJ, Bakx-Schotman JMT (2004) Formation of unreduced megaspores (diplospory) in apomictic dandelions (Taraxacum officinale, s.l.) is controlled by a sex-specific dominant locus. Genetics 166:483–492. https://doi.org/10.1534/genetics.166.1.483
3. Vijverberg K, Milanovic-Ivanovic S, Bakx-Schotman T, van Dijk PJ (2010) Genetic fine-mapping of DIPLOSPOROUS in Taraxacum (dandelion; Asteraceae) indicates a duplicated DIP-gene. BMC Plant Biol 10:154. https://doi.org/10.1186/1471-2229-10-154
4. Underwood CJ, Vijverberg K, Rigola D et al (2022) A PARTHENOGENESIS allele from apomictic dandelion can induce egg cell division without fertilization in lettuce. Nat Genet 54:84–93. https://doi.org/10.1038/s41588-021-00984-y
5. Ozias-Akins P, van Dijk PJ (2007) Mendelian genetics of apomixis in plants. Annu Rev Genet 41:509–537. https://doi.org/10.1146/annurev.genet.40.110405.090511
6. Conner JA, Mookkan M, Huo H et al (2015) A parthenogenesis gene of apomict origin elicits

embryo formation from unfertilized eggs in a sexual plant. Proc Natl Acad Sci USA 112: 11205–11210. https://doi.org/10.1073/pnas.1505856112

7. Boutilier K, Offringa R, Sharma VK et al (2002) Ectopic expression of BABY BOOM triggers a conversion from vegetative to embryonic growth. Plant Cell 14:1737–1749. https://doi.org/10.1105/tpc.001941
8. Jinek M, Chylinski K, Fonfara I et al (2012) A programmable dual-RNA–guided DNA endonuclease in adaptive bacterial immunity. Science 337:816–821. https://doi.org/10.1126/science.1225829
9. Ding Y, Li H, Chen L-L, Xie K (2016) Recent advances in genome editing using CRISPR/Cas9. Front Plant Sci 7:703. https://doi.org/10.3389/fpls.2016.00703
10. Karimi M, Depicker A, Hilson P (2007) Recombinational cloning with plant gateway vectors. Plant Physiol 145:1144–1154. https://doi.org/10.1104/pp.107.106989
11. Karimi M, Inzé D, Depicker A (2002) GATEWAY vectors for agrobacterium-mediated plant transformation. Trends Plant Sci 7:193–195. https://doi.org/10.1016/s1360-1385(02)02251-3
12. Doyle JJ, Doyle JL (1990) Isolation of plant DNA from fresh tissue. Focus 12:13–15
13. Clarke JD (2009) Cetyltrimethyl Ammonium Bromide (CTAB) DNA Miniprep for Plant DNA Isolation. Cold Spring Harb Protoc 2009:pdb.prot5177. https://doi.org/10.1101/pdb.prot5177
14. Lei Y, Lu L, Liu H-Y et al (2014) CRISPR-P: a web tool for synthetic single-guide RNA design of CRISPR-system in plants. Mol Plant 7:1494–1496. https://doi.org/10.1093/mp/ssu044
15. Dinkeloo K, Cantero AM, Paik I et al (2021) Genetic transformation technologies for the common dandelion, Taraxacum officinale. Plant Methods 17:59. https://doi.org/10.1186/s13007-021-00760-3
16. Whitton J, Sears CJ, Baack EJ, Otto SP (2008) The dynamic nature of Apomixis in the angiosperms. Int J Plant Sci 169:169–182. https://doi.org/10.1086/523369

Part V

Apomixis Model Species

Chapter 13

Map-Based Cloning of Apomixis Genes in Dandelion (*Taraxacum officinale s.l.*)

Peter J. van Dijk, Tatyana Radoeva, Rik H. M. Op den Camp, and Diana Rigola

Abstract

This chapter describes the erratic course of the research journey of our research group into the elucidation of the genetic basis of apomixis in the common dandelion, *Taraxacum officinale s. l.* Research is rarely straightforward; sometimes, it comes to a standstill, requiring a shift in approach. Our journey took years to complete, starting with the theoretical genetic models and culminating in the cloning of the parthenogenesis gene. Along the way, we employed a suite of genetic, cell biological, and molecular techniques.

Key words Apomixis, Diplospory, Parthenogenesis, Autonomous endosperm, Map-based cloning, Dandelion, *Taraxacum*

1 The Discovery of Apomixis and Sexual Reproduction in Dandelions

The common dandelion, *Taraxacum officinale*, is a cosmopolitan *Asteraceae* plant species of temperate and cooler regions. In 1903, Raunkiær discovered apomixis in *Taraxacum* through decapitation experiments [1]. The plants still produced seeds when the styles and stamens were removed at the bud stage. Since fertilization was thereby made impossible, the seeds had to have arisen parthenogenetically. Shortly before this, seed production had already been observed in isolated male-sterile *Taraxacum* plants. Through cytological studies, Murbeck (1904) and Juel (1905) confirmed diplospory and parthenogenesis in *Taraxacum* [2–4]. The first chromosome counts suggested 26 chromosomes, which was later corrected to 24. Through decapitation and crossing experiments, Ikeno (1910) demonstrated that some Japanese *Taraxacum* species were obligate outcrossing sexuals [5]. In 1913, Osawa showed that

Keygene N.V. owns patents and/or patent applications relating to the work described here.

Tatyana Radoeva and Peter J. van Dijk (eds.), *Apomixis: Methods and Protocols*, Methods in Molecular Biology, vol. 2987, https://doi.org/10.1007/978-1-0716-5001-1_13,

one of these sexual species, *T. platycarpum*, was diploid with 16 chromosomes [6].

Gustafsson (1932) conducted extensive decapitation experiments and chromosome counts on many European *Taraxacum* taxa [7]. He found that most apomicts were triploid ($2n = 3x = 24$), although tetraploids and pentaploids also occurred. Gustafsson further described three new sexual diploids ($2n = 2x = 16$), establishing the relationship between ploidy levels and apomixis.

Later studies mapped the geographical distribution of sexual and apomictic dandelions in Europe. Sexual species are more common in Central and Middle Europe, while apomicts occur throughout the continent, reaching as far as northern Scandinavia. This different distribution of apomicts and sexuals is a classic example of a general phenomenon, which Vandel (1928) dubbed *Geographic Parthenogenesis* [8].

Apomixis also had taxonomic implications. In Northern Europe, where dandelions occur only as apomicts, numerous distinguishable clones were described as microspecies, some with broad geographical distributions. In Central Europe, where sexuals occurred and the apomicts were less distinct, taxonomists used a broader species concept. Nowadays, microspecies are grouped into sections, which can be seen as comparable to Linnean species, although sexuals from different sections often can still hybridize.

Sørensen and Gudjónsson (1946) investigated spontaneous morphological aberrants (inflorescence and leaf characters) among thousands of cultivated offspring from various apomictic microspecies [9]. The same morphological aberrants were found in different microspecies. Abnormalities were found at low frequencies (e.g., for the mutant *tenuis* aberrant about 1 in 2000 plants) and mainly consisted of aneusomic plants ($2n = 23 = 3x–1$). All eight chromosomes could be morphologically distinguished, and a karyotype was constructed. A complete monosomic series ($2n = 3x–1$) for all eight chromosomes was isolated. Sørensen (1958) conducted further cytogenetic research on the offspring of these aneusomes [10]. Crossing experiments showed that some *tenuis* mutants, which had lost a satellite or nucleolar organizer region (NOR)-chromosome, seemed to lack diplospory. This suggested that a factor regulating diplospory (*DIP*) was located on the *H* chromosome. A second chromosome aberrant, *elegans*, also showed some return to sexuality, possibly lacking parthenogenesis. However, this type was backcrossed with triploids, and the resulting progeny was challenging to interpret in terms of maternal contribution. Loss of one of the other six morphologically distinguishable chromosomes did not result in a loss of apomictic reproduction.

2 What Causes Apomixis in Taraxacum?

Based on Sørensen's cytological observations, Richards (1970, 1973) proposed a two-locus simplex dominant genetic model, with separate loci for diplospory (*Ddd*) and parthenogenesis (*Ppp*) (Richards initially used the *Hhh* and *Ddd* respectively, following Sørensen's disomic nomenclature, but the use of D for parthenogenesis and not for diplospory is confusing) [11, 12]. According to Sørensen's findings, these factors were located on two different chromosomes and were, therefore, unlinked. However, this model by Richards was criticized by Mogie (1988, 1992) due to the small number of observations it was based on and doubts about the reliability of the identification of different chromosomes [13, 14].

Mogie argued that all known meiotic mutants were recessive and that diplospory was a shortened meiosis. He suggested that parthenogenesis was a pleiotropic effect of this shortened meiosis. Thus, Mogie proposed a single-locus recessive apomixis model with a dosage effect. According to Mogie, an apomict would have the genotype *Aaa*, while *Aa*, *AA*, *AAA*, and *AAa* were sexual. The homozygous recessive *aa* and *aaa* genotypes were lethal due to their harmful effects on mitotic divisions. Although ingenious, Mogie's model lacked empirical support from crossing experiments or chromosome loss studies.

3 The First Phase of Dandelion Apomixis-Genetics Research

In the mid-1990s, research on the cause of apomixis in *Taraxacum* was initiated at the Netherlands Institute of Ecology (NIOO-KNAW). First, it was examined whether apomixis could directly result from polyploidization, specifically, triploidy. To test this, colchicine-induced sexual autotetraploid lines were crossed with diploid sexual lines. However, among the large number of resulting 3x offspring, no apomixis was observed.

To determine whether apomixis might be epigenetically caused by DNA methylation, seeds from natural apomicts were treated with the demethylating agent 5-azacytidine (5AzaC). Yet, in the few dozen treated plants, no loss of apomixis occurred. Thus, there was no evidence that either polyploidization or epigenetics caused apomixis in *Taraxacum*.

Therefore, it was decided to study the inheritance of apomixis in crosses between sexual diploids and apomictic polyploids. To make this research possible, high-quality plant material and appropriate research tools must first be developed. Since dandelions typically require a cold period of at least six weeks (<6 °C) for floral induction and then flower only for a short period, apomicts and sexuals that did not require vernalization and had a long flowering time were selected from natural populations. Codominant, highly polymorphic short-sequence repeat (SSR) markers were developed to track the transmission of chromosome segments in crosses [15].

Phenotyping full apomixis is relatively straightforward, as *Taraxacum* apomicts form autonomous endosperm and, therefore, do not need to be pollinated for endosperm development. These plants produce large seed heads with approximately 100% fully developed, dark brown seeds. However, phenotyping plants that lack one of the components of apomixis (diplospory or parthenogenesis) is more complicated. The loss of diplospory (LoD) combined with the presence of parthenogenesis results in so-called "speckled" seed heads. Only seeds with a balanced chromosome number and carrying the PAR gene develop in these. These well-developed seeds are brown, while the aborted seeds remain white, creating the characteristic speckled seed head phenotype (Fig. 1c1, c2, c3). The brown seeds of LoD plants germinate and are triploid ($2n = 3x$) or hypotriploid ($2n = 3x$–1). Diplosporous triploid plants lacking parthenogenesis (loss of parthenogenesis [LoP]) produce small, compact seed heads when grown in isolation. However, when their inflorescences are pollinated with haploid pollen

Fig. 1 Seed heads of wild-type and apomixis loss deletion mutants. Panels show intact seed heads (1) and seed heads with seeds partially removed (2). (**a1**, **a2**) Wild-type; (**b1**, **b2**) *Loss-of-parthenogenesis* (*LoP*) mutant; (**c1**, **c2**) *Loss-of-diplospory* (*LoD*) mutant. (**c3**) Close-up view showing the speckled phenotype characteristic of the *LoD* mutant

from a diploid plant ($2n = 2x$), many offspring are produced, which are tetraploid ($2n = 4x$) based on flow cytometry. These unreduced egg cells need fertilization to develop into an embryo (Fig. 1b1, b2).

The initial crosses between sexual diploids and apomictic triploid pollen donors predominantly produced diploid offspring, all of which were sexual without exception. Additionally, several triploid hybrids were formed [16]. Aneuploid hybrids were completely absent despite the irregular size of the pollen from triploids, which indicated that many aneuploid pollen grains were formed. Apparently, either these unbalanced pollen grains or the embryos they produced were not viable.

Marker analysis (initially with allozymes, later with SSR markers) revealed that the diploids were predominantly the result of self-fertilization. This was remarkable because these plants were unable to produce seeds in isolation, which, in line with the literature, points to self-incompatibility (SI) in sexual diploids [17]. However, Morita et al. [18] had previously demonstrated that pollen from 3x plants could induce self-fertilization in diploids, likely due to a disruption of the sporophytic self-incompatibility (SSI) system, a phenomenon known as the "mentor pollen effect" [18].

Some triploid (3x) hybrids from the diploid (2x) x triploid crosses were apomictic, but most were not [19]. Test crosses with diploid pollen donors (*see* above) showed that some 3x hybrids produced 4x or near 4x offspring, suggesting that these 3x hybrids were diplosporous, but lacked the ability for parthenogenesis. This implied that parthenogenesis could not be a pleiotropic effect of diplospory, contrary to what Mogie [13] had assumed. Although the studies by Tas and Van Dijk [16] and Van Dijk et al. [19] suggested that these elements of apomixis were independent and could be uncoupled, the number of hybrids was too small for robust segregation analysis.

The tetraploid offspring resulting from the test crosses of diplosporous 3x hybrids lacking parthenogenesis were, like their 3x mothers, diplosporous, but produced evenly sized pollen, which, based on MTT (3-(4,5-Dimethylthiazol-2-yl)-2,5-Diphenyltetrazolium Bromide) viability staining, appeared to be viable and fertile. This suggested that, as expected, pollen meiosis in 4x offspring was much more balanced than in the 3x mother plants. In crosses with diploid seed donors, no induced self-fertilization occurred, allowing sufficient numbers of 3x hybrid offspring to be generated to study the inheritance of diplospory.

Van Dijk and Schotman [20] discovered that co-dominant SSR markers from a tetraploid diplosporous plant (coded as PAX) were inherited in a tetrasomic manner during pollen meiosis, indicating autopolyploidy. Although allopolyploidization and hybridization have often been proposed as possible causes of apomixis (e.g.,

[21, 22]), this explanation had to be rejected in the case of *Taraxacum*. Furthermore, Van Dijk and Schotman [20] found that for two SSR marker loci, one allele of PAX was associated with the occurrence of diplospory in the offspring, while the other three SSR alleles at these loci were not. From this, they concluded that diplospory was dominant and the genotype was simplex *Dddd*, consistent with Richards' model. They also found a weak, but significant, linkage with the rDNA locus, suggesting that diplospory was located on a satellite chromosome, as Sørensen had earlier proposed based on the *tenuis* mutant.

The number of available SSR markers, however, was too limited to generate a detailed genetic map. Since the *DIP* genotype was simplex dominant, it was also possible to use dominant markers, such as amplified fragment length polymorphism (AFLP) markers [23] for genetic mapping. Vijverberg et al. [24] focused on mapping the *DIP* locus using bulked segregant analysis. For this, a pool of diplosporous plants was compared to a pool of non-diplosporous offspring. These pools were screened for the specific presence or absence of AFLP fragments. This led to the identification of approximately 20 distinct AFLP markers, which were subsequently mapped onto 73 individual plants. These plants were individually phenotyped as DIP or non-DIP through test crosses and ploidy analysis of the offspring. Notably, there was no evidence of recombination suppression around the *Taraxacum DIP* locus, unlike what has been reported for apomeiosis loci in other apomictic species, such as *Erigeron* [25], *Pennisetum* [26], and *Pilosella* [27].

The next step was the fine mapping of the *DIP* locus. For this, several AFLP markers closely linked to *DIP* were converted into easily scorable sequence-characterized amplified region (SCAR) markers. These PCR markers were then used for the fine mapping of 2227 offspring from a 2x sexual × 4x PAX cross. Ultimately, 14 recombinants were identified between the two markers closest flanking the *DIP* locus [28].

A *Taraxacum* BAC genome library with a 10x coverage was created from a 3x apomictic plant A68 by the Arizona Genome Institute. The aim was to span the *DIP* locus using BAC walking based on BACs identified with PAX markers that, according to the genetic fine map, were closest to the *DIP* locus. The 3x apomictic plant A68 was chosen as the donor for the BAC library rather than the fine-mapped 4x diplosporous PAX plant because a 3x plant enables higher BAC genome coverage for the same cost (3.3-fold compared to 2.5-fold), with a lower risk of missing genomic sequences. Additionally, the PAX plant lacked the *PAR* allele at the *PAR* locus; thus, a BAC library of this plant could not be used for walking toward the PAR allele.

The most closely linked *DIP*-PAX SCAR markers were used to identify the starting BACs for BAC walking. For this, the filter-spotted BAC library was hybridized. Highly sensitive "Overgo" probes were synthesized from the sequences. The ends of the BACs were then sequenced, and new probes were designed for subsequent steps in the BAC walking process. Using this approach, several BACs were isolated, which were later utilized by Vašut et al. [29] for fluorescent in situ hybridization (FISH). Ribosomal DNA probes confirmed Sørensen's suggestion that the *DIP* locus was located on a satellite or nucleolar organizer region (NOR) chromosome and the weak linkage observed between the rDNA-NOR locus and the *DIP* allele linked SSR markers [20].

For the genetic mapping of both the *DIP* and the *PAR* locus, a cross was performed between a diploid male-sterile plant and the triploid A68 plant, the BAC library donor [30]. Since the diploid sexual seed plant produced no pollen, no diploid offspring could arise through induced self-pollination, and therefore, all offspring had to be hybrids. However, due to the low pollen fertility of the 3x A68 plant, many crosses (~100) were required to produce ~200 hybrid offspring.

Flow cytometry revealed that approximately half of the offspring were diploid, while the other half were triploid. Only one tetraploid offspring was found, likely resulting from an unreduced 3x pollen grain from the 3x plant. Analysis with SSR markers showed that the two SSR markers, which Van Dijk and Bakx-Schotman [20] had identified as linked to the *DIP* allele, were absent in the diploid hybrids. However, these markers were present in the triploid offspring with a frequency of about one-third. This implies that markers linked to the *DIP* allele are not transmitted via haploid pollen, which may explain the absence of diploid apomicts in *Taraxacum*. The question whether this phenomenon is a direct result of the *DIP* gene itself or due to a linked mutational load remains unanswered.

Through bulked segregant analysis (BSA), AFLP markers linked to the *PAR* locus were identified. The *DIP* and *PAR* loci were found to be unlinked, consistent with Sørensen's [10] suggestion that the controlling factors are located on different chromosomes. A simplex genotype (*Ppp*) was also observed in this case. However, genetic mapping of individual plants indicated that, unlike the *DIP* locus, the *PAR* locus might exhibit suppression of recombination, possibly frustrating map-based cloning.

The triploid hybrids segregated at a 1:2 ratio of apomicts to non-apomicts. All apomicts carried the closely linked SSR markers associated with the *DIP* locus. A second SSR marker was found to be tightly linked to the *PAR* locus [31]. However, a subset of plants carrying SSR markers linked to both apomixis loci were found to be non-apomictic. There was a significant deviation from the model with two unlinked loci and a simplex genotype, but not from a model with three loci. Therefore, it was possible that a third dominant factor, responsible for autonomous endosperm development, might be involved. This idea was later revisited by Van Dijk, Op den Camp and Schauer [31] (*see* below).

Unfortunately. when the SCAR markers from [32] were mapped in a cross involving the apomictic A68 plant, it was found that the linkage patterns at the *DIP* locus in A68 differed from those in the fine-mapped 4x *DIP* plant studied by Vijverberg et al. [28]. The two *DIP* loci differed due to the presence of an inversion. Since the BAC library had been constructed using A68, the marker relationships from the 4x PAX plant could not be used for BAC walking toward the *DIP* locus.

4 The Second Phase of Dandelion Apomixis—Genetics Research

In 2006, apomixis research moved from NIOO-KNAW to KeyGene. Due to the suspicion that the *PAR* locus might exhibit recombination suppression, it was decided to redirect the research. This decision was reinforced by findings from [33], who had achieved good results using gamma-deletion mutants in *Pilosella*. Radiation-induced deletions are independent of recombination suppression. An additional advantage of deletion mapping was that it could be conducted in the 3x apomictic A68 plant, which was also used as the basis for the BAC library. However, Asker and Jerling [34] reported (p. 92) that Von Hofsten [35] had not observed any loss of apomixis following X-ray irradiation of various apomictic *Taraxacum* clones. Upon reviewing the original study, it became clear that, although 3000 seeds had been irradiated, the reproductive system of only 21 plants had been examined—an entirely insufficient number, rendering Asker and Jerling's information misleading.

Therefore, about 5000 seeds of clone A68 were exposed to a dosage of 250–350 Gy gamma radiation from a ^{60}Co source. Ultimately, 3250 plants were phenotyped for the loss of apomixis [36, 37]. The aberrant types known from earlier crosses, corresponding to the loss of *DIP* and *PAR*, were observed again (Fig. 1). After careful phenotyping, including test crosses with diploid pollen donors and flow cytometry analysis of the offspring, 20–30 mutants of each type were identified.

Previously identified AFLP markers were used to map the positions and sizes of the deletions. Many new AFLP markers within the smallest *DIP* and *PAR* deletions were screened using bulked deletion analysis, a variant of bulked segregant analysis [37]. This yielded several new AFLP fragments located within the smallest deletions. These AFLP fragments were then used to identify new seed BACs. Meanwhile, *Taraxacum* transformation protocols for future functional gene testing were developed.

Instead of BAC walking, BAC assembly was accomplished via a DNA fingerprinting technique (Whole Genome Profiling; [38]).

However, the BAC clusters found by this approach did not fully cover the deletions. Besides gaps between the BACs, there were also sequence gaps within the BACs due to the short sequence read length of those days and the abundance of repetitive sequences. Among the genes in the *DIP* and *PAR* BACs that could be annotated, there were no outstanding candidate genes. Ultimately, sequencing with long reads, first with PacBio and later with Oxford Nanopore sequencing platforms, provided fully gap-free sequences of the three *DIP* and *PAR* loci haplotypes. The sizes of these two loci, as determined by the smallest deletions, were 900 and 1300 kb, respectively. These loci still had numerous coding sequences, and it was not yet feasible to functionally evaluate them all by targeted mutagenesis. Candidate genes in the *PAR* locus were ranked based on apomict allele-specific ovule expression (comparing sexual, *LoP*-mutant, and wild-type apomict plants) and by functional annotation. The top-listing candidate genes were targeted by gene editing to create targeted mutations in the apomictic alleles. A high-quality triploid genome sequence and the fully gap-free sequences of the three *PAR* loci haplotypes were instrumental for allele-specific gene editing target designs. After allele-specific gene editing of a single candidate gene, clear LoP phenotypes were obtained (*see* Chap. 12). Clones of these *LoP* mutant plants were transformed with genetic constructs of the candidate *PAR* gene. The *PAR* gene constructs could complement the LoP phenotype, and the transgenic lines reverted to apomictic reproduction. With these experiments, the *Taraxacum PAR* gene was identified as a small gene with a special zinc finger DNA-binding domain and a repressor EAR motif/domain (Fig. 2). The parthenogenesis function of the *PAR* gene has meanwhile been demonstrated in various crops (lettuce [37] rice [39], and foxtail millet [40]). Efforts to identify the *DIP* gene in the *DIP* locus by a similar approach are ongoing.

The genetic control of the third apomixis element in *Taraxacum*, autonomous endosperm development, has been studied less extensively. Most sexual flowering plants require a ratio of two maternal to one paternal genome for proper endosperm development. The double maternal genome dosage results from the fusion of two polar nuclei into the central cell nucleus. This 2:1 ratio is likely critical due to parental imprinting of the genes in the gametes [40, 41]. In many flowering plants, the degeneration of the endosperm causes a "triploid block" in crosses between diploid and tetraploid sexual plants (parental genome ratios 2:2 or 8:1, depending on the direction of the cross). In the case of triploid apomictic dandelions, the viable endosperm has a maternal-to-paternal genome of 6:0. How this is possible is a mystery according to [42]. However, when sexual diploid dandelions were crossed with sexual colchicine tetraploids (*see* above), many triploid progeny plants were obtained, suggesting the absence of a strong triploid

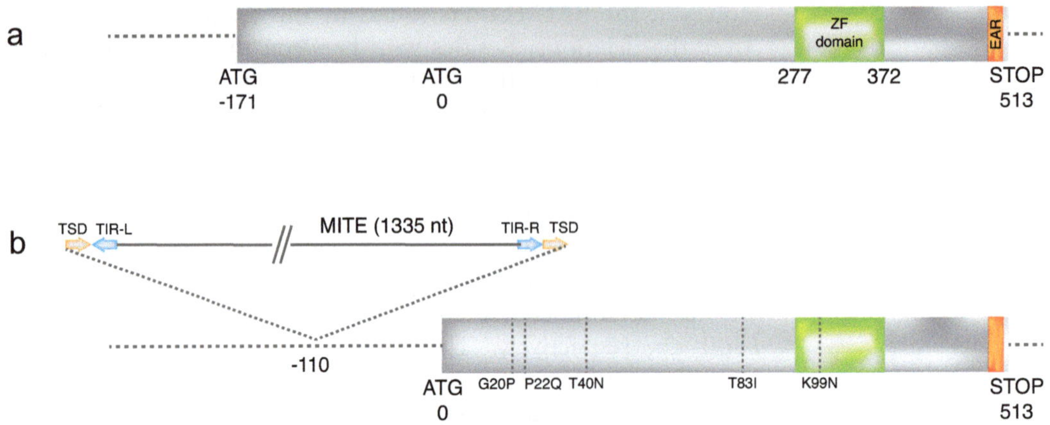

Fig. 2 The *PAR* gene of triploid apomictic dandelions. Panel (**a**) shows one of the two nearly identical sexual alleles; Panel (**b**) shows the apomictic allele. The *PAR* gene encodes a small 171-amino acid peptide containing a DNA-binding zinc finger motif (green) and an EAR repressor domain (orange). Single-nucleotide polymorphisms (SNPs) between the sexual and apomictic alleles lead to five amino acid substitutions. The major difference, however, is a transposable element insertion in the highly conserved promoter region of the apomictic allele. Transformation studies indicate that this insertion is responsible for parthenogenesis

block. Absence of a triploid block in diploid x tetraploid crosses was also observed in other *Asteraceae* species, such as *Tripleurospermum* [43] and *Pilosella* [44]. The normal development of autonomous endosperm, without a paternal genome contribution, may be related to the absence or weakness of a triploid block.

There are few studies on the genetic control of autonomous endosperm development. In *Erigeron*, it was shown that parthenogenesis and autonomous endosperm development were co-segregating, suggesting either a single gene, or two closely linked genes [45, 46]. Van Dijk et al. [31] investigated the segregation of autonomous endosperm in dandelion in the TJX3–20 (2x sexual, male sterile) x A68 (3x apomictic) cross. In this cross, in total, one-third of the triploid hybrid offspring were apomictic. All these apomicts carried the *DIP*- and *PAR*-linked SSR markers. Under a two-locus model, however, 44% of the plants were expected to be apomictic. This fraction of one-third apomicts does not deviate from a three-locus apomixis simplex model (*Ddd/Ppp/Aaa*), where '*A*' stands for autonomous endosperm. Not all the triploid hybrid offspring that carried the *DIP*- and *PAR*-linked SSR markers were apomictic; one-third were not. If this model were correct, then the one-third fraction that carried the *DIP*- and *PAR*-linked SSR markers would lack autonomous endosperm development. This was tested by Van Dijk et al. [31] using a flow cytometric seed screen (FCSS, *see* also Chap. 2) and Nomarski DIC microscopy. Unexpectedly, it was found that nearly all of these plants produced autonomous endosperm; the hexaploid FCSS peaks indicated that autonomous endosperm originated from the fusion of two triploid polar nuclei. This suggests that autonomous endosperm was not segregating as a third apomixis element in this cross. However, it should be noted that there were considerable differences in the hexaploid peak height between plants, suggesting that the amount of endosperm was highly variable between them. Since the endosperm nourishes the developing embryo, there may be a minimum threshold amount of autonomous endosperm required for embryo and viable seed development. It is possible that a modifier of autonomous endosperm development was segregating in this cross, converting a continuous trait into a binary trait. The genetic basis of autonomous endosperm in apomictic plants requires further investigation.

5 Conclusion

Progress in this field is often uneven over time, frequently relying on new technological advances. In this case, long-read sequencing and targeted mutagenesis were pivotal breakthroughs. The former enabled us to sequence all three haplotypes of the *DIP* and *PAR* loci completely and without gaps, while the latter provided functional evidence for the *PAR* gene. Ultimately, the research culminated in the publication of the *Taraxacum PAR* gene [37], 24 years after our research group's first publication on apomixis [15].

Today, reproducibility and credibility of research is an important issue. In the *Nature Genetics* article [37], we demonstrated, through collaboration with the research group of Ross Bicknell and Chris Winefield in Lincoln, New Zealand, that a homolog of the *PAR* gene is responsible for apomixis in *Pilosella*, including a similar, but not identical, transposon insertion in the gene's promoter (*see* Chap. 14). Nature itself had thus already performed the replication experiment, providing evidence that is difficult to surpass.

References

1. Raunkiær RC (1903) Kimdannelse uden Befrugtning hos Maelkebötte. Bot Tidsskr 25: 109–140
2. Murbeck S (1904) Parthenogenese bei den Gattungen *Taraxacum* und *Hieracium*. Bot Not 1904:285–296
3. Juel HO (1904) Die Tetradenteilung in die Samenanlage von *Taraxacum*. Arch f Bot 2:1–9
4. Juel HO (1905) Die Tetradenteilung bei *Taraxacum* und anderen Cichorieen. Kgl Sv Vet Ak Handl 39:1–21
5. Ikeno S (1910) Sind alle Arten der Gattung *Taraxacum* parthenogenetisch? Ber Deutsch Bot Ges 28:394–397
6. Osawa J (1913) Studies on the cytology of some species of *Taraxacum*. Arch f Zellforsch 10:450–469
7. Gustafsson Å (2010) Zytologische und experimentelle Studien in der Gattung *Taraxacum*. Hereditas 16:41–62. https://doi.org/10.1111/j.1601-5223.1932.tb02561.x
8. Vandel A (1928) La parthénogenèse géographique : Contribution à l'étude biologique et cytologique de la parthénogenèse naturelle. Bull Biol Fr Belg 62:164–281

9. Sørensen T, Gudjónsson G (1946) Spontaneous chromosome-aberrants in apomictic *Taraxaca*. Kong Danske Vid Selsk BioI 4:3–48
10. Sørensen T (1958) Sexual chromosome-aberrants in triploid apomictic *Taraxaca*. Bot Tidskr 54:1–22
11. Richards AJ (1970) Eutriploid facultative agamospermy in *Taraxacum*. New Phytol 69:761–774. https://doi.org/10.1111/j.1469-8137.1970.tb02461.x
12. Richards AJ (1973) The origin of *Taraxacum* agamospecies. Bot J Linn Soc 66:189–211. https://doi.org/10.1111/j.1095-8339.1973.tb02169.x
13. Mogie M (1988) A model for the evolution and control of generative apomixis. Biol J Linn Soc 35:127–153. https://doi.org/10.1111/j.1095-8312.1988.tb00462.x
14. Mogie M (1992) The evolution of asexual reproduction in plants, 1st edn. Chapman & Hall, London
15. Falque M, Keurentjes J, Bakx-Schotman JMT, Van Dijk PJ (1998) Development and characterization of microsatellite markers in the sexual-apomictic complex *Taraxacum officinale* (dandelion). Theor Appl Genet 97:283–292. https://doi.org/10.1007/s001220050897
16. Tas ICQ, Van Dijk PJ (1999) Crosses between sexual and apomictic dandelions (*Taraxacum*). I The inheritance of apomixis. Heredity 83: 707–714. https://doi.org/10.1046/j.1365-2540.1999.00619.x
17. Okabe S (1956) On the nature of incompatibility alleles *Taraxacum elatum*, *T. longeappendiculatum* (self-incompatibility studies in *Taraxacum* I). Bot Mag Tokyo 69: 592–597
18. Morita T, Menken SBJ, Sterk AA (1990) Hybridization between European and Asian dandelions (*Taraxacum* section *Ruderalia* and section *Mongolica*): 1. Crossability and breakdown of self-incompatibility. New Phytol 114:519–529. https://doi.org/10.1111/j.1469-8137.1990.tb00420.x
19. Van Dijk PJ, Tas ICQ, Falque M, Bakx-Schotman T (1999) Crosses between sexual and apomictic dandelions (*Taraxacum*). II The breakdown of apomixis. Heredity 83: 715–721. https://doi.org/10.1046/j.1365-2540.1999.00620.x
20. Van Dijk PJ, Bakx-Schotman JMT (2004) Formation of unreduced megaspores (diplospory) in apomictic dandelions (*Taraxacum officinale, s.l.*) is controlled by a sex-specific dominant locus. Genetics 166:483–492. https://doi.org/10.1534/genetics.166.1.483
21. Ernst A (1918) Bastardierung als Ursache der Apogamie im pflanzenreich: Eine Hypothese zur experimentellen Vererbung und Abstammungslehre. Gustaf Fischer, Jena
22. Carman JG (1997) Asynchronous expression of duplicate genes in angiosperms may cause apomixis, bispory, tetraspory, and polyembryony. Biol J Linn Soc 61:51–94. https://doi.org/10.1111/j.1095-8312.1997.tb01778.x
23. Vos P, Hogers R, Bleeker M et al (1995) AFLP: a new technique for DNA fingerprinting. Nucleic Acids Res 23:4407–4414. https://doi.org/10.1093/nar/23.21.4407
24. Michelmore RW, Paran I, Kesseli RV (1991) Identification of markers linked to disease-resistance genes by bulked segregant analysis: a rapid method to detect markers in specific genomic regions by using segregating populations. Proc Natl Acad Sci 88:9828–9832. https://doi.org/10.1073/pnas.88.21.9828
25. Noyes RD, Rieseberg LH (2000) Two independent loci control Agamospermy (Apomixis) in the triploid flowering plant *Erigeron annuus*. Genetics 155:379–390. https://doi.org/10.1093/genetics/155.1.379
26. Akiyama Y, Conner JA, Goel S et al (2004) High-resolution physical mapping in *Pennisetum squamulatum* reveals extensive chromosomal Heteromorphism of the genomic region associated with Apomixis. Plant Physiol 134:1733–1741. https://doi.org/10.1104/pp.103.033969
27. Okada T, Ito K, Johnson SD et al (2011) Chromosomes carrying meiotic avoidance loci in three apomictic eudicot *Hieracium* subgenus *Pilosella* species share structural features with two monocot apomicts. Plant Physiol 157:1327–1341. https://doi.org/10.1104/pp.111.181164
28. Vijverberg K, Milanovic-Ivanovic S, Bakx-Schotman T, Van Dijk PJ (2010) Genetic fine-mapping of *DIPLOSPOROUS* in *Taraxacum* (dandelion; *Asteraceae*) indicates a duplicated DIP-gene. BMC Plant Biol 10:154. https://doi.org/10.1186/1471-2229-10-154
29. Vašut RJ, Vijverberg K, Van Dijk PJ, De Jong H (2014) Fluorescent in situ hybridization shows *DIPLOSPOROUS* located on one of the NOR chromosomes in apomictic dandelions (*Taraxacum*) in the absence of a large hemizygous chromosomal region. Genome 57:609–620. https://doi.org/10.1139/gen-2014-0143
30. Van Dijk P, De Jong H, Vijverberg K, Biere A (2009) An Apomixis-gene's view on dandelions. In: Schön I, Martens K, Dijk P

(eds) Lost Sex. Springer, Netherlands, Dordrecht, pp 475–493

31. Van Dijk PJ, Op Den Camp R, Schauer SE (2020) Genetic dissection of spomixis in dandelions identifies a dominant parthenogenesis locus and highlights the complexity of autonomous endosperm formation. Genes 11:961. https://doi.org/10.3390/genes11090961
32. Vijverberg K, Van Der Hulst RGM, Lindhout P, Van Dijk PJ (2004) A genetic linkage map of the diplosporous chromosomal region in *Taraxacum officinale* (common dandelion; Asteraceae). Theor Appl Genet 108: 725–732. https://doi.org/10.1007/s00122-003-1474-y
33. Catanach AS, Erasmuson SK, Podivinsky E et al (2006) Deletion mapping of genetic regions associated with apomixis in *Hieracium*. Proc Natl Acad Sci 103:18650–18655. https://doi.org/10.1073/pnas.0605588103
34. Asker SE, Jerling L (1992) Apomixis in plants. CRC Press, Boca Raton. Ann Arbor London [etc.]
35. Von Hofsten CG (1954) Studies on the genus *Taraxacum* with special reference to the group *Vulgaria* DT in Scandinavia. LTs Förlag, Stockholm
36. Van Dijk PJ, Rigola D, Prins MW, Van Tunen AJ (2017) Diplospory gene, WO2017/039452 A1
37. Underwood CJ, Vijverberg K, Rigola D et al (2022) A *PARTHENOGENESIS* allele from apomictic dandelion can induce egg cell division without fertilization in lettuce. Nat Genet 54:84–93. https://doi.org/10.1038/s41588-021-00984-y
38. Van Oeveren J, De Ruiter M, Jesse T et al (2011) Sequence-based physical mapping of complex genomes by whole genome profiling. Genome Res 21:618–625. https://doi.org/10.1101/gr.112094.110
39. Song M, Wang W, Ji C et al (2024) Simultaneous production of high-frequency synthetic apomixis with high fertility and improved agronomic traits in hybrid rice. Mol Plant 17:4–7. https://doi.org/10.1016/j.molp.2023.11.007
40. Köhler C, Grossniklaus U (2005) Seed development and genomic imprinting in plants. In: Jeanteur P (ed) Epigenetics and chromatin. Springer, Berlin/Heidelberg, pp 237–262
41. Gehring M, Satyaki PR (2017) Endosperm and imprinting, inextricably linked. Plant Physiol 173:143–154. https://doi.org/10.1104/pp.16.01353
42. Haig D, Westoby M (1991) Genomic imprinting in endosperm: its effect on seed development in crosses between species, and between different ploidies of the same species, and its implications for the evolution of apomixis. Philos Trans Biol Sci 333:1–13
43. Kay QON (1969) The origin and distribution of diploid and tetraploid *Tripleurospermum inodorum* (L.) Schultz Bip. Watsonia 7:130–141
44. Chrtek J, Herben T, Rosenbaumová R et al (2017) Cytotype coexistence in the field cannot be explained by inter-cytotype hybridization alone: linking experiments and computer simulations in the sexual species *Pilosella echioides* (*Asteraceae*). BMC Evol Biol 17:87. https://doi.org/10.1186/s12862-017-0934-y
45. Noyes RD, Wagner JD (2014) Dihaploidy yields diploid apomicts and parthenogens in *Erigeron* (Asteraceae). Am J Bot 101:865–874
46. Noyes RD (2022) Mendelian segregation for parthenogenetic embryo development at the diploid level in the flowering plant *Erigeron*. Am J Bot 109:1641–1651. https://doi.org/10.1002/ajb2.16071

Chapter 14

Pilosella: A Dicotyledonous Model for Studying Aposporous, Autonomous Apomixis

Ross Bicknell and Anna M. G. Koltunow

Abstract

Pilosella, a member of the Compositae, is a model system used to study the molecular genetics of aposporous apomixis. These plants are small, rapidly growing perennials that are easy to cultivate both in the greenhouse and in tissue culture. Apomixis in *Pilosella* occurs by apospory where mitotically derived embryo sacs arise adjacent to cells undergoing female gamete meiosis in the ovule. Seed initiation is autonomous, where both embryo and endosperm form without fertilization in the aposporous embryo sac. Apomixis is not fully penetrant in *Pilosella*. Instead, plants are facultatively apomictic, and apomixis can be easily scored through the simple decapitation of the immature capitulum bud. Natural sexual and facultatively apomictic forms are readily cross-compatible, facilitating comparative studies of inheritance and allele function. A wide range of experimental methods have been described for these plants, including histological techniques for studying the cytological aspects of apomixis, an efficient *Agrobacterium*-mediated transformation system, CRISPR/*Cas9* mutagenesis, and mapping approaches that use deletion mutation and segregation in polyhaploid populations. Freely available online resources include a genome assembly, a molecular map based on cDNA markers and a transcriptome database. Collectively, these resources make *Pilosella* a highly tractable experimental system for studying the genetic control of native apomixis.

Key words *Pilosella*, *Hieracium*, Apospory, Model system

1 Introduction

The genus *Pilosella* is a versatile model system used to study the molecular genetics of apospory, a form of gametophytic apomixis where cells in the ovule adjacent to those undergoing the typical events of meiosis enlarge and form a mitotically derived (maternal) embryo sac. In *Pilosella,* both embryo and endosperm formation is fertilization-independent in aposporous embryo sacs. Consequently, seed formation is often precocious, typically being well advanced before the flower opens [1]. The meiotically derived embryo sac may or may not survive; thus, apomixis is facultative and its penetrance is variety dependent [2–6].

Tatyana Radoeva and Peter J. van Dijk (eds.), *Apomixis: Methods and Protocols*, Methods in Molecular Biology, vol. 2987, https://doi.org/10.1007/978-1-0716-5001-1_14,

Intriguingly, the first to study inheritance in this plant was Gregor Mendel [7, 8]. From his publications and personal correspondence, it is apparent that Mendel had difficulty reconciling the differences he observed when studying inheritance in *Pisum* (pea) and *Hieracium* (syn. *Pilosella*) [9]. The reasons for this are now obvious, but more than 30 years were to elapse before Ostenfeld and Rosenberg demonstrated that most *Pilosella* species are apomicts, including all of the plants used by Mendel in his inheritance studies [10–12].

Apomixis in *Pilosella* is conferred by the inheritance of dominant determinants at two (probably three) loci: one associated with the avoidance of meiosis (the *LOSS OF APOMEIOSIS* locus: *LOA*) one with spontaneous embryo formation (the *LOSS OF PARTHENOGENESIS* locus: *LOP*) [2, 4, 13] and a probable third associated with *AUTONOMOUS ENDOSPERM* formation (*AutE*). Catanach et al. [2] noted that autonomous endospermy in *Pilosella piloselloides* var. *praealta* (syn. *H. praealtum*; syn. *Hieracium caespitosum*) was lost along with parthenogenesis following the deletion of the *LOP* locus, suggesting that the two developmental events may be associated either via genetic linkage, physiological epistasis, or be controlled by the same gene. Ogawa et al. [14] described two segregants from among the progeny of a related *Pilosella* species in which autonomous endospermy was expressed without parthenogenesis, indicating the presence of a separate "*AUTONOMOUS ENDOSPERM*" locus (*AutE*) in *Pilosella*. The *AutE* locus has since been located to a unique linkage group in *Pilosella* [15]; however, the identity of *AutE* remains unknown.

The dominant actions of the *LOA* and *LOP* loci are inferred by their influence when present as simplex copies. In *Pilosella piloselloides* var. *piloselloides* (syn. *P. piloselloides; Hieracium piloselloides*) and *Pilosella piloselloides* var. *praealta* (syn. *Hieracium praealtum*), the loss of *LOA* by deletion resulted in plants exhibiting a meiotic phenotype, while the loss of *LOP* resulted in plants requiring fertilization to initiate both embryogenesis and endosperm formation [2, 4]. The loss of both loci resulted in a "mutant" phenotype with the reappearance of sexual reproduction in this otherwise asexually reproducing plant. The uncovering of sexual reproduction following the removal of these loci supports the proposal that sexual reproduction is the default state, while apomixis is superimposed, and that genes controlling apomixis lead to the avoidance of meiosis and eliminate the requirement for fertilization to induce embryo and endosperm formation [4]. Further support for this hypothesis comes from studies of gene expression in apomictic and sexual biotypes of *Pilosella*. Tucker et al. [16] reported that the same genes appear to act in both biotypes throughout most of gametogenesis, indicating that the sexual and apomictic pathways of reproduction in *Pilosella* differ in the expression of only a small percentage of the genes and/or alleles necessary to achieve the formation of a new asexually derived sporophytic generation. Bräuning et al. [6] further noted that the transcriptome of a *LOP* mutant of *Pilosella piloselloides* var. *praealta* surveyed over the period of parthenogenesis was substantially similar to that of the corresponding wild type. At least some alleles associated with apomixis in *Pilosella* appear to be haploid gamete lethal [17], restricting their transfer to diploid or polyploid gametes. The molecular mechanism underlying this effect is unknown, but it potentially explains the almost universal association between apomixis and polyploidy in this system [17, 18].

Shirasawa et al. [13] constructed genetic linkage maps for two apomictic species, *Pilosella piloselloides* var. *piloselloides* (syn. *Hieracium piloselloides*) and *Pilosella piloselloides* var. *praealta* (syn. *Hieracium praealtum*), using SSR markers associated with expressed transcripts. In *Pilosella piloselloides* var. *praealta,* the *LOA* and *LOP* loci were mapped to linkage groups LGR15 and LGR08, respectively, supporting previous findings that the two loci are unlinked in this plant. The partial structure of the *LOA* locus was described by Kotani et al. [19]. It was found to be a hemizygous region of low recombination, containing extensive, often complex repeat sequences. Kotani et al. [19] addressed the potential role of repeat sequence at the *LOA* locus, noting that *LOA* remained functional when it was separated from the majority of the surrounding repeat sequence through recombination. They concluded that the association between hemizygosity, repeat sequence, and apomixis loci is more likely to be a consequence of the action of these loci in suppressing recombination, rather than a requirement for their function.

Bicknell et al. [17] mapped the *LOP* locus to a 654 kb nonrecombinant interval in *P. piloselloides* var. *praealta,* which includes the gene *PpPAR*. Underwood et al. [5] reported the structure and function of *PAR* in the related apomicts *Taraxacum officinale* (*ToPAR*) and *Pilosella piloselloides* var. *praealta* (*PpPAR*). In *Taraxacum, PAR* acts as an initiator of parthenogenesis, activating the egg cell to begin the first mitotic division of embryogenesis. Underwood et al. further noted that in *Taraxacum,* a unique MITE is positioned in the promoter of the dominant *ToPAR* allele, which appears to be important for its function [5]. Curiously, transposons of similar size (1233–1335 bp) were also seen in the dominant alleles of *PAR* in both *Pilosella piloselloides* and *Hieracium policheae* [17]. In each case, the transposon was inserted a similar distance upstream of the *PAR* start codon (−110 to −154 bp); however, the internal sequences of the elements and even their terminal repeat sequences were different. This finding suggests that similarly modified *PAR* alleles play a critical role in the expression of parthenogenesis in the three genera. It is notable that, although all three genera are related members of the Cichorioideae, *Taraxacum* employs meiotic diplospory for apomeiosis, whereas

Hieracium employs mitotic diplospory and *Pilosella* employs apospory. The role of *PAR* in these plants, therefore, does not appear to be associated with the mechanism of apomeiosis and, furthermore, the observed differences in transposon sequence suggest that the *PAR* gene has been similarly modified on at least three separate occasions to enable the expression of parthenogenesis [17]. These findings, together with observations on haploid gamete lethality (discussed above) and the importance of perennialism in native apomicts, were used to formulate a model for the evolution of apomixis in the Cichorioideae [17]. Interestingly, in apomictic *Citrus* species, which form embryos directly from maternal ovule tissues surrounding a sexually derived embryo sac, a MITE transposon is also evident in the promoter region of an identified gene and absent in sexually reproducing *Citrus* species [20]. In *Citrus*, the modified gene is an RWP-RK-domain protein, different in both structure and function to PAR, which is a K2-2 zinc finger EAR-domain protein [5].

2 Using the *Pilosella* System

2.1 The Plant

Pilosella is a genus within the tribe Cichorioideae of the family Asteraceae. It is native to Eurasia with some species found throughout the temperate world as adventive colonists [21]. The genus currently includes 248 accepted species, but the classification of *Pilosella/Hieracium* has been revised many times, resulting in different estimates of its size [22]. At the time of Linnaeus, *Pilosella* species were included in the genus *Hieracium*. Later, *Pilosella* was reclassified as a subgenus of *Hieracium*, before it was finally elevated to the status of a genus [22]. As a result of these changes, the names of the plants used to study apomixis have occasionally changed. In particular, early publications, including that of Gregor Mendel [7], use the generic name *Hieracium* and occasionally also use different specific names, to accommodate the rules of Latin-based nomenclature. The main species (and varieties) currently used to study the molecular genetics of apomixis are *Pilosella piloselloides* var. *praealta*, *Pilosella piloselloides* var. *piloselloides*, *Pilosella aurantiaca, and Pilosella officinarum* (Fig. 1). Plants of *Pilosella* are typically self-incompatible, utilizing a form of sporophytic self-incompatibility (SI) mediated at the stigmatic surface (Fig. 2). Cross-compatibility between species within the genus, however, is common and interspecific hybridization is frequently reported in the field [22, 23]. Wild accessions range in ploidy from diploid to octoploid [24]. Native apomicts of *Pilosella* have always been recorded as polyploid [18]; however, diploid apomicts can be experimentally derived, and these have been used extensively for genome assembly and molecular mapping (*see* below). Bicknell et al. [17] recently demonstrated the action of tetrasomic

Fig. 1 *Pilosella aurantiaca*. Left: plant in flower. Middle: wild-type apomictic seed formation; Right: EMS-induced mutant with sexual seed formation

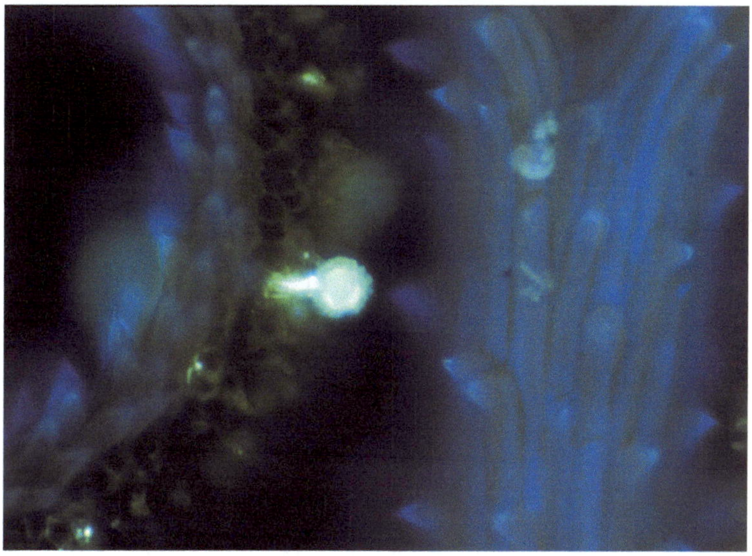

Fig. 2 Pollen germination on the stigma of *Pilosella piloselloides* var. *praealta* visualized with aniline blue staining

inheritance in the apomict *P. piloselloides* var. *praealta*, confirming earlier speculation that this plant acts as an autopolyploid [13, 25].

Plants of *Pilosella* are either obligately sexual or facultatively apomictic [26, 27]. During apospory (currently said to be of the "*Hieracium* type") [28], a somatic cell adjacent to the megaspore mother cell undergoes differentiation into an aposporous initial, a unique cell type that mimics the form and developmental function of a selected megaspore. The egg cell in the aposporous embryo sac has a 2n maternal ploidy, which contrasts with the 1n maternal ploidy of the egg in the sexually formed embryo sac. In sexual forms of *Pilosella*, endosperm formation requires the fertilization of the two fused, reduced polar nuclei and results in a maternal: paternal genome ratio of 2:1. In apomicts, the endosperm forms autonomously within the aposporous embryo sac from the fusion of unreduced polar nuclei and results in a maternal: paternal genome ratio of 4:0, and the embryo has a 2:0 maternal to paternal genome ratio [29]. Unlike in diplosporous plants, sexually derived embryo sacs in *Pilosella* often initiate, and may complete development alongside asexually derived embryo sacs within the same ovule. If a sexually derived embryo sac is fertilized, then both sexually derived and asexually derived embryos may form in the same seed. Such polyembryonic seeds are a common feature of aposporous apomicts and they result in progeny populations consisting of a mix of hybrid and maternal seedlings, typically with asexually derived (maternal) seedlings predominating. Estimates of the levels of maternal seed formation in *Pilosella* vary between 97% [30] and 99% [18], with the remainder arising from combinations of hybridity and parthenogenesis. This plasticity is one of the great strengths of the *Pilosella* system, as it can be used to quantify factors influencing apomixis, such as Genotype x Environment effects [31], it facilitates the screening of mutant populations [2], and it enables the production of polyhaploid populations for genetic analyses [17, 19].

To facilitate the study of ovule and seed development, in sexual and apomictic varieties, a developmental calendar was created to assist staging of material for cytological and molecular characterizations (Fig. 3) [1].

2.2 Cultivation

Plants of *Pilosella* grow rapidly under greenhouse conditions, completing a generation (seed to seed) in approximately 3 months. They appear to be well adapted to high light conditions and grow significantly slower under artificial lighting. Vegetative plants develop as closely appressed rosettes until induced to flower by a photoperiod exceeding 14 hours of light per day. Once induced, the terminal meristem transitions into floral bud development and an inflorescence quickly rises above the level of the rosette leaves.

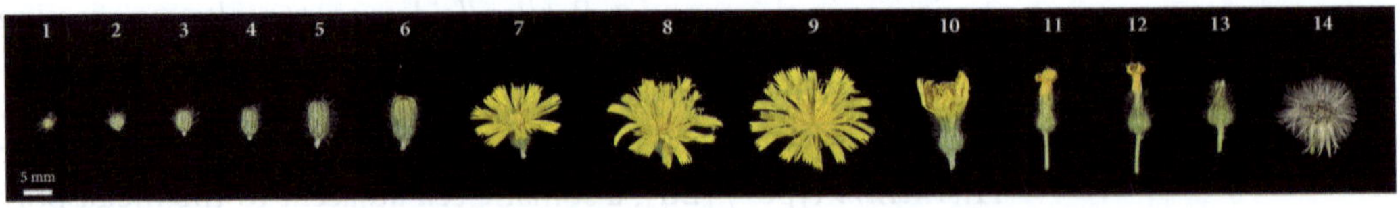

Fig. 3 Developmental staging of flower and seed formation

With the conversion of the terminal bud to flowering, lateral bud-break gives rise to stolons that quickly extend to establish new rosettes. Removing these rosettes and allowing them to root in moist sand provides a simple, efficient method for the vegetative increase of valuable genotypes. The inflorescence of *Pilosella* is a cyme, typically bearing 5–20 capitula. Individual capitula include a variable number of florets, depending on the species and variety, but commonly between 30 and 50. Floret development on a capitulum is centripetal, with development being slightly more advanced in the outer rows than the inner rows, facilitating analyses of developmental events within a capitulum. In the capitulum of *Pilosella*, the individual flowers are all ray florets, so a capitulum can be considered as a population of flowers of similar form and genotype. The histological stages of seed development are well documented for both sexual and apomictic biotypes [1] and correlated to bud development using a numeric scoring system [32]. As *Pilosella* is an autonomous gametophytic apomict, seed forms without fertilization, and therefore, also without pollination. This forms the basis of a simple test for apomixis. Severing the top of a floral bud at stages 5–6 removes both the stigma and anther tissues prior to pollen dehiscence. Apomictic varieties continue to form seeds in decapitated capitula while sexual plants do not.

To facilitate cultivation, in vitro methods are available for the micropropagation [33], anther culture [34], and Agrobacterium-mediated transformation [35] of *Pilosella*. Plantlets can be easily regenerated from severed leaf, shoot, or root segments [33], using a basal medium supplemented with 2.5 μM IBA and 2 μM BAP. Roots readily form on shoots placed onto hormone-free medium. To germinate seeds in vitro, seed should be treated with a 1% sodium hypochlorite solution for 50 minutes at room temperature, then rinsed in sterile water, and allowed to dry in the laminar flow hood before sowing onto agar-solidified hormone-free medium. Stainless steel sieves are ideal for this purpose, and they can be stored sterile under ethanol until required.

2.3 Histology

Histological techniques are documented for wax- and resin-based sectioning [1], callose deposition [36], whole-mount clearing using methyl salicylate [4], and confocal microscopy (*see* Fig. 4 and Chap. 3). It is critical with whole-mount techniques to include a mild acid treatment to dissolve oxalate crystals. These commonly occur in the ovary tissues of *Pilosella* and are highly refractive. Pollen tube germination on the stigmatic surface (Fig. 2) can be visualized using aniline blue. Pollen grains should be stained at least 5 hours after controlled pollination to allow sufficient time for the pollen tubes to germinate. As *Pilosella* utilizes sporophytic self-incompatibility, pollen germination can be used to score compatibility.

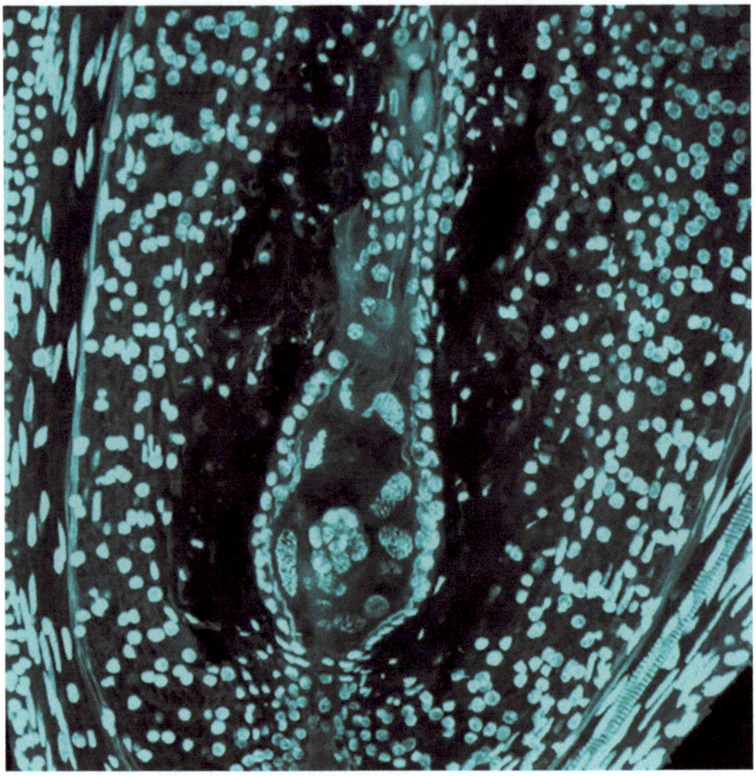

Fig. 4 Confocal image of early embryogenesis (stage 7) in *Pilosella piloselloides* var. *praealta* with free nuclear division in the endosperm

2.4 Mutagenesis

Mutants of *Pilosella* have been produced following exposure to gamma irradiation [2] and to the chemical mutagen ethyl methane sulfonate (EMS) (Fig. 1). For gamma-induced mutation, dry seeds were exposed to a dose of 400 Gy using a ^{60}Co source. This dose was found in a test series to result in an approximate 50% lethality (LD_{50}) among the seedlings. For EMS mutagenesis, the LD_{50} dose was estimated to be exposure to a 0.8% solution for 12 hours. Both forms of mutation led to chimeric M0 plants. To retain and purify mutant sectors away from other tissues of the chimeric plant, shoots were regenerated from excised tissue sections using the methods described above. Henderson et al. [37] recently reported an efficient CRISPR/*Cas9*-mediated mutagenesis protocol for *P. piloselloides* var. *piloselloides* (syn. *H. piloselloides*), including multiallelic deletion. This advance presents exciting new opportunities for targeted mutagenesis in this system.

2.5 Molecular Mapping

Details of the tools used to conduct molecular mapping in *Pilosella* are described elsewhere in this book. Briefly, mapping has progressed using segregation analysis [13, 14], deletion mapping [2], and polyhaploid mapping [13, 17] with the plants analyzed using AFLP, cDNA, SSR, SCAR, and CAPs markers. Lists of the relevant

Fig. 5 Wild-type (left) and polyhaploid (right) of *Pilosella piloselloides* var. *praealta*

molecular markers are available in these publications. Apomixis only occurs naturally in the polyploids of this genus, but apomictic, diploid "polyhaploids" (Figs. 5 and 6) are easily isolated using either a negative selection system or a mutant stock deficient in apomeiosis (*see* Chap. 10 on molecular mapping in apomicts for details). Diploid polyhaploids were used in these studies, as they greatly simplify both segregation analysis and genome assembly.

2.6 Online Resources

A draft genome assembly for PH70, a diploid polyhaploid of *P. piloselloides* var. *praealta*, is available under the BioProject accession code: PRJNA982544: *Pilosella piloselloides* Genome sequencing and assembly (TaxID: 1628030). The assembly was constructed by the MaSuRCA hybrid genome assembly tool [38, 39] using both long- and short-read data. It resolved into

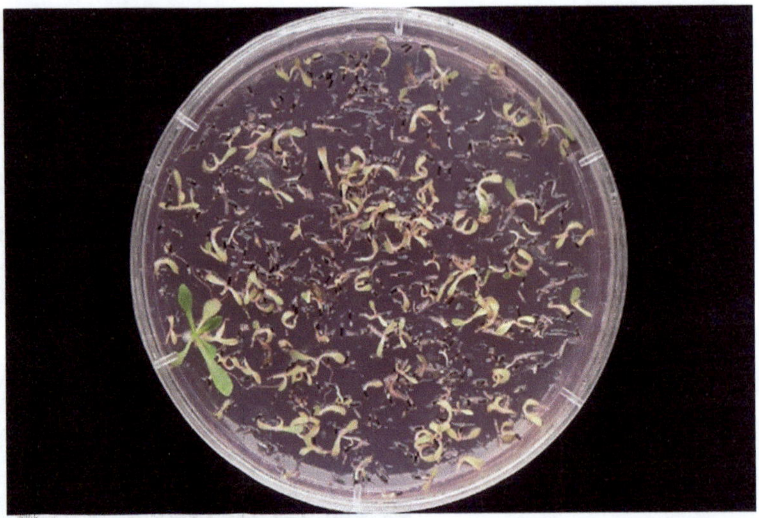

Fig. 6 Selection of polyhaploid seedlings on 5-fluorocytosine medium using *codA* a negative selection marker

21,000 contigs with a BUSCO score of 94%. The genome of PH70 includes dominant alleles at both the *LOA* and *LOP* loci critical for the inheritance of apomixis in *Pilosella* [17]. Molecular maps for *P. piloselloides* var. *praealta* and *P. piloselloides* var. *piloselloides* are available [13] positioning 302 and 533 cDNA-derived markers, respectively, across the 9 linkage groups of these species. Similarly, a reference transcriptome is available at the Shotgun Assembly project DDBJ/EMBL/GenBank under the accession number GEEH00000000. This contains 147,632 unique transcripts belonging to 74,424 loci [6].

3 Summary

The genus *Pilosella* is a tractable experimental system for studying the molecular genetics of aposporous, autonomous apomixis. The plants are easily cultivated and easily scored for apomixis. Cross-compatible sexual and apomictic biotypes are available, and self-incompatibility simplifies the formation of hybrid populations. A wide range of methods have been described for the experimental use of these plants, including histological techniques, molecular mapping approaches, and targeted mutagenesis using CRISPR/*Cas9*-mediated deletion. A growing number of online data resources are also available and freely accessible. Collectively, these resources facilitate the study of apomixis in this system. Unlinked loci are known to separately control apomeiosis (*LOA*) and parthenogenesis (*LOP*) and the likely presence of another locus critical for autonomous endospermy has been detected (*AutE*). Genes at these loci are now under study.

Acknowledgement

The author would like to thank the former New Zealand Foundation for Science and Technology (now the New Zealand Ministry of Business Innovation and Employment) for their financial support (grant number C10X0404 to R.B.).

Conflicts of Interest The authors have no conflicts of interest to declare that are relevant to the content of this chapter.

References

1. Koltunow A, Johnson SD, Bicknell RA (1998) Sexual and apomictic development in *Hieracium*. Sex Plant Reprod 11(4):213–230
2. Catanach AS, Erasmuson SK, Podivinski E, Jordan BR, Bicknell RA (2006) Deletion mapping of genetic elements associated with apomixis in *Hieracium*. Proc Natl Acad Sci U S A 103(49):18650–18655
3. Rodrigues JCM, Tucker MR, Johnson SD, Hrmova M, Koltunow AMG (2008) Sexual and apomictic seed formation in *Hieracium* requires the plant Polycomb-group gene *FERTILIZATION INDEPENDENT ENDOSPERM*. Plant Cell 20(9):2372–2386
4. Koltunow AM, Johnson SD, Rodrigues J, Okada T, Hu Y, Tsuchiya T, Wilson S, Fletcher P, Ito K, Suzuki G (2011) Sexual reproduction is the default mode in apomictic *Hieracium* subgenus *Pilosella*, in which two dominant loci function to enable apomixis. Plant J 66(5):890–902
5. Underwood C, Vijverberg K, Rigola D, Okamoto S, Oplaat C, Camp R, Radoeva T, Schauer S, Fierens J, Jansen K, Mansveld S, Busscher M, Xiong W, Datema E, Nijbroek K, Blom E-J, Bicknell R, Catanach A, Erasmuson S, Winefield C, van Tunen A, Prins M, Schranz E, van Dijk P (2022) A *PARTHENOGENESIS* allele from apomictic dandelion can induce egg cell division without fertilization in lettuce. Nat Genet 54:84–93. https://doi.org/10.1038/s41588-021-00984-y
6. Bräuning S, Catanach A, Lord JM, Bicknell R, Macknight RC (2018) Comparative transcriptome analysis of the wild-type model apomict *Hieracium praealtum* and its loss of parthenogenesis (*lop*) mutant. BMC Plant Biol 18(1):206
7. Mendel G (1869) Ueber einige aus künstlicher Befruchtung gewonnene *Hieracium*-Bastarde. Verhandlungen des naturforchenden Vereins in Brünn 8:26–31
8. Nogler G (2006) The lesser-known Mendel: his experiments on *Hieracium*. Genetics 172(1):1–6
9. Bicknell R, Catanach A, Hand M, Koltunow A (2016) Seeds of doubt: Mendel's choice of *Hieracium* to study inheritance, a case of right plant, wrong trait. Theor Appl Genet 129(12):2253–2266
10. Ostenfeld CH, Rosenberg O (1906) Experimental and cytological studies in the Hieracia. I. Castration and hybridization experiments with some species of Hieracia. Botanisk Tidsskrift 27:225–248
11. Ostenfeld C (1910) Further studies on the apogamy and hybridization of the Hieracia. Mol Gen Genet 3(1):241–285
12. Rosenberg O (1907) Experimental and cytological studies in Hieracia: II cytological studies on the Apogamy in *Hieracium*. Botanisk Tidsskrift 28:143–170
13. Shirasawa K, Hand ML, Henderson ST, Okada T, Johnson SD, Taylor JM, Spriggs A, Siddons H, Hirakawa H, Isobe S, Tabata S, Koltunow AMG (2014) A reference genetic linkage map of apomictic *Hieracium* species based on expressed markers derived from developing ovule transcripts. Ann Bot 115(4):567–580
14. Ogawa D, Johnson SD, Henderson ST, Koltunow AMG (2013) Genetic separation of autonomous endosperm formation (*AutE*) from the two other components of apomixis in *Hieracium*. Plant Reprod 26(2):113–123
15. Henderson ST, Johnson SD, Eichmann J, Koltunow AM (2017) Genetic analyses of the inheritance and expressivity of autonomous endosperm formation in *Hieracium* with different modes of embryo sac and seed formation. Ann Bot 119(6):1001–1010

16. Tucker MR, Araujo ACG, Paech NA, Hecht V, Schmidt EDL, Rossell JB, de Vries SC, Koltunow AMG (2003) Sexual and apomictic reproduction in *Hieracium* subgenus *Pilosella* are closely interrelated developmental pathways. Plant Cell 15(7):1524–1537
17. Bicknell R, Gaillard M, Catanach A, McGee R, Erasmuson S, Fulton B, Winefield C (2023) Genetic mapping of the *LOSS OF PARTHENOGENESIS* locus in *Pilosella piloselloides* and the evolution of apomixis in the Lactuceae. Front Plant Sci 14. https://doi.org/10.3389/fpls.2023.1239191
18. Mráz P, Zdvořák P (2019) Reproductive pathways in *Hieracium* (Asteraceae): strict sexuality in diploids and apomixis in polyploids. Ann Bot 123(2):391–403
19. Kotani Y, Henderson ST, Suzuki G, Johnson SD, Okada T, Siddons H, Mukai Y, Koltunow AM (2014) The *LOSS OF APOMEIOSIS* (*LOA*) locus in *Hieracium praealtum* can function independently of the associated large-scale repetitive chromosomal structure. New Phytol 201(3):973–981
20. Shimada T, Endo T, Fujii H, Nakano M, Sugiyama A, Daido G, Ohta S, Yoshioka T, Omura M (2018) MITE insertion-dependent expression of CitRKD1 with a RWP-RK domain regulates somatic embryogenesis in citrus nucellar tissues. BMC Plant Biol 18:1–19
21. Kew: Plants of the world online (2024) https://powo.science.kew.org/taxon/urn:lsid:ipni.org:names:30065694-2
22. Bräutigam S, Greuter W (2007) A new treatment of *Pilosella* for the Euro-Mediterranean flora. Willdenowia 37:123–137
23. Krahulcova A, Chrtek J, Krahulec F (1999) Autogamy in *Hieracium* subgen. *Pilosella*. Folia Geobot 34(3):373–376. https://doi.org/10.1007/Bf02912821
24. Krahulcová A, Krahulec F (2000) Offspring diversity in *Hieracium* subgen. *Pilosella* (Asteraceae): new cytotypes from hybridization experiments and from open pollination. Fragm Flor Geobot 45:239–255
25. Krahulcová A, Krahulec F, Chrtek J (2001) Chromosome numbers and reproductive systems in selected representatives of *Hieracium* subgen. *Pilosella* in the Krkonose Mts (the Sudeten Mts)-2. Preslia-Praha 73(3):193–212
26. Hand ML, Koltunow AMG (2014) The genetic control of apomixis: asexual seed formation. Genetics 197(2):441–450
27. Hojsgaard D, Hörandl E (2019) The rise of apomixis in natural plant populations. Front Plant Sci 10:358
28. Crane CF (2001) Classification of apomictic mechanisms. In: Savidan Y, Carman JG, Dresselhaus T (eds) Flowering of apomixis: from mechanisms to genetic engineering. CIMMYT, IRD, European Commission DG VI, Mexico, pp 24–34
29. Matzk F, Meister A, Schubert I (2000) An efficient screen for reproductive pathways using mature seeds of monocots and dicots. Plant J 21(1):97–108. https://doi.org/10.1046/j.1365-313x.2000.00647.x
30. Bicknell RA, Lambie SC, Butler RC (2003) Quantification of progeny classes in two facultatively apomictic accessions of *Hieracium*. Hereditas 138:11–20
31. Houliston GJ, Chapman HM, Bicknell RA (2006) The influence of genotype and environment on the fecundity and facultative expression of apomixis in *Hieracium pilosella* L. Folia Geobot 41:165–181
32. Tucker MR (2003) Molecular & cytological aspects of seed development in sexual and apomictic *Hieracium*. PhD Thesis, The University of Adelaide, Adelaide
33. Bicknell RA (1994) Micropropagation of *Hieracium aurantiacum*. Plant Cell Tissue Organ Cult 37:197–199
34. Bicknell RA, Borst NK (1995) Isolation of reduced genotypes of *Hieracium pilosella* using anther culture. Plant Cell Tissue Organ Cult 45:37–41
35. Bicknell RA, Borst NK (1994) Agrobacterium-mediated transformation of *Hieracium aurantiacum*. Int J Plant Sci 155(4):467–470
36. Tucker MR, Paech NA, Willemse MTM, Koltunow AMG (2001) Dynamics of callose deposition and beta-1,3-glucanase expression during reproductive events in sexual and apomictic *Hieracium*. Planta 212(4):487–498. https://doi.org/10.1007/s004250000445
37. Henderson SW, Henderson ST, Goetz M, Koltunow AM (2020) Efficient CRISPR/Cas9-mediated knockout of an endogenous phytoene desaturase gene in t1 progeny of apomictic hieracium enables new strategies for apomixis gene identification. Genes 11(9):1064
38. Zimin AV, Marçais G, Puiu D, Roberts M, Salzberg SL, Yorke JA (2013) The MaSuRCA genome assembler. Bioinformatics 29(21):2669–2677
39. Luo MC, Gu YQ, Puiu D, Wang H, Twardziok SO, Deal KR, Huo N, Zhu T, Wang L, Wang Y, McGuire PE (2017) Genome sequence of the progenitor of the wheat D genome *Aegilops tauschii*. Nat Biotechnol 551:498–502

Chapter 15

The Genus *Boechera* as a Model for Apomixis Research

Terezie Mandáková

Abstract

The genus *Boechera*, a prominent member of the tribe Boechereae within the family Brassicaceae, has emerged as an exceptional model for apomixis research, owing to its unique evolutionary, reproductive, and genomic characteristics. With over 480 genetically distinct taxa and a distribution spanning North America, Greenland, and parts of Asia, *Boechera* exemplifies remarkable ecological and genetic diversity. A hallmark of the genus is its high frequency of gametophytic apomixis, including diplospory and apospory, which, combined with hybridization, drive genetic innovation and environmental adaptability. Moreover, *Boechera* is notable for its diploid expression of apomixis, independent of polyploidy, providing an unparalleled system for dissecting asexual reproduction. Genomic adaptations, such as heterochromatic supernumerary chromosomes, further highlight its evolutionary complexity. Recent advances, including the discovery of critical regulatory genes like *APOLLO* and *UPGRADE2*, have deepened our understanding of the genetic basis of apomixis. These insights position *Boechera* as a cornerstone for elucidating apomictic pathways and leveraging these mechanisms for crop improvement. The genus not only enhances our understanding of apomixis but also offers transformative potential for agriculture.

Key words *Boechera*, *Brassicaceae*, Apomixis, Gametophytic reproduction, Hybridization, Chromosome dynamics, Heterochromatic chromosomes, Evolutionary biology, Genomic architecture

1 The Tribe Boechereae (Brassicaceae)

The Brassicaceae family, commonly known as the mustard family, ranks among the largest and most diverse plant families, encompassing nearly 4000 species across 351 genera [1]. Originating during the late Eocene to Oligocene epochs, this family holds considerable ecological and agricultural significance, thriving across a wide range of ecosystems worldwide [2]. Brassicaceae is classified into two subfamilies: the smaller and less diverse Aethionemoideae, and the expansive Brassicoideae, which comprises five supertribes and 58 tribes, including the tribe Boechereae [2].

The tribe Boechereae stands out for its distinctive reproductive strategies, particularly the extensive presence of apomixis [3]. This tribe encompasses several genera that vary in diversity and

Tatyana Radoeva and Peter J. van Dijk (eds.), *Apomixis: Methods and Protocols*, Methods in Molecular Biology, vol. 2987,
https://doi.org/10.1007/978-1-0716-5001-1_15,

geographic distribution. Notably, seven genera—*Anelsonia*, *Cusickiella*, *Nevada*, *Phoenicaulis*, *Polyctenium*, *Sandbergia*, and *Yosemitea*—are either mono- or bispecific and predominantly restricted to the western United States. In contrast, the genus *Borodinia* comprises eight species, with seven native to eastern North America and one extending into Siberia and northeastern coastal Asia [4].

The genus *Boechera* is the most diverse and geographically extensive member of the tribe Boechereae. It includes approximately 480 genetically distinct taxa [5], with a distribution that spans from Alaska across much of North America, and extends to Greenland and Siberia for certain species [4]. Recent advancements in phylogenetic analyses utilizing high-resolution genomic tools have significantly enhanced our understanding of the evolutionary relationships within the tribe. These studies have effectively distinguished genera and resolved clades within *Boechera*, providing a robust framework for exploring the evolutionary dynamics and diversification processes of the Boechereae [2, 4, 6].

2 Apomixis and Reproductive Diversity in Boechereae

The tribe Boechereae serves as an evolutionary hotspot for the origin and diversification of apomictic taxa. Within the genus *Boechera*, gametophytic apomixis is observed in over 100 taxa, significantly contributing to genotype and phenotype diversity by stabilizing the genetic outcomes of hybridization and reticulate evolutionary processes [7]. Detailed embryological investigations have identified three distinct mechanisms of gametophytic apomixis in *Boechera*: *Antennaria*-type diplospory, *Taraxacum*-type diplospory, and *Hieracium*-type apospory [7, 8]. Among these, *Antennaria*-type diplospory is notably rare and primarily documented in individuals that predominantly reproduce via *Taraxacum*-type diplospory. *Taraxacum*-type diplospory and apospory, in contrast, are more prevalent and occur at higher frequencies than sexual reproduction in natural populations. Remarkably, both apospory and diplospory are equally represented among *Boechera* taxa, and in seven species, these two modes of apomixis co-occur within the same individual, each expressed at exceptionally high frequencies [7].

Apospory, while extensively studied in *Boechera*, has also been identified in other genera within the tribe Boechereae, including *Borodinia laevigata*, *Phoenicaulis cheiranthoides* (in its di-, tri-, and tetraploid forms), *Polyctenium fremontii* (tetraploid), and *Sandbergia whitedii* (triploid) [3, 7, 9]. These discoveries highlight the reproductive and evolutionary adaptability across Boechereae, emphasizing its significance as a model system for investigating apomixis in angiosperms.

3 Chromosome Dynamics and Hybridization in *Boechera*

The genus *Boechera* is characterized by a basic chromosome number of $x = 7$, with sexual species predominantly diploid ($2n = 14$). Although primarily self-pollinating (autogamous), interspecific hybridization among sexual diploids frequently results in the formation of allodiploids ($2n = 14, 15$) [3, 10–12]. These hybrids often adopt apomictic reproduction, a strategy that mitigates sterility typically associated with allodiploidy, thereby enhancing their viability and persistence within natural populations. Remarkably, apomixis in allodiploid *Boechera*—arising from hybridization between sexual diploids—represents a rare phenomenon among flowering plants [7].

A distinctive trait of many apomictic *Boechera* species is their ability to produce unreduced ($2n$) pollen, a phenomenon rarely observed in other apomictic angiosperms [13]. This unique capability allows $2n$ sperm from apomictic diploids to fertilize haploid ($1n$) eggs of sexually reproducing taxa, resulting in the formation of novel triploid apomicts with chromosome numbers of $2n = 21$ or 22 [5, 12, 14]. Tetraploid apomicts can also emerge through this process, although they occur much less frequently [15, 16]. These apomictic hybrids demonstrate enhanced ecological adaptability compared to their sexual progenitors, enabling them to thrive across diverse environmental niches [17, 18]. However, frequent introgression between sexual and apomictic taxa often obscures morphological and genetic distinctions, complicating the taxonomy of *Boechera* species [5].

4 Chromosome Structure and Evolution

The genomic architecture of the tribe Boechereae has been elucidated through studies on species representing seven genera: *Boechera*, *Borodinia*, *Cusickiella*, *Phoenicaulis*, *Polyctenium*, *Nevada*, and *Sandbergia* [3, 5, 9, 19]. These investigations revealed a conserved chromosome base number of $x = 7$ across all analyzed taxa. Comparative analyses with the sister tribe Halimolobeae, which retains a base number of $x = 8$, suggest that the ancestral genome of Boechereae ($n = 7$) originated via descending dysploidy from an $n = 8$ genome (*see* Fig. 1a, b). This genomic reorganization is estimated to have occurred approximately eight million years ago, coinciding with the divergence of extant Boechereae lineages [3].

Of the seven chromosomes present in Boechereae taxa, three (Boe4, Boe6, and Boe7) remain structurally intact and resemble their ancestral configuration. However, the other chromosomes have undergone extensive reshuffling, including an end-to-end translocation, two reciprocal translocations, and a pericentric

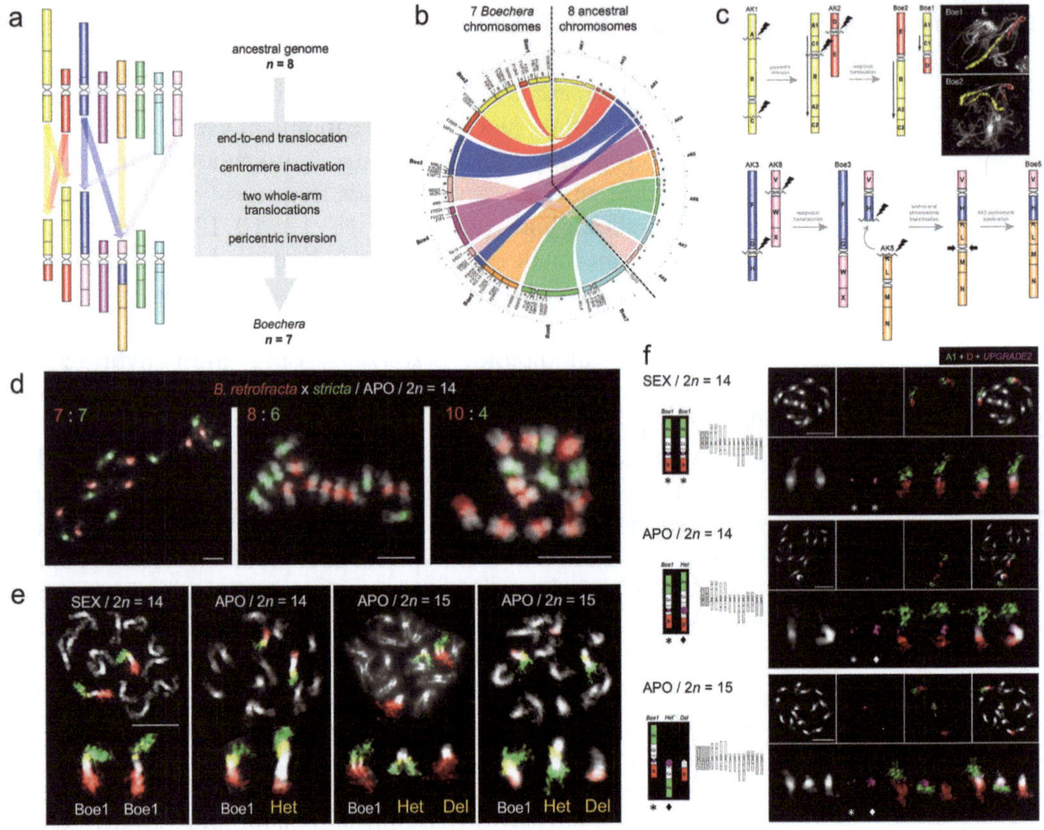

Fig. 1 Structure and evolution of *Boechera* genomes and apomixis-associated chromosomes. (**a**) Origin of the seven *Boechera* chromosomes from an eight-chromosome ancestral genome, modified from [3]. (**b**) Collinearity between the seven *Boechera* chromosomes and the eight ancestral chromosomes, modified from [19]. (**c**) Origin of the reshuffled chromosomes Boe1, Boe2, Boe3, and Boe5, modified from [3]. (**d**) Genomic in situ hybridization showing variable numbers of parental chromosomes in different apomictic *B. retrofracta* x *stricta* lines (unpublished data). (**e**) Structure of chromosomes Boe1, *Het*, and *Del* revealed by comparative chromosome painting using probes corresponding to genomic blocks A1 (green), C1 (yellow), and D (red) (unpublished data). (**f**) Fluorescence in situ hybridization of the *UPGRADE2* gene modified from [27]. All chromosomes are counterstained with DAPI. Scale bars, 10 μm

inversion, culminating in the formation of chromosomes Boe1, Boe2, Boe3, and Boe5 (*see* Fig. 1c) [3, 19]. While Boechereae generally display genomic stability, specific chromosomal rearrangements—particularly inversions—have been linked to diversification within the tribe [3].

One prominent example is a significant pericentric inversion identified in *Boechera stricta*, which has been associated with ecologically important traits and the differentiation of populations [20]. These findings highlight the role of structural genomic changes in reproductive isolation and the speciation processes that have shaped the evolutionary trajectory of the tribe Boechereae.

The genus *Boechera* has garnered significant scientific attention due to its compact genome size (~250 Mb), its evolutionary proximity to *Arabidopsis thaliana* [2, 21], and the coexistence of diploid sexual and diploid apomictic lineages. Notably, the presence of diploid apomictic lines in *Boechera* highlights that polyploidy is not a prerequisite for the expression of apomixis in this genus. Genomic in situ hybridization studies have shown that apomictic *Boechera* are interspecific hybrids, with variability in the number of parental chromosomes resulting from the substitution of homeologous parental chromosomes (*see* Fig. 1d) [10].

A defining feature of apomictic *Boechera* genomes is the presence of heterochromatic supernumerary chromosomes. These additional chromosomes were first documented in aneuploid apomicts with 15 chromosomes as early as 1951 [14]. Subsequent research identified these smaller, heterochromatic chromosomes, referred to as B-like chromosomes, which are absent in sexual diploids, but consistently found in apomictic diploids ($2n = 15$) and triploids ($2n = 22$). These B-like chromosomes were hypothesized to carry genetic elements essential for apomixis [22, 23].

Further cytogenetic investigations revealed the ubiquitous presence of a highly heterochromatic chromosome known as *Het* in all diploid apomicts. Additionally, a smaller derivative chromosome, termed *Del* (as presumed, a deletion chromosome), was detected in diploid and triploid apomictic aneuploids [10]. Studies on 14-chromosomal apomicts identified the *Het* chromosome as a modified Boe1 homeolog. This chromosome contains genomic blocks A1, C1, and D, characterized by expanded pericentromeric heterochromatin (*see* Fig. 1e) [19]. In 15-chromosomal aneuploid apomicts, the centric fission of the *Het* chromosome resulted in two new derivatives: a telocentric *Het* chromosome (comprising blocks A1 and C1) and a *Del* chromosome (block D) (*see* Fig. 1e) [19]. These fission-derived chromosomes are stably inherited, a process facilitated by the apomictic reproductive mode, which ensures their persistence in populations.

Interestingly, centric fission in *Boechera* represents the only documented case of this chromosomal rearrangement, leading to ascending dysploidy, within the *Brassicaceae* family. This unique genomic feature further underscores the genus's value as a model for studying chromosomal evolution in plants.

5 Genetic Mechanisms Underlying Apomixis in *Boechera*

The regulation of apomixis in *Boechera* is orchestrated by multiple genetic factors, each associated with distinct components of the reproductive pathway. Prominent among these are the *APOLLO* gene, implicated in female apomeiosis [24], and *UPGRADE2*, which is linked to male apomeiosis [25]. Genome-wide studies investigating gene expression during ovule development have identified significant alterations in regulatory patterns that are characteristic of apomictic reproduction [26]. The functional elements of apomixis—namely apomeiosis, parthenogenesis, and pseudogamy—exhibit variable inheritance patterns within and between *Boechera* species, as demonstrated through segregation and transmission studies [27]. These findings support the hypothesis that apomixis is governed by the interplay of multiple genetic factors rather than a single genetic determinant.

Further insights into the genomic basis of apomixis have emerged from studies on the chromosomal localization of apomixis-associated genes. The *UPGRADE2* gene has been mapped to the Boe1 hom(e)ologous chromosomes, where an increased copy number has been identified on the *Het* chromosomes in apomictic diploids ($2n = 14$) and aneuploids ($2n = 15$), respectively (*see* Fig. 1f) [27]. This amplification correlates with the expansion of pericentromeric heterochromatin on the *Het* and *Het'* chromosomes, suggesting a functional link between chromatin structure and gene dosage. The consistent presence of *Het* chromosomes containing *UPGRADE2* in all apomictic lineages underscores its pivotal role in controlling apomictic reproduction. Collectively, these findings highlight the critical contributions of chromosomal rearrangements and gene copy number variations to the evolution and persistence of apomixis in *Boechera*.

6 Conclusion

The genus *Boechera* serves as an unparalleled model for studying apomixis, particularly its expression at the diploid level, independent of the confounding effects of polyploidy. Its diverse mechanisms of gametophytic apomixis, genomic adaptations such as heterochromatic chromosomes, and the identification of pivotal regulatory genes like *APOLLO* and *UPGRADE2* make *Boechera* a critical system for exploring the genetic, evolutionary, and ecological dynamics of asexual reproduction. This genus not only advances our understanding of apomixis but also provides a foundation for translating these insights into practical applications.

7 Future Directions

Although apomixis in *Boechera* has been the subject of extensive research for decades, significant gaps remain, including the absence of a complete genome sequence for an apomict and a limited understanding of the epigenetic mechanisms underlying this phenomenon. However, recent technological advancements have set the stage for transformative progress. Long-read sequencing technologies, paired with sophisticated algorithms for genome assembly, now enable the generation of telomere-to-telomere genome assemblies, providing unprecedented opportunities to decode the complex genomic architecture of apomicts. Advances in cytogenetics, integrating chromosome painting and immunolocalization, further allow for validation of chromosome-level assemblies and enable detailed investigations into the three-dimensional structure and spatial organization of chromosomes and subgenomes. These tools promise to illuminate the links between hybridization, chromosome reorganization, and the evolution of apomixis.

Future research should prioritize leveraging these breakthroughs to unravel the intricate genetic and regulatory networks governing apomixis, with a focus on the mechanistic roles of key genes. Investigating the environmental and ecological factors that influence apomixis expression will enhance our understanding of its adaptive significance, while detailed studies of heterochromatic chromosomes could shed light on their evolutionary and reproductive functions. Comparative analyses across apomictic systems may uncover conserved pathways, facilitating broader generalizations.

Furthermore, cutting-edge gene-editing tools such as CRISPR/Cas provide an exciting avenue to experimentally reconstruct plant chromosomes and replay evolutionary trajectories, offering unique insights into the interplay between chromosomal dynamics and reproductive strategies. Ultimately, applying synthetic biology and gene-editing technologies to transfer apomictic traits into crop species holds transformative potential for agriculture, paving the way for the development of high-yield, genetically uniform plants. This progress could address critical challenges in global food security and biodiversity conservation, marking a new era for both basic and applied research in plant biology.

Acknowledgments

This work was supported by the Czech Science Foundation (project 24-11371S) and Masaryk University Grant Agency (project MUNI/R/1268/2022).

References

1. German DA, Hendriks KP, Koch MA, Lens F, Lysak MA, Bailey CD, Mummenhoff K, Al-Shehbaz IA (2023) An updated classification of the Brassicaceae (Cruciferae). PhytoKeys 220:127–144
2. Hendriks KP, Kiefer C, Al-Shehbaz IA, Bailey CD, van Huysduynen AH, Nikolov LA, Nauheimer L, Zuntini AR, German DA, Franzke A, Koch MA, Lysak MA, Toro-Núñez O, Özüdoğru B, Invernón VR, Walden N, Maurin O, Hay NM, Shushkov P, Mandáková T, Thulin M, Windham MD, Rešetnik I, Španiel S, Ly E, Pires JC, Harkess A, Neuffer B, Vogt R, Bräuchler C, Rainer H, Janssens SB, Schmull M, Forrest A, Guggisberg A, Zmartzy S, Lepschi BJ,

Scarlett N, Stauffer FW, Schönberger I, Heenan P, Baker WJ, Forest F, Mummenhoff K, Lens F (2023) Less is more: global Brassicaceae phylogeny based on filtering of 1,000 gene dataset. Curr Biol 33:1–17

3. Mandáková T, Price B, Ashby K, Hloušková P, Windham MD, Mitchell-Olds T, Carman J, Lysak MA (2020) The origin of the Boechereae (Brassicaceae) was preceded by descending dysploidy, whereas intra-tribal cladogenesis was marked by overall genome stasis. Front Plant Sci 11:514
4. Alexander PJ, Windham MD, Beck JB, Al-Shehbaz IA, Allphin L, Bailey CD (2013) Molecular phylogenetics and taxonomy of the genus Boechera and related genera (Brassicaceae: Boechereae). Syst Bot 38:192–209
5. Li F W, Rushworth CA, Beck J B, Windham M D (2017) Boechera microsatellite website: an online portal for species identification and determination of hybrid parentage. Database 2017: baw169
6. Hay N, Windham MD, Mandáková T, Lysák MA, Hendriks KP, Mummenhoff K, Lens F, Pryer KM, Bailey CD (2023) A Hyb-Seq phylogeny of Boechera and related genera using a combination of Angiosperms353 and Brassicaceae-specific bait sets. Am J Bot e16226
7. Carman JG, Mateo de Arias M, Gao L, Zhao X, Kowallis B, Sherwood DA et al (2019) Apospory in addition to diplospory is common in Boechera where it may facilitate speciation by recombination-driven apomixis-to-sex reversals. Front Plant Sci 10:724
8. Windham MD, Beck JB, Li F-W, Allphin L, Carman JG, Sherwood DA et al (2015) Searching for diamonds in the apomictic rough: a case study involving Boechera lignifera (Brassicaceae). Syst Bot 40:1031–1044
9. Mandáková T, Ashby K, Price B, Windham M, Carman J, Lysak M (2021) Genome structure and apomixis in Phoenicaulis. J Syst Evol 59: 83–92
10. Kantama L, Sharbel T, Schranz ME, Mitchell-Olds T, de Vries S, de Jong H (2007) Diploid apomicts of the Boechera holboellii complex display large-scale chromosome substitutions and aberrant chromosomes. Proc Natl Acad Sci USA 104:14026–14031
11. Beck JB, Alexander PJ, Allphin L, Al-Shehbaz IA, Rushworth C, Bailey CD et al (2012) Does hybridization drive the transition to asexuality in diploid Boechera (Brassicaceae)? Evolution 66:985–995
12. Alexander PJ, Windham MD, Beck JB, Al-Shehbaz IA, Allphin L, Bailey CD (2015) Weaving a tangled web: divergent and reticulate speciation in Boechera fendleri sensu lato (Brassicaceae; Boechereae). Syst Bot 40:572–596
13. Asker SE, Jerling L (1992) Apomixis in plants. CRC Press, Boca Raton
14. Böcher TW (1951) Cytological and embryological studies in the amphiapomictic Arabis holboellii complex. Det Kongelige Danske Videnskabernes Selskab. Biol Skrif 6:1–59
15. Schranz ME, Dobes C, Koch MA, Mitchell-Olds T (2005) Sexual reproduction, hybridization, apomixis, and polyploidization in the genus Boechera (Brassicaceae). Am J Bot 92: 1797–1810
16. Aliyu OM, Schranz ME, Sharbel TF (2010) Quantitative variation for apomictic reproduction in the genus Boechera (Brassicaceae). Am J Bot 97:1719–1731
17. Windham MD, Al-Shehbaz IA (2006) New and noteworth species of Boechera (Brassicaceae) I: sexual diploids. Harvard Papers Bot 11:61–88
18. Shah JN, Kirioukhova O, Pawar P, Tayyab M, Mateo JL, Johnston AJ (2016) Depletion of key meiotic genes and transcriptome-wide abiotic stress reprogramming mark early preparatory events ahead of apomeiotic transition. Front Plant Sci 7:1539
19. Mandáková T, Schranz ME, Sharbel TF, de Jong H, Lysak MA (2015) Karyotype evolution in apomictic Boechera and the origin of the aberrant chromosomes. Plant J 82:785–793
20. Lee CR, Wang B, Mojica JP, Mandáková T, Prasad KVSK, Goicoechea JL et al (2017) Young inversion with multiple linked QTLs under selection in a hybrid zone. Nat Ecol Evol 1:119
21. Song B-H, Mitchell-Olds T (2007) High genetic diversity and population differentiation in Boechera fecunda, a rare relative of Arabidopsis. Mol Ecol 16:4079–4088
22. Sharbel TF, Voigt ML, Mitchell-Olds T, Kantama L, de Jong H (2004) Is the aneuploid chromosome in an apomictic Boechera holboellii a genuine B chromosome? Cytogenet Genome Res 106:173–183
23. Sharbel TF, Mitchell-Olds T, Dobes C, Kantama L, de Jong H (2005) Biogeographic distribution of polyploidy and B chromosomes in the apomictic Boechera holboellii complex. Cytogenet Genome Res 109:283–292
24. Corral JM, Vogel H, Aliyu OM, Hensel G, Thiel T, Kumlehn J et al (2013) A conserved apomixis-specific polymorphism is correlated with exclusive DEDDh exonuclease expression

in premeiotic ovules of apomictic Boechera. Plant Physiol 163:1660–1672

25. Mau M, Corral JM, Vogel H, Melzer M, Fuchs J, Kuhlmann M et al (2013) The conserved chimeric transcript UPGRADE-2 is associated with unreduced pollen formation and is exclusively found in apomictic Boechera species. Plant Physiol 163:1640–1659
26. Sharbel TF, Voigt M-L, Corral JM, Galla G, Kumlehn J, Klukas C et al (2010) Apomictic and sexual ovules of Boechera display heterochronic global gene expression patterns. Plant Cell 22:655–671
27. Mau M, Mandáková T, Ma X, Ebersbach J, Zou L, Lysak MA, Sharbel TF (2022) Convergent evolution of an apomixis-specific allele class of the tapetum-active UPGRADE2 gene in supernumerary chromatin of apomicts. Front Plant Sci 13:890038

Chapter 16

How Apomictic and Sexual *Cenchrus* Species Can Help Elucidate the Mechanisms Leading to Apospory

Joann A. Conner, Yuji Ke, and Peggy Ozias-Akins

Abstract

Reproduction of apomictic and sexual plants within *Cenchrus*, previously classified as both *Cenchrus* and *Pennisetum*, has been studied since the 1950s. Seventeen of approximately 108–120 *Cenchrus* species contain apomictic cytotypes. Apomictic cytotypes are polyploids and all are classified as developing via apospory. While most apomictic species in *Cenchrus* produce viable offspring that are highly maternally clonal, the ability for sexual reproduction has been retained. With advancement in molecular technologies, understanding the genetic composition, physical location, and genes associated with apospory has been forthcoming. Additional gains including whole-genome assembly of *Cenchrus* species to study the evolutionary mechanisms underlying apomixis are on the horizon.

Key words Apomixis, *Cenchrus*, *Pennisetum*, Reproduction, Apomeiosis, Parthenogenesis, Apospory

1 An Overview of Apomixis

The term apomixis was coined to describe all types of asexual reproduction, which can act as a substitute for sexual reproduction. Winkler is credited with describing terminology for apomixis and asserting because the normal sexual pathway in higher plants contains two essential steps, meiosis and fertilization, separated by time and gamete development, that apomicts are influenced by genetic factor(s) controlling both steps. Further published work by authors such as Stebbins, Gustafsson, and Asker has solidified the current developmental terms used in apomixis research [1–4]. In plants, apomixis leads to maternal clonal propagation through seed. While this process is biologically and evolutionarily fascinating, a major advantage of engineering apomixis into agricultural crops would be the ability to fix superior hybrid genotypes over multiple generations without requiring repeated hybridizations, which are both labor and resource intensive.

Tatyana Radoeva and Peter J. van Dijk (eds.), *Apomixis: Methods and Protocols*, Methods in Molecular Biology, vol. 2987,
https://doi.org/10.1007/978-1-0716-5001-1_16,

In plants, sexual reproduction is considered the default pathway to create offspring. Through sexual reproduction, offspring are a mixture of the genetic makeup of both parents. However, numerous types of developmental pathways can lead to apomictic offspring in plants. Adventitious embryony (sporophytic apomixis) occurs when maternal somatic cell(s) in the ovule change developmental fate to become an embryo. Within the ovule, a functional sexual embryo sac must be fertilized for endosperm formation and seed creation. Adventitious embryony, common in citrus, can lead to seed containing both apomictic and sexual embryos. Gametophytic apomicts have a developmental pathway that generates an embryo sac within the ovule containing cells that are chromosomally unreduced and nonrecombined (apomeiosis). The unreduced egg cell develops, without fertilization, into an embryo (parthenogenesis). The central cell of the gametophytic embryo sac can either develop autonomously or be fertilized by pollen sperm nuclei (pseudogamy) to produce endosperm required for seed development. Gametophytic embryo sacs can be derived from two different types of cells within the ovule. In diplospory, the unreduced embryo sac develops from a megaspore mother cell, which, in a sexual plant, would normally undergo meiosis. In apospory, the unreduced embryo sac forms from nucellar cell(s) within the ovule that change developmental fate to become aposporous initials. In *Cenchrus*, an aposporous initial undergoes two mitotic divisions to form an aposporous embryo sac. The term facultative apomixis describes plants that reproduce both sexual and clonal seeds within the same plant. For facultative apomicts, the range of sexual reproduction can vary from plants that almost exclusively produce clonal seed to plants that mostly produce sexual seed.

2 Phylogeny

Various scientific names have been given to apomictic and sexual species currently designated as *Cenchrus*. *Cenchrus* species are members of the Poaceae (Gramineae) family of monocotyledonous flowering plants, collectively known as the grasses, and reside as part of the tribe Paniceae, subtribe Cenchrinae. Cenchrinae, also referred to as the bristle clade, comprises two genera: *Cenchrus* and *Setaria* [5]. The bristle clade is a monophyletic group characterized by bristle formation in the inflorescence [6]. Nuclear, plastid, and morphological data were used to combine the previously designated *Pennisetum*, *Cenchrus*, and *Odontelytrum* species to *Cenchrus* [7]. *Cenchrus* currently includes 108 (Plants of the World Online managed by the Royal Botanic Gardens, Kew: https://powo.science.kew.org/) to 120 species [8]. Adoption of the new scientific names in literature and websites has been sporadic. Table 1

Table 1
***Cenchrus* species**

Current *Cenchrus* classification	Synonyms	Common name	Base chromosome number	Ploidy levels
C. americanus	*Pennisetum glaucum* & *Pennisetum typhoides*	Pearl millet	7	2x, 4x[a]
C. purpureus	*Pennisetum purpureum*	Napier grass; Elephant grass	7	4x
C. *squamulatus*	*Pennisetum squamulatum*	–	7	8x
C. *mezianus*	*Pennisetum mezianum*	Bamboo grass	8	2x,4x
C. *ciliaris*	*Pennisetum ciliare*	Buffel grass Anjan grass	9	4x, 5x, 6x, 7x, anueploid
C. *setigerus*	*Pennisetum setigerum* & *Cenchrus ciliaris var. setigerus*	Birdwood grass	9	4x
C. *flaccidus*	*Pennisetum flaccidum*	Himalayan fountaingrass	9	2x,4x,5x
C. *massaicus*	*Pennisetum massaicum*	–	9	4x
C. *orientalis*	*Pennisetum orientale*	Oriental pennisetum	9	2x, 3x, 4x, 5x, 6x
C. *setosus*	*Pennisetum polystachion*	Mission grass	9	4x, 6x, 7x
C. *setaceus*	*Pennisetum setaceum*	Fountain grass	9	3x, 6x
C. *longisetus*	*Pennisetum villosum*	Feathertop grass	9	2x, 3x, 4x, 5x, 6x

[a]Induced with colchicine

denotes the current scientific name, previous scientific name(s), and common names found within apomixis literature.

Species within *Cenchrus* can be annual or perennial in nature and are utilized as food, forage, livestock feed, ornamentals, and erosion control. All identified apomictic species within the genus develop by apospory and have pseudogamous endosperm development. The protogynous nature of *Cenchrus* carpels promotes outcrossing, making both the apomictic and sexual species highly heterozygous in genetic makeup. With the ability of *Cenchrus* to thrive on little rain fall and in nutrient-poor and sandy soil, five *Cenchrus* species have been identified as invasive according to the Global Invasive Species Database (https://www.iucngisd.org/gisd/). Four of the five species have apomictic cytotypes.

3 Identification of Apomixis Within *Cenchrus*

Early classification of apomictic reproduction was based on combinations of cytological observation of mature embryo sacs, meiotic behavior of chromosomes, and visual phenotyping of self and/or crossed offspring to gauge uniformity and similarity to the maternal parent. Apospory in *Cenchrus* is relatively easy to identify based on the morphology of the aposporous embryo sacs. A sexual polygonum-type embryo sac is composed of seven cells and eight nuclei. There is one egg cell, two synergids, one central cell with two nuclei, and three antipodal cells. Embryo sac cells are oriented such that the egg and synergid cells are at the micropylar end of the ovule and antipodals are at the chalazal end (Fig. 1a). Aposporous embryo sacs in *Cenchrus* have four nuclei, which are formed after two mitotic divisions of an aposporous initial (panicum-type embryo sac). The unreduced embryo sac lacks antipodals and has an egg cell, one or two synergids, and one or two central cell nuclei. Aposporous embryo sacs can, but will not always, be oriented with the egg cell toward the micropylar end (Fig. 1b). The number of aposporous embryo sacs per ovule is variable and differs between apomictic cytotypes and within plant introduction lines of the same species. This variation suggests that factors (genetic and environmental) outside of the apomixis locus can quantitatively modify the number of aposporous initials within an ovule. The complexity of multiple embryo sacs within an ovule can make the identification of ovules with sexual and aposporous embryo sacs difficult, although accessions with lower numbers of aposporous embryo sacs or ovules with only sexual embryo sacs within an apomictic plant are easier to evaluate (Fig. 1c) [9]. Identification of apospory in

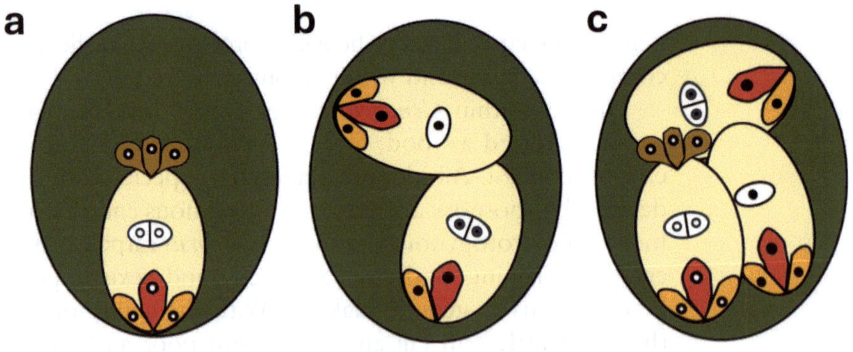

Fig. 1 Graphic representation of varying types of reproductive phenotypes within ovules of *Cenchrus* species. (**a**) A sexual embryo sac. (**b**) Apomixis-derived 4-celled, 4-nucleate embryo sacs. (**c**) Ovules from apomictic reproducing maternal plants can contain both sexual and apomixis derived embryo sacs. Red represents egg cells, orange represents synergid cells, dark tan represents antipodal cells, and light tan represents the central cell with unfused or single nucleus (white). Open circles within cells denote meiotically reduced nuclear content, while filled circles denote parental nuclear content

Cenchrus was noted in the early-to-mid-1950s for *P. setaceum*, *P. orientale*, *P. villosum*, *P. clandestinum*, *P. ciliare*, and *C. setigerus* with a total of 17 apomictic species being noted within the genera [10–14].

4 Ploidy Level(s) in Apomictic *Cenchrus* Species

Accessions within *Cenchrus* apomictic species can have multiple ploidy levels and aneuploidy. Both chromosomal counts and flow cytometry have been used to describe the chromosomal variations found within *C. ciliaris*. The largest flow cytometry screen evaluated 568 *C. ciliaris* accessions from the USDA National Plant Germplasm collection to determine DNA content and, by inference, chromosome number. This screen identified over half of the accessions (54%) with a base tetraploid chromosome number of $2n = 4x = 36$, 24% as pentaploids, 4% as hexaploids, 1% septaploids, and 17% as probable aneuploids [15]. Ploidy levels of *C. ciliaris* using flow cytometry and chromosome counts in a study of 28 populations collected from Tunisia identified only one tetraploid population; 24 were hexaploid and one was hyper-aneuploid with 55 chromosomes [16]. Nuclear DNA content based on flow cytometry and chromosomal spreads was also validated for *C. ciliaris* genotypes within Argentina [17]. High correlations for chromosome number were found between flow cytometry and chromosomal counts. Ploidy levels identified in other *Cenchrus* apomicts are found in Table 1.

Ploidy differences in apomicts can be generated by different mechanisms. A BIII ($2n + n$) hybrid is produced when an unreduced egg is fertilized by pollen prior to the initiation of parthenogenesis. BIII hybrids were produced experimentally through crossing a pentaploid *C. ciliaris* ($2n = 5x = 45$) with apomictic *C. ciliaris* or *C. setigerus* (both ($2n = 4x = 36$)) to yield heptaploid offspring ($2n = 7x = 63$). In other examples, crossing a facultative apomict *P. flaccidum* ($2n = 4x = 36$) 2–3 days prior to anthesis with *P. mezianum* ($2n = 4x = 32$) yielded BIII hybrids ($2n = 52$) while self-pollinating apomictic *C. ciliaris* ($2n = 4x = 36$) up to 3 days prior to anthesis yielded hexaploids ($2n = 6x = 54$) [9, 18–20]. Genetic inheritance of amplified fragment length polymorphism (AFLP) markers in different populations and between ploidy levels of *C. ciliaris* in Tunisia suggested that pentaploids could also have evolved via unbalanced meiosis of a hexaploid with parthenogenic development [21]. BIII hybrids were also obtained when a rare recombination event genetically separated the apomeiosis and parthenogenesis components of apomixis in *C. ciliaris* leading to fertilization of an unreduced but nonparthenogenetic egg [22]. The parthenogenetic development of a meiotically derived egg cell within an apomictic plant can also lead to a reduction of ploidy level. This phenomenon has been noted in production of a polyhaploid plant from a pearl millet x *P. squamulatum* hybrid [23].

5 Genetic Mapping of the Apomixis-Associated Region in *Cenchrus*

Research on genetic inheritance and the generation of linkage maps using molecular markers such as restriction fragment length polymorphisms (RFLPs), random amplification of polymorphic DNA (RAPDs), sequence characterized amplified regions (SCARs), AFLPs, transposon display, duplex-specific nuclease (DSN) and expressed sequence tag-simple sequence repeats (EST-SSRs) have been undertaken. These studies have been limited to *C. ciliaris* and *P. squamulatum* due to a lack of sexual cytotypes or a lack of seed production of interspecific hybridizations using pollen from apomictic males crossed onto sexual cytotypes as females.

Genetic inheritance of apomixis in *C. ciliaris* relied on the identification of a highly sexual *C. ciliaris* plant "SB" [24] that was clonally maintained and registered as TAM-CRD B-1s germplasm [25]. Genetic inheritance based on self-pollinated SB/B-1s seed obtained from an isolated field nursery generated sexual and apomictic offspring, and crosses of SB with apomictic *C. ciliaris* lines "blue" and "common" suggested an SB genotype of "AaBb" where B conditions sexuality and is epistatic to A, which controls apomixis [26]. A later study used a sexual plant B-2s (pedigree unknown) and progeny from open-pollinated B-2s plants. The germinated open-pollinated B-2s seed produced both sexual and apomictic offspring. The original B-2s plant, propagated via tillers, and the sexually derived offspring became the female parents in crosses with apomictic offspring from B-2s and the apomictic cultivar "Higgins." The F_1 hybrids and self-pollinated sexual plants were classified for reproductive phenotype based on the examination of ~20 pistils per plant using a cleared pistil technique [27]. This method replaced the time-consuming sectioning steps and allowed better reproductive characterization by increasing the number of ovules per plant that could be analyzed. All progeny derived from self-pollination or crossing between sexual plants yielded only sexual phenotypes. This is in contrast with the results of self-pollinated SB/B-1s, which showed apomictic offspring. Either B-2s lost the SB/B-1s "sexual B" genetic factor, SB/B-1s was a highly sexual facultative apomict or the isolated nursery where SB/B-1s seed was harvested had a low level of apomictic pollen available for fertilization of B-1s. Models for the B-2s study suggested the following genotypes as sexual (aaaa), and apomictic (Aaaa or AAaa), depending on the apomictic pollen source [28]. Additional obligately sexual *C. ciliaris* have been reported such as IGFRI-CcSx-08/1 [29, 30], IG-96-443 [31], x136-1-7-11, and x153-1-9-1 [32].

While multiple linkage maps have been published for *C. ciliaris* populations, the use of different parental genotypes and molecular markers not in common to all parental genotypes still leaves open questions regarding the exact genomic region and allelic composition required for inheritance of apomixis [33–40]. The genetic inheritance and marker conflicts could be due in part to the classification of mode of female reproduction based on aposporous embryo sac formation exclusive of parthenogenesis, different penetrance of the apospory phenotype, or structural differences within the locus between the different male apomictic parents used. Long-read genome sequencing or whole-genome assembly of the various apomictic parents could help address this conflict in data. While not yet completed, a limited number of apomictic genomes including species from *Cenchrus*, *Megathyrsus* (Guinea grass), and *Urochloa* (signalgrass) are being sequenced and assembled (https://jgi.doe.gov/csp-2021-apomictic-relatives-of-switchgrass/). In addition, a haplotype-resolved genome assembly of apomictic *Urochloa decumbens* cv. Basilisk was recently submitted to the European Nucleotide Archive (ENA) and NCBI as assembly: GCA_964030465.3. A haplotype-resolved *U. humidicola* cv. Tully genome assembly has also been submitted to NCBI as BioProject 1276850.

Genetic studies of apomixis in *P. squamulatum* were facilitated by the ability to introgress apomixis from this wild apomictic species into *P. glaucum* (pearl millet). Pearl millet is the sixth most important food grain for human consumption worldwide in addition to having desirable forage qualities. *P. glaucum* has a diploid genome ($2n = 2x = 14$) that can be maintained after chromosome doubling as a tetraploid ($2n = 4x = 28$) [41, 42]. Multiple apomictic species including *C. ciliaris* ($2n = 36$), *P. mezianum* ($2n = 32$), *P. macrourum* ($2n = 36$), *P. orientale* ($2n = 36$), *P. pedicellatum* ($2n = 54$), *P. polystachion* ($2n = 54$), *P. setaceum* ($2n = 27$), and *P. squamulatum* ($2n = 56$ [43]; previously $2n = 54$) were used as pollen donors for crossing with various lines of diploid and tetraploid *P. glaucum* to create hybrid seed. Limited or no hybrid offspring were produced using pollen from all pollen donors except for *P. squamulatum*. The few hybrids recovered, excluding those from the *P. glaucum* x *P. squamulatum* crosses, displayed male sterility and poor aposporous female fertility [44–46]. Relatively large numbers of fertile apomictic hybrid lines could be produced when *P. squamulatum* was used as the pollen donor. The *P. glaucum x P. squamulatum* hybrids shared characteristics associated with both parents and segregated for apomictic and sexual female reproduction. Most lines retained partial male fertility [41, 42].

Genetic mapping of the *P. glaucum* x *P. squamulatum* hybrids is based on a single pollen donor line, Ps26. A common theme between the mapping techniques for the *P. glaucum* x *P squamulatum* hybrids was the identification of multiple markers, many of which were hemizygous (only associated with apomictic individuals) and tightly linked to the aposporous phenotype. The term ASGR (Apospory-Specific Genomic Region) was coined to denote the locus in *Cenchrus* [47–52]. Conservation of ASGR molecular markers between *P. squamulatum* and *C. ciliaris* was observed [40].

6 Sequence and Chromosomal Characterization of the ASGR

Bacterial artificial chromosome (BAC) libraries were developed from *C. ciliaris* (apomict B12-9) and an apomictic polyhaploid ($2n = 21$) generated via parthenogenesis from a reduced egg of an apomictic *P. glaucum* x *P. squamulatum* hybrid. Fifty-three ASGR-linked BAC clones were identified from both libraries, although BAC clones for each ASGR marker screened were not always identified in both libraries. The BAC clones verified that the ASGR in these genotypes was physically large as no BAC clones shared multiple ASGR markers. However, there was good conservation of the ASGR between *C. ciliaris* and *P. squamulatum* including highly similar duplicate regions. Genomic walking across the ASGR with BAC clones was attempted but achieved limited success due to repetitive elements [53].

The ASGR-linked BAC clones were used to study the location and physical size of the ASGR in apomictic *Cenchrus* species. The physical location of the ASGR in *P. squamulatum* is at the telomeric end of a *P. squamulatum* chromosome with physical synteny outside the ASGR to chromosome 2 in both *Sorghum bicolor* and *Setaria italica* [54]. The hemizygous region of the *P. squamulatum* ASGR leads to disturbed chromosome pairing with other *P. squamulatum* chromosomes during pollen meiosis and contributes to the low level of marker recombination. Additional physical characteristics of the *P. squamulatum* ASGR include an ~15 MB of limited repetitive DNA (the low-copy region), flanked on both sides with blocks of *Opie-2-like* repetitive DNA. The *Opie-2-like* fluorescence in situ hybridization (FISH) signal in *P. squamulatum* is unique to the ASGR, i.e., there is no signal on other *P. squamulatum* chromosomes. A physical size of ~50 Mb was estimated using the distance between a BAC signal derived from a clone containing a proximal recombinant AFLP marker and the distal end of the ASGR [55].

In contrast, the *C. ciliaris* ASGR is located near the centromere on a chromosome associated with an 18S rDNA FISH signal. The number of 18S rDNA signals in tetraploid ($2n = 4x$), pentaploid ($2n = 5x$), and hexaploid ($2n = 6x$) *C. ciliaris* populations are 4, 5, and 6, respectively, with the signal at the telomeric end of the chromosome, suggesting genome stability at the three ploidy levels [56]. While the low-copy ASGR-linked BAC probes generate distinct FISH signals for the ASGR, the repetitive *Opie-2-like* probe generates signals on every *C. ciliaris* chromosome. Like the *P. squamulatum* ASGR, the signal flanks the low-copy ASGR-linked BAC signals [57]. Additional high-copy repetitive elements such as Ty3-gypsy and Ty1-copia were also identified within *C. ciliaris* chromosomes [58].

The ASGR-linked BAC clones identified by the highly conserved UGT197 SCAR were shown to have synteny to rice chromosome 11 and currently are known to contain the parthenogenetic genetic factor, the *ASGR-BabyBooM-Like (ASGR-BBML)* gene [59]. Sample sequencing of other ASGR-linked BAC clones identified additional potential coding regions, an abundance of repetitive elements and a lack of continued macro-syntenic regions of the ASGR to the genomes of rice and sorghum [60, 61]. While a strong candidate for the apomeiosis factor based on protein similarity was not found within the sample BAC sequencing, the identity of multiple potential coding regions conserved between *C. ciliaris* and *P. squamulatum* ASGRs illustrates the need to limit the physical regions containing the apomeiosis genetic factors and to have genome sequence across the ASGR. A recombinant plant, denoted A8, was identified that contained the apomeiosis genetic factor while losing the ability for parthenogenesis [22].

The ASGR-linked BAC probes were also used to look at the ASGR in species without genetic mapping data. Three low-copy ASGR-linked BAC clones, including the UGT197/*ASGR-BBML* BAC clone, along with a BAC clone containing large regions of the repetitive *Opie-2-like* sequence were used for FISH. The low-copy BAC signals were identified on a single chromosome in all apomictic species/cytotypes tested, except for *P. orientale* ($2n = 6x$) that had two signals. None of the sexual genotypes used in the study showed chromosomal signal with low-copy BACs. These signals confirm the hemizygosity of the low-copy region of the ASGR in apomictic *Cenchrus*. The high-copy *Opie-2-like* signal, however, was not always associated with or flanking the UGT197 BAC signal as had been observed in *P. squamulatum* and *C. ciliaris*. Apomictic *P. setaceum*, *P. villosum*, and *P. massaicum* do not have the associated high-copy *Opie-2-like* signals flanking the UGT197 signal [62]. The differing chromosomal locations of the ASGR in multiple apomictic species and potential changes of DNA content of the ASGR suggest a complex evolution of the region between *Cenchrus* species.

7 Characterization of the Parthenogenesis Gene, *ASGR-BBML*

The *ASGR-BBML* gene was a good candidate for the parthenogenesis factor based on many features. The gene had protein similarity to the *Brassica napus BABY BOOM (BnBBM)* gene, which when overexpressed in Arabidopsis leaf tissue caused the production of somatic embryos [63]. The gene is located within the ASGR, shows high sequence similarity across apomictic *Cenchrus* species and displays expression 1 day prior to anthesis in ovaries of all tested apomictic *Cenchrus* species. In addition, an RNAi study to knock down *ASGR-BBML* expression in apomictic F_1 hybrids (RNAi containing *P. glaucum* x *P. squamulatum*) showed a correlation between lower levels of *ASGR-BBML* transcription and reduced parthenogenesis/embryo development. Sexual tetraploid pearl millet, rice, maize, and tobacco lines transformed with various constructs of the *P. squamulatum ASGR-BBML* gene have been shown to produce haploid offspring at varying frequencies. Seed set within pearl millet, rice, and maize transgenic lines with *PsASGR-BBML* and which generated haploid seed was reduced [59, 64–66].

Multiple factors for reduced seed set can be postulated. *PsASGR-BBML* constructs containing the endogenous promoter or other egg cell-specific promoters (DD45/*EGG CELL 1.2* (At2g21740) [67]) may not have an expression level identical to that of a highly fertile natural *Cenchrus* apomict. PsASGR-BBML may not efficiently interact with the transformed species interacting proteins such as the HOMEODOMAIN GLABROUS (HDG) transcription factors, which were identified in Arabidopsis to act antagonistically to AtBBM [68]. However, parthenogenesis driven by endogenous BBM transgenes under egg cell-specific promoters can also display reduced fertility [69]. Fertilization of the central cell and maturation of endosperm and embryo may not be optimal.

To investigate endosperm development, homozygous transgenic plants from the R488-6E lineage, containing a single transgene integration on rice chromosome 4 [70], and whole-mount confocal imaging of Eosin B stained and cleared rice ovaries were used to monitor embryo and endosperm development [71]. Homozygous siblings from the R488-6E lineage displayed 17–35% seed set (based on mature seed recovered per number of florets from a minimum of 3 panicles), wild-type pollen fertility (based on Alexander staining), 50–90% haploid flow cytometry signals from ungerminated seed (25–30 seed tested per plant), and ~50% haploid and diploid seedling survival after germination. Self-pollinated branches from wild-type and homozygous R488-6E lineage were fixed 4 days after emergence of anthers on that branch and displayed a range of development. From the collected florets, only ovaries displaying expansion were analyzed using confocal microscopy to eliminate ovaries that may not have been pollinated

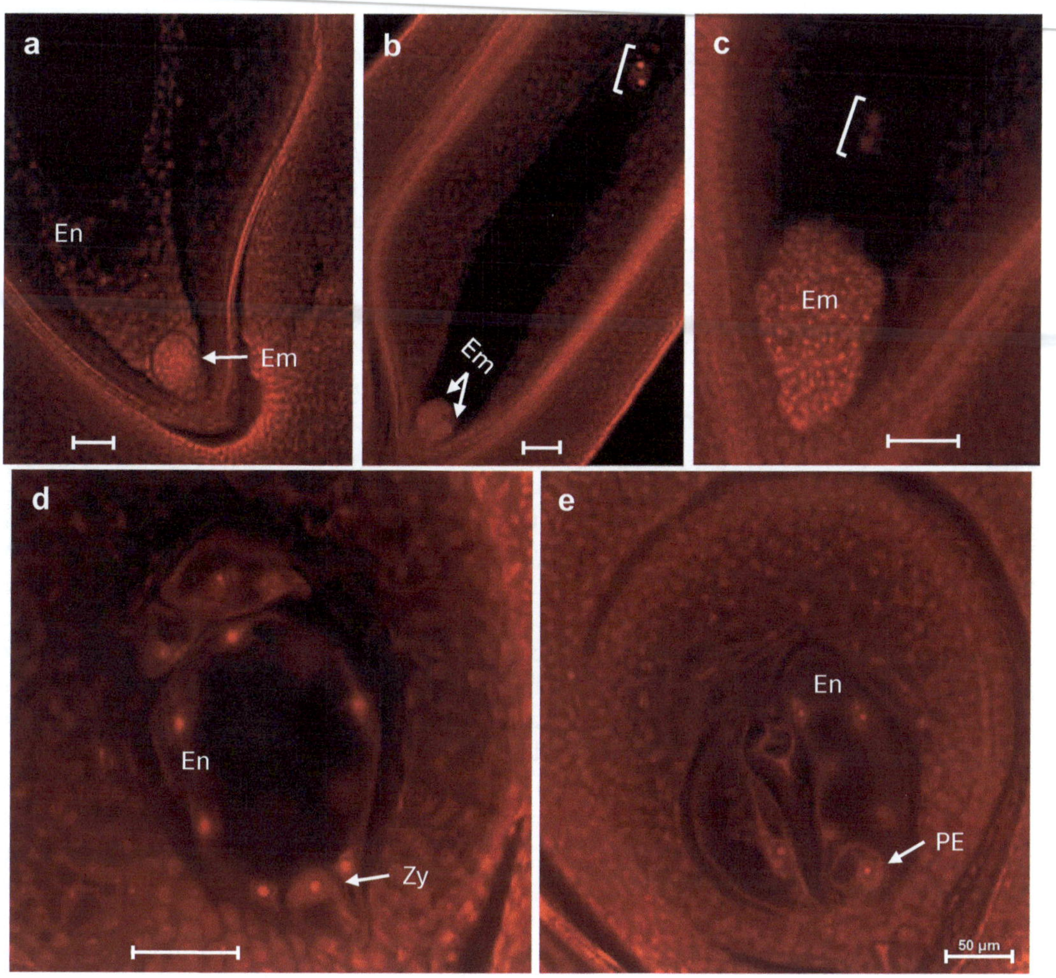

Fig. 2 Confocal images of fertilized rice and *C. ciliaris* ovaries. (**a**) Wild-type rice with developing embryo (Em) and endosperm (En). (**b**, **c**) A lack of central cell fertilization (central cell nuclei denoted with brackets) and parthenogenetic embryo development in 448-6E rice lines. Twin embryos are indicated in (**b**). (**d**) Endosperm development has progressed in a fertilized sexual *C. ciliaris* embryo sac prior to the first cell division of the zygote (Zy). Two antipodal cells are starred. (**e**) Endosperm development in one aposporous embryo sac of B12-9 prior to the first cell division of a parthenogenic egg (PE). To the left, a second aposporous embryo sac (functionality unknown) with two central cell nuclei but without a prominent egg cell. (Scale bars = 50 μM)

or where embryo development did not occur. Wild-type ovaries displayed both endosperm and characteristic wild-type embryo development (Fig. 2a). Within the 488-6E lines, ~60% of the ovaries (of 47 analyzed) had clearly defined unfused central cell nuclei and parthenogenic embryos consisting of variable cell number (Fig. 2b, c). Twin embryos were also identified (Fig. 2b). The remaining ovaries (~40%) displayed clear endosperm development with embryo development ranging from wild type in character to "off-types." This analysis could not distinguish in the 448-6E lines, the embryos that are haploid (derived through parthenogenesis) and embryos that are diploid (derived through sexual reproduction).

To compare endosperm development in transgenic rice to ovaries from open-pollinated *C. ciliaris,* B-2s (sexual) and B12-9 (apomictic) were analyzed with confocal imaging at ~24 h post pollination. Similar to the rice study, only healthy ovules within the fixed inflorescence (~80%) were analyzed. For B-2s, the ovaries analyzed all showed endosperm development. Approximately ~50% were still in the zygote stage (with fertilization of the egg cell based on endosperm development) (Fig. 2d), while the remaining 50% contained 2–4-celled embryos. For B12-9, the ovaries analyzed contained a range of 1–4 aposporous embryo sacs. Given the number of aposporous embryo sacs within a given ovule, multiple developmental outcomes could be observed within one ovule. However, ~75% of the B12-9 ovaries had at least one aposporous embryo sac displaying endosperm development, with the majority still displaying an enlarged parthenogenetic egg prior to the first cell division (Fig. 2e). While the analysis of development in rice and *C. ciliaris* took place on different days after open pollination, the data suggest that some of the reduction in fertility in one transgenic rice line is due to a lack of endosperm development, potentially due to parthenogenic development of the embryo in rice inhibiting pollen sperm nuclei from fusing with the central cell nuclei. Apomictic *C. ciliaris* may avoid this issue by postponing the first division of the parthenogenetic egg until after the central cell has been fertilized.

Transformations of rice, maize, and Arabidopsis with a variety of *BBM* genes derived from sexual plants expressed from an egg-specific promoter also induce parthenogenesis [69, 72–75]. Within monocots, seven sexual BBM genes have been used to induce parthenogenesis including two from rice (*OsBBM1*: LOC_Os11g19060 and *OsBBM4*: LOC_Os04g42570), two from maize (*ZmBBM1*: Zm00001eb247080 and *ZmBBM2*: Zm00001eb144510), and three from foxtail millet (*SiBBM1*: Seita.8G107100, *SiBBM2*: Seita.5G415800, and *SiBBM3*: Seita.1G232200). Including PsASGR-BBML, the BBML proteins range in size from 545 to 709 amino acids. Alignment of these proteins shows that the two AP2 domains are highly conserved between all BBMs, as are the *bbm-1*, *-2,* and *-3* domains identified previously as unique to the BBM lineage of AP2 proteins (Fig. 3) [76, 77]. With these data, we theorize that parthenogenesis in apomictic *Cenchrus* resulted from an expression pattern change and transposition/neofunctionalization of a sexual *BBM* gene to the apomixis locus.

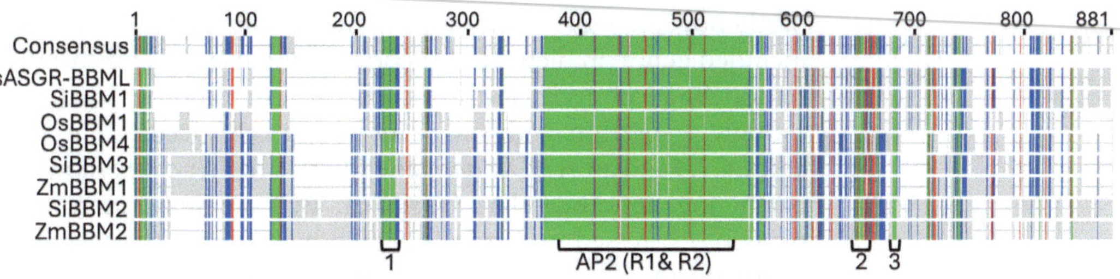

Fig. 3 Expresso protein alignment of monocot BBMs that promote parthenogenesis. BBMs show high conservation within the two AP2 domains and bbm-1, -2, and -3 domains. Colors are based on percentage of amino acid similarity shared between the BBM proteins [green (100%), red (80–99%), blue (60–79%), gray (<60%)]

8 Transcriptomic Data from Reproductive Tissues in *Cenchrus*

Gene expression studies in *Cenchrus* to identify transcripts unique to apospory are challenging due to a lack of assembled genomes and the complex transcriptional changes that take place during apomeiosis and parthenogenesis. Early research in apomictic *Cenchrus* included differential display analysis and suppression subtractive hybridization (SSH) between sexual and apomictic ovaries of *C. ciliaris* [78, 79] and differential screening by microarray analysis after SSH comparing sexual *P. glaucum* to a BC_1 *P. glaucum* x *P. orientale* line ($2n = 23$) displaying aposporous embryo sac formation [80]. Comparison of 454 Life Sciences-generated transcripts from the aposporous initial stage of ovaries from *P. squamulatum* and an apomictic pearl millet BC_8 line was used to identify transcripts in common between the two samples. Of the 61 transcripts identified as identical between the two samples, only one mapped to the ASGR region. The others mapped to the *P. squamulatum* ASGR carrier chromosome. The ASGR transcript contained a transposase domain [81]. A transcriptomic study using RNA from laser capture microdissected egg cells from emasculated sexual and apomictic *C. ciliaris* genotypes B-2s and B12-9 at day of anthesis and NextSeq PE150 sequencing was done to identify differentially expressed genes/pathways up- or downregulated during parthenogenetic embryo development prior to the first egg cell division. From this study, a set of high-quality sexual and apomictic *C. ciliaris* ovule transcriptomic data were generated [82]. As a genomic reference was not available for *C. ciliaris*, transcripts from both genotypes were de novo assembled using Trinity and 4625 transcripts were identified as differentially expressed at $\log2FC > 2$, false discovery rate <0.05 using DESeq2 with 2571 transcripts upregulated in the parthenogenic egg cell and 2054 transcripts upregulated in the sexual egg cell. Many transcription factors showed reproductive mode-specific expression such as *BBMs* (including *CcASGR-BBML*), *YABBY*, and *WOX* genes in parthenogenic eggs. Additional differentially expressed pathways included genes predicted to function during histone and chromatin remodeling, cell cycle, and auxin and calcium responses [83]. Interestingly, transcripts with similarity to egg cell-specific gene *EC1.3* and *EC1.4* and an *RKD2-like* gene showed higher expression in the parthenogenetic egg than in sexual. ECs are small cysteine-rich proteins expressed in egg cells, which help trigger sperm cell activation during double fertilization. In Arabidopsis, *EC1* RNA is not detectable after fertilization [84]. *RKD2* is a transcription factor, expressed in egg cells and promotes egg cell identity in Arabidopsis [85]. In Arabidopsis, another member of the *RKD* family, *AtRKD5*, has been shown to help repress expression of *AtBBM* prior to fertilization [86]. The potential role of *RKD5* in controlling transcription of *BBM* in apomictic egg cells of monocots will need further investigation.

Single-cell and spatial transcriptomics technologies could help potentially identify genes associated with spatially restricted phenotypes, such as apomeiosis, where changes within a nucellar cell lead to an aposporous initial and to help determine differences between an aposporous egg cell prior to parthenogenesis and sexually derived egg cells.

9 Introgression of Apomixis into *P. glaucum* (Pearl Millet)

While the creation of F_1 hybrids between *P. glaucum* and *P. squamulatum* was successful to produce mapping populations for apomixis, the transfer of apomixis into pearl millet and maintenance of a high level of penetrance necessary for use in an agricultural setting proved to be challenging. Apomictic BC_1 plants were derived by crossing sexual tetraploid *P. glaucum* with an apomictic double-cross hybrid of [*P. glaucum x P. purpureum*] and [*P. glaucum x P. squamulatum*], which helped to increase male fertility [87]. Additional back-crosses to tetraploid *P. glaucum* yielded a strongly apomictic, male fertile BC_3 line with 29 chromosomes, albeit with low seed set [88]. Back-crosses, focusing on lines with high aposporous embryo sac formation and male fertility, continued to the BC_8 stage. FISH/GISH studies analyzed numbers of *P. glaucum* and *P. squamulatum* chromosomes in the back-crossed lines. BC_3 contained 29 chromosomes, three of which were derived from the *P. squamulatum* genome. An apomictic BC_7 line 58 showing strong aposporous embryo sac formation (88–100%), high ASGR inheritance (98%), and reduced seed set (6.5 seed per cm head) contained 28 chromosomes with two derived from the *P. squamulatum* genome. Three BC_8 lines displaying lower aposporous embryo sac formation (63–94%) and more ovaries with sexual embryo sac formation, lower ASGR inheritance (73–85%), and higher seed set (31.5–46.2 seed per cm head) contained

28 chromosomes with one derived from the *P. squamulatum* genome [89, 90]. The loss of a strong apomictic phenotype with increased seed set remains the key issue for integrating apomixis from wild species into agronomically acceptable pearl millet by traditional breeding. Reduced seed set in introgression lines likely is due to endosperm imbalance caused by a decrease in the proportion of uninucleate central cells in aposporous embryo sacs of the back-cross compared with the apomictic parent [91]. A uninucleate central cell in an unreduced panicum-type embryo sac is the only condition that would maintain a 2:1 maternal: paternal genome ratio in the endosperm. The only surviving seeds in introgression lines showed triploid endosperm.

Another approach for introducing desired crop wild relative (CWR) traits, including apomixis, from *P. squamulatum* into a synthetic cultivar with improved domesticated traits is the apomixis-mediated genome addition (AMGA) strategy. This method generated F_1 hybrids through a *P. glaucum* ($2n = 4x = 28$) x *P. squamulatum* ($2n = 8x = 56$) cross. The hybrids were evaluated for multiple phenotypic traits, including fodder and yield, perenniality, and apomixis. This evaluation reduced the number of candidates that were given a two-year station trial containing other rangeland perennial grasses. Then a multilocation and multiyear evaluation was done for adaptability, fodder traits, and nutritional quality, leading to the release of one variety for commercial cultivation [92].

10 Seed Set in Natural Apomicts

Studies on seed set in apomictic species have shown wide variation. Pollen fertility and percent open-pollinated seed set from five inflorescences were analyzed for three apomictic species, *P. macrourum*, *P. polystachion*, and *P. squamulatum*. Pollen fertility, based on I_2KI staining, was above 60% with seed set at <1% for *P. macrourum*, 80% for *P. polystachion*, and 34% for *P. squamulatum* [93]. In *P. setaceum*, open-pollinated plants yielded 10% seed set, while seed set of self-pollinated apomictic *P. ciliare* was 18.6% [94]. Other accessions of apomictic *C. ciliaris* showed a wide range of seed set from 32% to 91% [9, 13]. *C. setigerus* showed ~60% pollen viability based on I_2KI staining and 65–81% seed set when selfed or cross-pollinated [95].

Variation in open pollinated seed set and germination rates for greenhouse-grown *Cenchrus* species was also identified (Table 2). Pollen fertility ranged from 34% to 76%, seed set ranged from 5% to 92%, and germination rates ranged from 16% to 52%. The higher rate of seed set in *C. ciliaris* apomict B12-9, as calculated in this experiment, is at least partially due to an increase in the number of ovules per fascicle in this genotype. The wide range of results is not

Table 2
Fertility of greenhouse-grown *Cenchrus*

	PI/Cultivar	Reproduction	Pollen fertility[a] (%)	Fascicle # threshed	Seed set[b] (%)	Seed germination[c] (%)
C. ciliaris	B2s	Sexual	74	1087	62	16
C. ciliaris	B12-9	Apomictic	55	990	92	36
C. massaicus	87A11508[d]	Apomictic	34	687	44	30
C. mezianus	PI 365021	Apomictic	76	871	5	20
C. orientalis	PI 315867	Apomictic	36	356	24	26
C. setaceus	PI 364994	Apomictic	34	566	36	52
C. squamulatus	PI 319196/ Ps26	Apomictic	40	776	15	36

[a]Based on Alexander staining
[b]Percent seed set based on the number seed threshed from counted fascicles removed from 5 self-pollinated inflorescences after maturation. The number of ovaries per fascicle can vary from 1 to 3 in the various species studied. *C. ciliaris* B12-9 had the largest number of ovaries per fascicle
[c]Based on 25 (*C. setaceus*), 40 (*C. massaicus*), and 50 (remaining species) seeds sown on soil. Seedlings were counted after 2 weeks of growth under lights at room temperature
[d]Seed originally obtained from Joseph Karl Wipff

unexpected due to the undomesticated, natural wildness of the apomictic species tested, and the complexity of seed maturation of apomicts. Given these results and previous work on apomixis in *Cenchrus* species, *C. ciliaris* is the best "model" system choice.

11 Transformation of *Cenchrus*

With the increasing use of *Cenchrus* species for food and forage and the advent of CRISPR/Cas9 technologies that allow the production of transgene-free, DNA-edited plants, having a robust transformation system for *Cenchrus* would be very beneficial. Currently, the ability to transform *Cenchrus* remains low. Transgenic lines from sexual *P. glaucum* have been created using particle gun bombardment of embryogenic callus induced via multiple tissue types [96, 97], through Agrobacterium transformation of shoot apices [98], and through a leaf transformation system using morphogenic genes, *WUSCHEL2* and *BBM* [99]. Transgenic sexual *C. pupureus* has been generated using bombardment of embryogenic calli generated from immature inflorescences [100].

Plant regeneration of five *C. ciliaris* genotypes using shoot induction has been reported [101] along with the regeneration of three *C. ciliaris* genotypes via somatic embryogenesis initiated from anther tissue [102]. An ~0.2% transformation success rate was reported from the bombardment of calli derived from immature inflorescences of a single genotype of *C. ciliaris* [103]. A preprint research article *Agrobacterium–mediated genetic engineering system for the C4 apomictic grass model species Cenchrus ciliaris L* by Shashi [https://doi.org/10.21203/rs.3.rs-3892291/v1] has developed a shoot induction protocol using shoot apices and *Agrobacterium*, which gives a transformation rate of ~1.4%.

12 Conclusion

The study of apospory in *Cenchrus* species has demonstrated that apospory is a complex developmental system that enables the generation of maternal offspring through seed. Apospory in *C. ciliaris* and *C. squamulatus* is controlled by the ASGR, a physically large, dominant genetic locus. The *ASGR-BBML* gene was identified in *Cenchrus* and shown to promote parthenogenesis. The discovery of *ASGR-BBML* gene function has become one basis for the use of *BBM* transgenes within sexual agronomic crops to promote parthenogenesis. Understanding the complex evolution of the ASGR in *Cenchrus* species will be facilitated by comparing whole-genome assemblies. Identifying the genetic component at the ASGR controlling apomeiosis is actively being pursued and can be validated by either creating (i.e., through CRISPR) or identifying a mutation that leads to the loss of apomeiosis in a natural apomict or through the creation of the apomeiosis phenotype within a sexual species. Additionally, research to increase the transformation efficiency of *Cenchrus* is needed to open new avenues of gene discovery for both apomixis and agronomic *Cenchrus* crops.

Funding

This study was funded in part by USDA-NIFA Award 2018-67013-27419.

References

1. Stebbins GL (1950) Apomixis in relation to variation and evolution. In: Stebbins GL (ed) Variation and evolution in plants. Columbia University Press, New York, pp 380–419
2. Asker S (1979) Progress in apomixis research. Hereditas 91(2):231–240
3. Asker S, Jerling L (1992) Apomixis in plants. CRC Press, Boca Raton
4. Gustafsson Å (1939) The interrelation of meiosis and mitosis: I. The mechanism of agamospermy. Hereditas 25(3):289–322
5. Delfini C, Aliscioni SS, Acosta JM, Pensiero JF, Zuloaga FO (2023) An update of the Cenchrinae (Poaceae, Panicoideae, Paniceae) and a new genus for the subtribe to clarify the dubious position of a species of Panicum L. Plants 12(4):749

6. Doust AN, Kellogg EA (2002) Inflorescence diversification in the panicoid "bristle grass" clade (Paniceae, Poaceae): evidence from molecular phylogenies and developmental morphology. Am J Bot 89(8):1203–1222
7. Chemisquy MA, Giussani LM, Scataglini MA, Kellogg EA, Morrone O (2010) Phylogenetic studies favour the unification of Pennisetum, Cenchrus and Odontelytrum (Poaceae): a combined nuclear, plastid and morphological analysis, and nomenclatural combinations in Cenchrus. Ann Bot 106(1):107–130
8. Gallaher TJ, Peterson PM, Soreng RJ, Zuloaga FO, Li DZ, Clark LG, Tyrrell CD, Welker CA, Kellogg EA, Teisher JK (2022) Grasses through space and time: an overview of the biogeographical and macroevolutionary history of Poaceae. J Syst Evol 60(3): 522–569
9. Hignight K, Bashaw E, Hussey M (1991) Cytological and morphological diversity of native apomictic buffelgrass, Pennisetum ciliare (L.) Link. Bot Gaz 152(2):214–218
10. Narayan K (1955) Cytogenetic studies of apomixis in Pennisetum: Pennisetum clandestinum Hochst. In: Proceedings/Indian Academy of Sciences, vol 5. Springer, pp 196–208
11. Narayan KN (1951) Cytogenetic studies of apomixis in Pennisetum. University of California, Davis
12. Fisher W, Bashaw E, Holt E (1954) Evidence for apomixis in Pennisetum ciliare and Cenchrus setigerus. Agron J 46:401–404
13. Snyder L, Hernandez AR, Warmke H (1955) The mechanism of apomixis in Pennisetum ciliare. Bot Gaz 116(3):209–221
14. Ozias-Akins P, Akiyama Y, Hanna WW (2003) Molecular characterization of the genomic region linked with apomixis in Pennisetum/Cenchrus. Funct Integr Genomics 3(3):94–104
15. Burson BL, Actkinson JM, Hussey MA, Jessup RW (2012) Ploidy determination of buffel grass accessions in the USDA National Plant Germplasm System collection by flow cytometry. S Afr J Bot 79:91–95
16. Kharrat-Souissi A, Siljak-Yakovlev S, Brown SC, Chaieb M (2013) Cytogeography of Cenchrus ciliaris (Poaceae) in Tunisia. Folia Geobot 48:95–113
17. Carloni-Jarrys EJ, Acosta-Bragato MC, Grunberg-Fraga K (2018) Nuclear DNA content and ploidy level in apomictic buffelgrass genotypes. Rev Fitotec Mex 41(1):23–29
18. Bashaw E, Hignight K (1990) Gene transfer in apomictic buffelgrass through fertilization of an unreduced egg. Crop Sci 30(3): 571–575
19. Burson BL, Hussey MA, Actkinson JM, Shafer GS (2002) Effect of pollination time on the frequency of 2n+ n fertilization in apomictic buffelgrass. Crop Sci 42(4):1075–1080
20. Bashaw E, Hussey M, Hignight K (1992) Hybridization (n+ n and 2n+ n) of facultative apomictic species in the Pennisetum agamic complex. Int J Plant Sci 153(3, Part 1): 466–470
21. Kharrat-Souissi A, Baumel A, Torre F, Juin M, Siljak-Yakovlev S, Roig A, Chaieb M (2011) New insights into the polyploid complex Cenchrus ciliaris L.(Poaceae) show its capacity for gene flow and recombination processes despite its apomictic nature. Aust J Bot 59(6):543–553
22. Conner JA, Gunawan G, Ozias-Akins P (2013) Recombination within the apospory specific genomic region leads to the uncoupling of apomixis components in Cenchrus ciliaris. Planta 238(1):51–63
23. Dujardin M, Hanna W (1986) An apomictic polyhaploid obtained from a pearl millet x Pennisetum squamulatum apomictic interspecific hybrid. Theor Appl Genet 72:33–36
24. Bashaw E (1962) Apomixis and sexuality in Buffelgrass. Crop Sci 2:412–415
25. Bashaw E (1969) Registration of buffelgrass germplasm. Crop Sci 9:396
26. Taliaferro C, Bashaw E (1966) Inheritance and control of obligate apomixis in breeding buffelgrass, Pennisetum ciliare 1. Crop Sci 6(5):473–476
27. Young B, Sherwood R, Bashaw E (1979) Cleared-pistil and thick-sectioning techniques for detecting aposporous apomixis in grasses. Can J Bot 57(15):1668–1672
28. Sherwood R, Berg C, Young B (1994) Inheritance of apospory in buffelgrass. Crop Sci 34(6):1490–1494
29. Kumar S, Chandra A, Gupta M, Shukla G (2010) Molecular and embryological analyses of rare sexual plant in buffelgrass (Cenchrus ciliaris L.). Range Manag Agrofor 31(1): 36–40
30. Kumar S, Chandra A, Gupta M, Shukla G (2012) IGFRI-CcSx-08/1 (IC0590889; INGR11062), an Anjan grass (Cenchrus ciliaris L.) germplasm with a rare obligate sexual plant. Indian J Plant Genet Resour 26(1): 99–100
31. Gupta S, Gupta M, Bhat B, Bhat V (2001) Status of apomixis and sexuality in four species of Cenchrus. J Plant Biol-New Delhi 28(2): 153–160

32. Quiroga M, Grunberg K, Ribotta A, Colomba EL, Carloni E, Tommasino E, Luna C, Griffa S (2013) Obtaining sexual genotypes for breeding in buffel grass. S Afr J Bot 88:118–123
33. Jessup R, Burson B, Burow G, Wang Y-W, Chang C, Li Z, Paterson A, Hussey M (2002) Disomic inheritance, suppressed recombination, and allelic interactions govern apospory in buffelgrass as revealed by genome mapping. Crop Sci 42(5):1688–1694
34. Gustine D, Sherwood R, Huff D (1997) Apospory-linked molecular markers in buffelgrass. Crop Sci 37(3):947–951
35. Dwivedi K, Bhat S, Bhat V, Bhat B, Gupta M (2007) Identification of a SCAR marker linked to apomixis in buffelgrass (Cenchrus ciliaris L.). Plant Sci 172(4):788–795
36. Bhan Yadav C, Anuj KS, Gupta M, Bhat V (2012) Genetic linkage maps of the chromosomal regions associated with apomictic and sexual modes of reproduction in Cenchrus ciliaris. Mol Breed 30:239–250
37. Kumar S, Saxena S (2016) Sequence characterized amplified regions linked with apomictic mode of reproduction in four different apomictic Cenchrus species. Mol Plant Breed 7:1–14
38. Kumar S, Saxena S, Gupta MG (2017) Marker-assisted screening of breeding populations of an apomictic grass Cenchrus ciliaris L. segregating for the mode of reproduction. Crop Breed Appl Biotechnol 17:10–17
39. Abdi S, Dwivedi A, Shashi, Kumar S, Bhat V (2019) Development of EST-SSR markers in Cenchrus ciliaris and their applicability in studying the genetic diversity and cross-species transferability. J Genet 98:1–16
40. Roche D, Cong P, Chen Z, Hanna WW, Gustine DL, Sherwood RT, Ozias-Akins P (1999) An apospory-specific genomic region is conserved between Buffelgrass (Cenchrus ciliaris L.) and Pennisetum squamulatum Fresen. Plant J 19(2):203–208
41. Dujardin M, Hanna WW (1989) Crossability of pearl millet with wild Pennisetum species. Crop Sci 29(1):77–80
42. Dujardin M, Hanna WW (1983) Apomictic and sexual pearl millet x Pennisetum squamulatum hybrids. J Hered 74(4):277–279
43. Akiyama Y, Goel S, Chen Z, Hanna WW, Ozias-Akins P (2006) Pennisetum squamulatum: is the predominant cytotype hexaploid or octaploid? J Hered 97(5):521–524
44. Read J, Bashaw E (1974) Intergeneric hybrid between pearl millet and buffelgrass 1. Crop Sci 14(3):401–403
45. Hanna WW (1979) Interspecific hybrids between pearl millet and fountaingrass. J Hered 70(6):425–427
46. Hanna WW, Dujardin M (1982) Apomictic interspecific hybrids between pearl millet and Pennisetum orientale LC Rich 1. Crop Sci 22(4):857–859
47. Ozias-Akins P, Roche D, Hanna WW (1998) Tight clustering and hemizygosity of apomixis-linked molecular markers in Pennisetum squamulatum implies genetic control of apospory by a divergent locus that may have no allelic form in sexual genotypes. Proc Natl Acad Sci 95(9):5127–5132
48. Huo H, Conner JA, Ozias-Akins P (2009) Genetic mapping of the apospory-specific genomic region in Pennisetum squamulatum using retrotransposon-based molecular markers. Theor Appl Genet 119(2):199–212
49. Lubbers E, Arthur L, Hanna W, Ozias-Akins P (1994) Molecular markers shared by diverse apomictic Pennisetum species. Theor Appl Genet 89:636–642
50. Ozias-Akins P, Lubbers E, Hanna W, McNay J (1993) Transmission of the apomictic mode of reproduction in Pennisetum: co-inheritance of the trait and molecular markers. TAG Theor Appl Genet 85(5):632–638
51. Goel S, Chen Z, Akiyama Y, Conner JA, Basu M, Gualtieri G, Hanna WW, Ozias-Akins P (2006) Comparative physical mapping of the apospory-specific genomic region in two apomictic grasses: Pennisetum squamulatum and Cenchrus ciliaris. Genetics 173(1):389–400
52. Conner J, Sapkota S, Deschamps S, Fengler K, Cigan M, Ozias-Akins P (2016) Targeted sequencing of a complex locus within a polyploid genome using reduced representation libraries. Mol Breed 36:1–11
53. Roche D, Conner J, Budiman M, Frisch D, Wing R, Hanna W, Ozias-Akins P (2002) Construction of BAC libraries from two apomictic grasses to study the microcolinearity of their apospory-specific genomic regions. Theor Appl Genet 104(5):804–812
54. Sapkota S, Conner JA, Hanna WW, Simon B, Fengler K, Deschamps S, Cigan M, Ozias-Akins P (2016) In silico and fluorescence in situ hybridization mapping reveals collinearity between the Pennisetum squamulatum apomixis carrier-chromosome and chromosome 2 of sorghum and foxtail millet. PLoS One 11(3):e0152411
55. Akiyama Y, Conner JA, Goel S, Morishige DT, Mullet JE, Hanna WW, Ozias-Akins P (2004) High-resolution physical mapping in

Pennisetum squamulatum reveals extensive chromosomal heteromorphism of the genomic region associated with apomixis. Plant Physiol 134(4):1733–1741

56. Kharrat-Souissi A, Siljak-Yakovlev S, Pustahija F, Chaieb M (2012) Physical mapping of 5S and 18S-5.8 S-26S RNA gene families in polyploid series of Cenchrus ciliaris Linnaeus, 1771 (Poaceae). Comp Cytogenet 6(3):273
57. Akiyama Y, Hanna WW, Ozias-Akins P (2005) High-resolution physical mapping reveals that the apospory-specific genomic region (ASGR) in Cenchrus ciliaris is located on a heterochromatic and hemizygous region of a single chromosome. Theor Appl Genet 111(6):1042–1051
58. Rathore P, Schwarzacher T, Heslop-Harrison-J, Bhat V, Tomaszewska P (2022) The repetitive DNA sequence landscape and DNA methylation in chromosomes of an apomictic tropical forage grass, Cenchrus ciliaris. Front Plant Sci 13:952968
59. Conner JA, Mookkan M, Huo H, Chae K, Ozias-Akins P (2015) A parthenogenesis gene of apomict origin elicits embryo formation from unfertilized eggs in a sexual plant. Proc Natl Acad Sci 112(36):11205–11210
60. Gualtieri G, Conner JA, Morishige DT, Moore LD, Mullet JE, Ozias-Akins P (2006) A segment of the apospory-specific genomic region is highly microsyntenic not only between the apomicts Pennisetum squamulatum and buffelgrass, but also with a rice chromosome 11 centromeric-proximal genomic region. Plant Physiol 140(3):963–971
61. Conner JA, Goel S, Gunawan G, Cordonnier-Pratt M-M, Johnson VE, Liang C, Wang H, Pratt LH, Mullet JE, DeBarry J (2008) Sequence analysis of bacterial artificial chromosome clones from the apospory-specific genomic region of Pennisetum and Cenchrus. Plant Physiol 147(3):1396–1411
62. Akiyama Y, Goel S, Conner JA, Hanna WW, Yamada-Akiyama H, Ozias-Akins P (2011) Evolution of the apomixis transmitting chromosome in Pennisetum. BMC Evol Biol 11(1):1
63. Boutilier K, Offringa R, Sharma VK, Kieft H, Ouellet T, Zhang L, Hattori J, Liu C-M, van Lammeren AA, Miki BL (2002) Ectopic expression of BABY BOOM triggers a conversion from vegetative to embryonic growth. Plant Cell 14(8):1737–1749
64. Conner JA, Podio M, Ozias-Akins P (2017) Haploid embryo production in rice and maize induced by PsASGR-BBML transgenes. Plant Reprod 30(1):41–52
65. Zhang Z, Conner J, Guo Y, Ozias-Akins P (2020) Haploidy in tobacco induced by PsASGR-BBML transgenes via parthenogenesis. Genes 11(9):1072
66. Sidhu GS, Conner JA, Ozias-Akins P (2022) Controlled induction of parthenogenesis in transgenic rice via post-translational activation of PsASGR-BBML. Front Plant Sci 13: 925467
67. Steffen JG, Kang IH, Macfarlane J, Drews GN (2007) Identification of genes expressed in the Arabidopsis female gametophyte. Plant J 51(2):281–292
68. Horstman A, Fukuoka H, Muino JM, Nitsch L, Guo C, Passarinho P, Sanchez-Perez G, Immink R, Angenent G, Boutilier K (2015) AIL and HDG proteins act antagonistically to control cell proliferation. Development 142(3):454–464
69. Khanday I, Skinner D, Yang B, Mercier R, Sundaresan V (2019) A male-expressed rice embryogenic trigger redirected for asexual propagation through seeds. Nature 565(7737):91–95
70. Sidhu GS (2021) Controlling the induction of parthenogenesis in transgenic rice via post-translational activation of PsAsgr-Bbml and determining the effect of egg cell ploidy level on penetrance of parthenogenesis. University of Georgia
71. Huang X, Peng X, Sun MX (2017) OsGCD1 is essential for rice fertility and required for embryo dorsal-ventral pattern formation and endosperm development. New Phytol 215(3):1039–1058
72. Skinner DJ, Mallari MD, Zafar K, Cho M-J, Sundaresan V (2023) Efficient parthenogenesis via egg cell expression of maize BABY BOOM 1: a step toward synthetic apomixis. Plant Physiol 193(4):2278–2281
73. Chen B, Maas L, Figueiredo D, Zhong Y, Reis R, Li M, Horstman A, Riksen T, Weemen M, Liu H (2022) BABY BOOM regulates early embryo and endosperm development. Proc Natl Acad Sci 119(25): e2201761119
74. Chahal LS, Conner JA, Ozias-Akins P (2022) Phylogenetically distant BABY BOOM genes from Setaria italica induce parthenogenesis in rice. Front Plant Sci 13:863908
75. Qi X, Gao H, Lv R, Mao W, Zhu J, Liu C, Mao L, Li X, Xie C (2023) CRISPR/dCas-mediated gene activation toolkit development and its application for parthenogenesis induction in maize. Plant Commun 4(2):100449
76. El Ouakfaoui S, Schnell J, Abdeen A, Colville A, Labbé H, Han S, Baum B,

Laberge S, Miki B (2010) Control of somatic embryogenesis and embryo development by AP2 transcription factors. Plant Mol Biol 74(4–5):313–326

77. Armougom F, Moretti S, Poirot O, Audic S, Dumas P, Schaeli B, Keduas V, Notredame C (2006) Expresso: automatic incorporation of structural information in multiple sequence alignments using 3D-Coffee. Nucleic Acids Res 34(suppl_2):W604–W608
78. Vielle-Calzada J-P, Nuccio ML, Budiman MA, Thomas TL, Burson BL, Hussey MA, Wing RA (1996) Comparative gene expression in sexual and apomictic ovaries of Pennisetum ciliare (L.) Link. Plant Mol Biol 32(6): 1085–1092
79. Singh M, Burson BL, Finlayson SA (2007) Isolation of candidate genes for apomictic development in buffelgrass (Pennisetum ciliare). Plant Mol Biol 64:673–682
80. Sahu PP, Gupta S, Malaviya D, Roy AK, Kaushal P, Prasad M (2012) Transcriptome analysis of differentially expressed genes during embryo sac development in apomeiotic non-parthenogenetic interspecific hybrid of Pennisetum glaucum. Mol Biotechnol 51: 262–271
81. Zeng Y, Conner J, Ozias-Akins P (2011) Identification of ovule transcripts from the Apospory-Specific Genomic Region (ASGR)-carrier chromosome. BMC Genomics 12(1):1
82. Ke Y (2019) Novel insights into the parthenogenetic potential of natural apomictic buffelgrass (Cenchrus ciliaris) from the perspective of egg apparatus transcriptome. University of Georgia
83. Ke Y, Podio M, Conner J, Ozias-Akins P (2021) Single-cell transcriptome profiling of buffelgrass (Cenchrus ciliaris) eggs unveils apomictic parthenogenesis signatures. Sci Rep 11(1):1–17
84. Sprunck S, Rademacher S, Vogler F, Gheyselinck J, Grossniklaus U, Dresselhaus T (2012) Egg cell–secreted EC1 triggers sperm cell activation during double fertilization. Science 338(6110):1093–1097
85. Kőszegi D, Johnston AJ, Rutten T, Czihal A, Altschmied L, Kumlehn J, Wüst SE, Kirioukhova O, Gheyselinck J, Grossniklaus U (2011) Members of the RKD transcription factor family induce an egg cell-like gene expression program. Plant J 67(2):280–291
86. Liu Q, Han D, Cheng D, Chen J, Tian S, Wang J, Liu M, Yuan L (2024) AtRKD5 inhibits the parthenogenic potential mediated by AtBBM. J Integr Plant Biol 66:1517
87. Dujardin M, Hanna W (1984) Cytogenetics of double cross hybrids between Pennisetum americanum—P. purpureum amphiploids and P. americanum X Pennisetum squamulatum interspecific hybrids. Theor Appl Genet 69: 97–100
88. Dujardin M, Hanna W (1989) Developing apomictic pearl millet-characterization of a BC3 plant. J Genet Breed 43:145–151
89. Goel S, Chen Z, Conner JA, Akiyama Y, Hanna WW, Ozias-Akins P (2003) Delineation by fluorescence in situ hybridization of a single hemizygous chromosomal region associated with aposporous embryo sac formation in Pennisetum squamulatum and Cenchrus ciliaris. Genetics 163(3):1069–1082
90. Singh M, Conner J, Zeng YJ, Hanna W, Johnson V, Ozias-Akins P (2010) Characterization of apomictic BC7 and BC8 pearl millet: meiotic chromosome behavior and construction of an ASGR-carrier chromosome-specific library. Crop Sci 50(3):892–902
91. Morgan R, Ozias-Akins P, Hanna W (1998) Seed set in an apomictic BC3 pearl millet. Int J Plant Sci 159:89–97
92. Roy A, Chakraborti M, Radhakrishna A, Dwivedi K, Srivastava M, Saxena S, Paul S, Khare A, Malaviya D, Kaushal P (2022) Alien genome mobilization and fixation utilizing an apomixis mediated genome addition (AMGA) strategy in Pennisetum to improve domestication traits of P. squamulatum. Theor Appl Genet 135(7):2555–2575
93. Dujardin M, Hanna W (1984) Microsporogenesis, reproductive behavior, and fertility in five Pennisetum species. Theor Appl Genet 67:197–201
94. Simpson C, Bashaw E (1969) Cytology and reproductive characteristics in Pennisetum setaceum. Am J Bot 56(1):31–36
95. Read JC, Bashaw E (1969) Cytotaxonomic relationships and the role of apomixis in speciation in buffelgrass and birdwoodgrass 1. Crop Sci 9(6):805–806
96. Goldman J, Hanna W, Fleming G, Ozias-Akins P (2003) Fertile transgenic pearl millet [Pennisetum glaucum (L.) R. Br.] plants recovered through microprojectile bombardment and phosphinothricin selection of apical meristem-, inflorescence-, and immature embryo-derived embryogenic tissues. Plant Cell Rep 21:999–1009
97. O'Kennedy MM, Burger J, Botha F (2004) Pearl millet transformation system using the positive selectable marker gene phosphomannose isomerase. Plant Cell Rep 22:684–690

98. Jha P, Shashi, Rustagi A, Agnihotri PK, Kulkarni VM, Bhat V (2011) Efficient Agrobacterium-mediated transformation of Pennisetum glaucum (L.) R. Br. using shoot apices as explant source. Plant Cell Tissue Organ Cult 107:501–512

99. Wang N, Ryan L, Sardesai N, Wu E, Lenderts B, Lowe K, Che P, Anand A, Worden A, Van Dyk D (2023) Leaf transformation for efficient random integration and targeted genome modification in maize and sorghum. Nat Plants 9(2):255–270

100. Faleiro F, Kim J, Parikh A, Altpeter F (2021) Generation of transgenic Napier grass (Cenchrus purpureus Schum.) plants by biolistic gene transfer of a minimal expression cassette. Trop Plant Biol 14:371–380

101. Kumar S, Bhat V (2012) High-frequency direct plant regeneration via multiple shoot induction in the apomictic forage grass Cenchrus ciliaris L. In Vitro Cell Dev Biol Plant 48:241–248

102. Carloni E, Ribotta A, Lopez Colomba E, Griffa S, Quiroga M, Tommasino E, Grunberg K (2014) Somatic embryogenesis from in vitro anther culture of apomictic buffel grass genotypes and analysis of regenerated plants using flow cytometry. Plant Cell Tissue Organ Cult 117:311–322

103. Laishram SD, Goyal S, Shashi, Kulkarni VM, Kumar S, Bhat V (2020) Assessment of biolistic and agrobacterium-mediated genetic transformation methods in Cenchrus ciliaris. Nucleus 63:303–312

Part VI

Engineering Apomixis

Chapter 17

Synthetic Apomeiosis in Tomato via the Mutagenesis of *SlSPO11-1*, *SlREC8*, and *SlTAM* to Engineer *Mitosis Instead of Meiosis*

Yazhong Wang, Iris E. Zahn, Bas Heidemann, Roven Rommel Fuentes, and Charles J. Underwood

Abstract

Heterosis, or hybrid vigor, is a crucial driver of yield and robustness in modern hybrid breeding. However, F1 hybrid plants give rise to genetically variable F2 offspring due to the segregation of genetic material in male and female meiosis. Bioengineered clonal gametes, containing identical genetic material as the parental plant, offer an innovative route to maintain genetic heterozygosity as a component of synthetic apomixis, or as a tool in breeding to carry out polyploid genome design. In this protocol, we present methods to generate clonal gametes in hybrid tomato plants by inducing synthetic apomeiosis through the *Mitosis instead of Meiosis* (*MiMe*) system. This protocol can be used to implement *MiMe* in tomato plants of choice, and describes the methods for comprehensive assessment of plant fertility and ploidy.

Key words Meiosis, Tomato, Ploidy, Apomeiosis, *Mitosis instead of Meiosis* (*MiMe*), Apomixis, Clonal gametes, Plant hybridization

1 Introduction

The tomato (*Solanum lycopersicum*), once referred to as the "wolf peach," is an important food crop and genetic model system. As one of the most widely consumed vegetables globally, tomatoes are economically important and are consumed in fresh and processed forms. The tomato genome is relatively small (around 900 Mbp) and more than 1000 cultivated and wild accessions have been sequenced, which has been an important basis for understanding the genetic basis of disease resistance, fruit quality, and yield [1, 2].

Hybrid seed production is a major challenge in crop breeding, especially in crops, such as tomato, which require manual emasculation, and pollination is required to obtain hybrids. Hybrid seeds and plants offer advantages in terms of doubled genetic map for gene introgression, higher yields, and increased robustness.

Tatyana Radoeva and Peter J. van Dijk (eds.), *Apomixis: Methods and Protocols*, Methods in Molecular Biology, vol. 2987, https://doi.org/10.1007/978-1-0716-5001-1_17,

However, the cost and complexity of producing hybrid seeds are substantial, as it requires controlled pollination and careful management to prevent contamination from self-pollination. Meanwhile, hybrid vigor cannot pass to subsequent generations due to meiotic recombination and random segregation of homologous chromosomes during sexual reproduction [3–5]. This means young plant producers and farmers must purchase new hybrid seeds each season to maintain uniform phenotypes and high yields. The need to ensure hybrid seed purity and seed quality further complicate the hybrid seed production process, making it an expensive and labor-intensive endeavor.

The application of clonal gametes in synthetic apomixis—clonal seed production in crops—offers an innovative solution to some of the challenges associated with hybrid seed production [6, 7]. Clonal eggs combined with embryogenesis without fertilization (parthenogenesis) allow for the production of offspring that are genetically identical to the parent plant [6, 7]. This enables the fixation of desirable traits, such as hybrid vigor, without the need for annual hybrid seed production. In addition, clonal seed production has the potential to significantly reduce the costs of breeding and seed production, as well as to create crops with uniform traits, which is crucial for large-scale commercial farming.

Abolishing recombination and genome reduction to generate clonal gametes involves modifying different elements of meiosis and it is therefore not straightforward to engineer in every plant species [3, 5]. Clonal egg and sperm cells have been engineered by modifying three crucial steps in meiosis through a process known as *Mitosis instead of Meiosis* (*MiMe*) [8]. *MiMe* is a process in which first the homologous chromosome pairing and recombination is abolished, second the sister chromatids separate in advance during meiosis I, and third, the second cell division is skipped to mimic a mitotic-like cell division [8]. *MiMe* was first established in *Arabidopsis* [9] and later rice [9], and has recently also been accomplished in tomato [10]. Tomato is the first dicot crop species in which the bioengineering of clonal gametes has been demonstrated [10]. Apomeiosis through *MiMe* is a promising method in plant breeding, as it results in clonal gametes and supports the fixing of favorable genotypes over generations. *MiMe* could be especially useful to maintain heterozygous genotypes displaying hybrid vigor.

This protocol describes how to generate tomato hybrid *MiMe* plants producing clonal gametes and how to assess the genomic composition of their progeny. In short, this involves the generation of triple knock-out mutants via mutation of *SPORULATION 11-1* (*SlSPO11-1*) to remove meiotic recombination initiation, mutation of *RECOMBINATION 8* (*SlREC8*) to ablate sister chromatid cohesion in meiosis I, and mutation of *TARDY ASYNCHRONOUS MEIOSIS* (*SlTAM*) to avoid the second meiotic division. Gene mutagenesis is carried out here using a Clustered Regularly

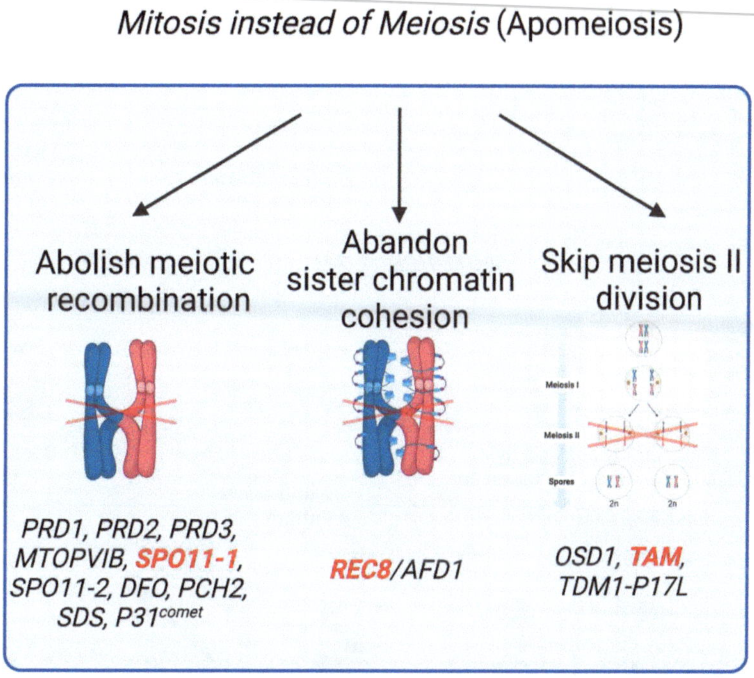

Fig. 1 A schematic display of processes and genes that must be modified to establish a *Mitosis instead of Meiosis* (*MiMe*) system. *MiMe* leads to unreduced clonal male and female gametes in plants. It consists of modifying three crucial pathways (removing all three of (1) meiotic recombination initiation, (2) sister chromatid cohesion during meiosis I, and (3) the second meiotic division). Potential gene targets to mutate and establish *MiMe* are listed at the bottom of each process. Those genes highlighted in red are the targets for the tomato *MiMe* system focused on in this protocol. (Created with Biorender.com)

Interspaced Short Palindromic Repeats/CRISPR-associated protein 9 (CRISPR/Cas9) genome editing system, but could be equally achieved using other genome editing systems or conventional approaches (e.g., by chemical mutagenesis). *MiMe* triple mutants are analyzed by visualizing chromosome behaviour by meiotic cytology, pollen viability and size phenotyping, and offspring ploidy determination through flow cytometry. Finally, whole-genome sequencing is applied to check for the presence/absence of meiotic crossovers in the complete genomes of offspring plants.

So far, tomato *MiMe* has been used to generate clonal gametes (Fig. 1), which produce tetraploid offspring after fertilization of the diploid egg cell by the diploid sperm cell. In the future, tomato *MiMe* may be used to engineer clonal reproduction/synthetic apomixis [6, 7, 10, 11], which combines apomeiosis with parthenogenesis (embryogenesis without fertilization of the egg cell) or haploid induction/chromosome elimination, resulting in diploid

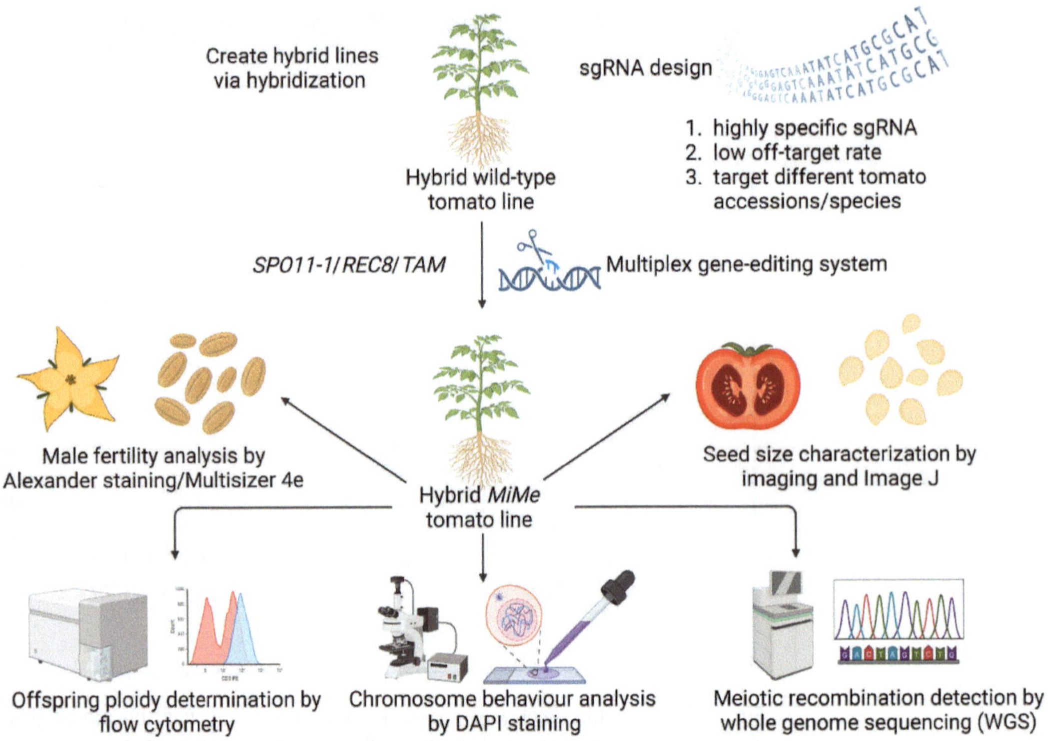

Fig. 2 A schematic of the design, implementation, and downstream analysis related to *Mitosis instead of Meiosis* (*MiMe*) in hybrid tomato. First, specific sgRNAs targeting *SPO11-1*, *REC8*, and *TAM* were designed by checking the gene sequences in different tomato accessions/species. The predicted on-target cutting efficiency and off-target rate of sgRNAs was checked. Through a multiplex gene-editing system, triple gene *MiMe* mutants can be obtained in hybrid tomato plants, validated via genotyping and then analyzed through several phenotyping approaches. (Created with Biorender.com)

progeny that is genetically identical to the mother plant. In this protocol, we describe the detailed approach for setting up *MiMe* in hybrid tomato plants and the methods for the downstream analysis (Figs. 2, 3, 4, and 5).

2 Materials

2.1 Plant Material and Growth Conditions

Tomato plants (*Solanum lycopersicum*) are grown under long-day conditions of 16-h light (25 °C) and 8-h dark (20 °C). Hybrid plants should be used to facilitate analysis of presence/absence of meiotic crossovers by analysis of genetic marker segregation in offspring plants. In the protocol described here, Moneyberg-TMV x Micro-Tom hybrid plants are used.

2.2 Seed Germination

1. 2 mL Eppendorf tubes.
2. Sterile water.

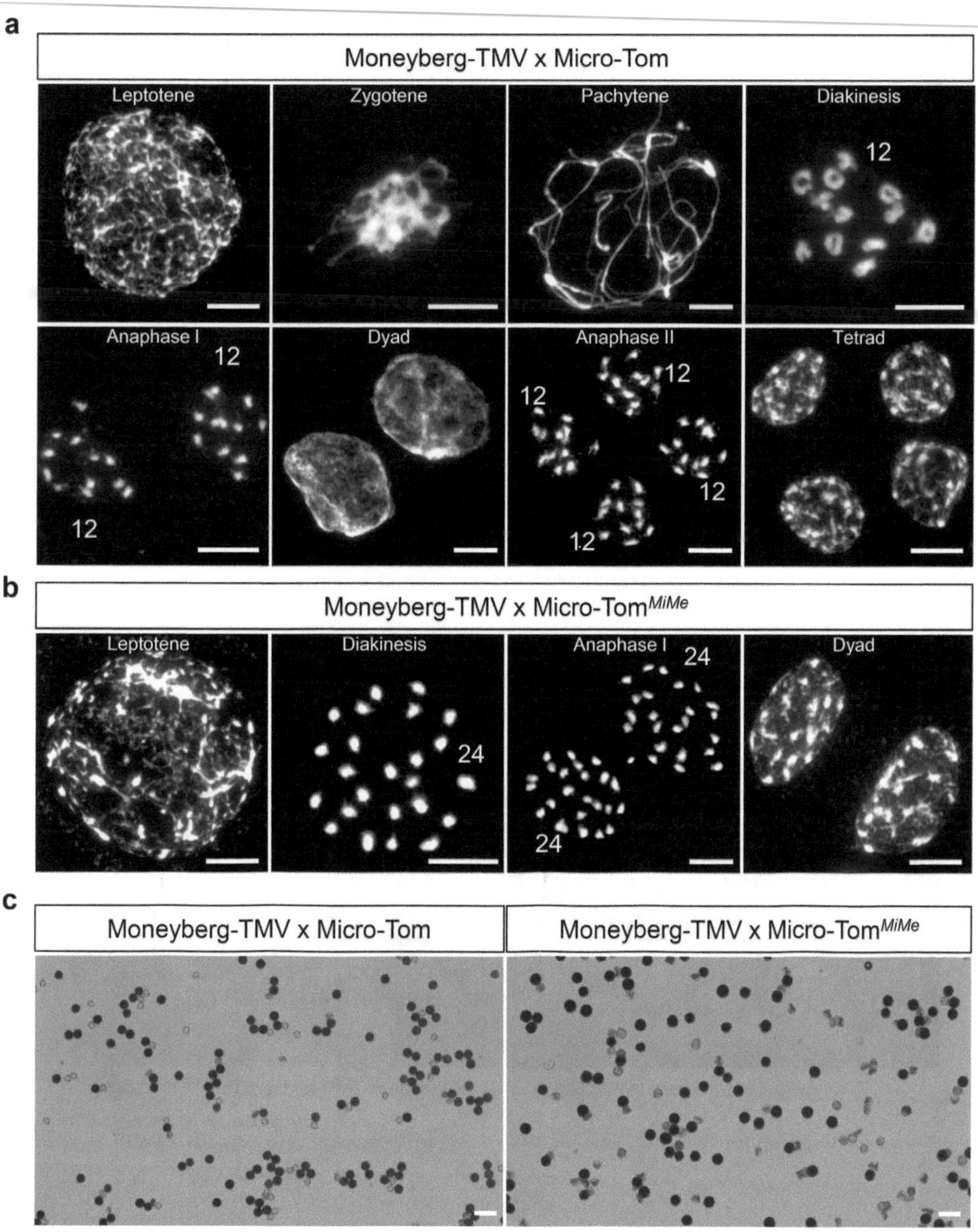

Fig. 3 The chromosome behavior of male meiocytes and pollen viability from control and *MiMe* mutant plants. (**a**) Meiotic atlas of male meiocytes via DAPI staining from wild-type hybrid Moneyberg-TMV × Micro-Tom. Scale bars = 10 μm. (**b**) Meiotic atlas of male meiocytes via DAPI staining from hybrid Moneyberg-TMV × Micro-Tom^{MiMe}. Scale bars = 10 μm. (**c**) Alexander staining results from hybrid Moneyberg-TMV × Micro-Tom and hybrid Moneyberg-TMV × Micro-Tom^{MiMe} mutants. Scale bars = 50 μm

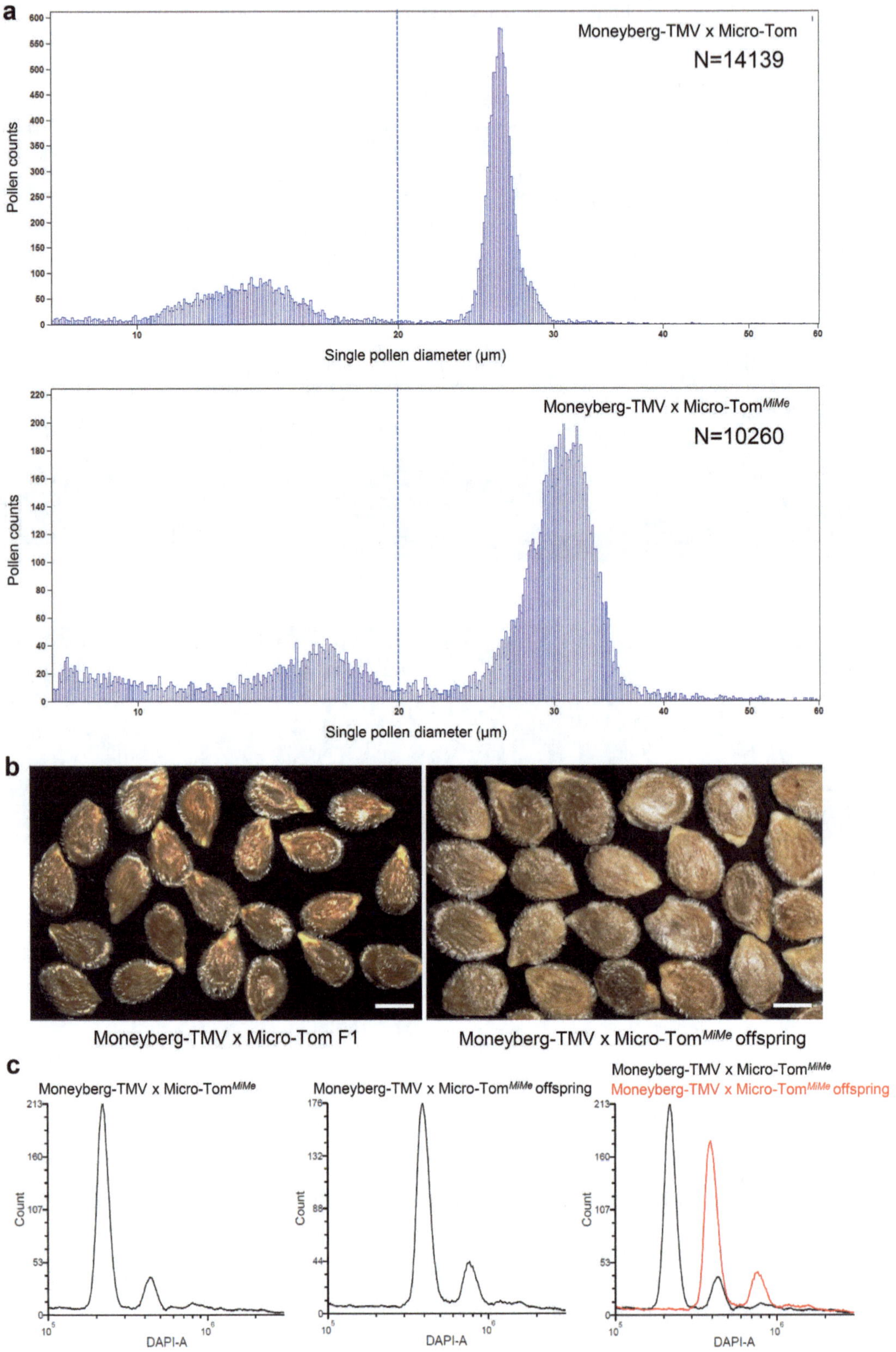

Fig. 4 Phenotyping of control, hybrid *MiMe* mutant plants and their offspring. (**a**) High-throughput single pollen diameter distribution in hybrid Moneyberg-TMV × Micro-Tom (control) and hybrid Moneyberg-TMV × Micro-Tom^{MiMe} mutants. The blue dotted line at 20 μm represents a dividing line between normal and abnormal

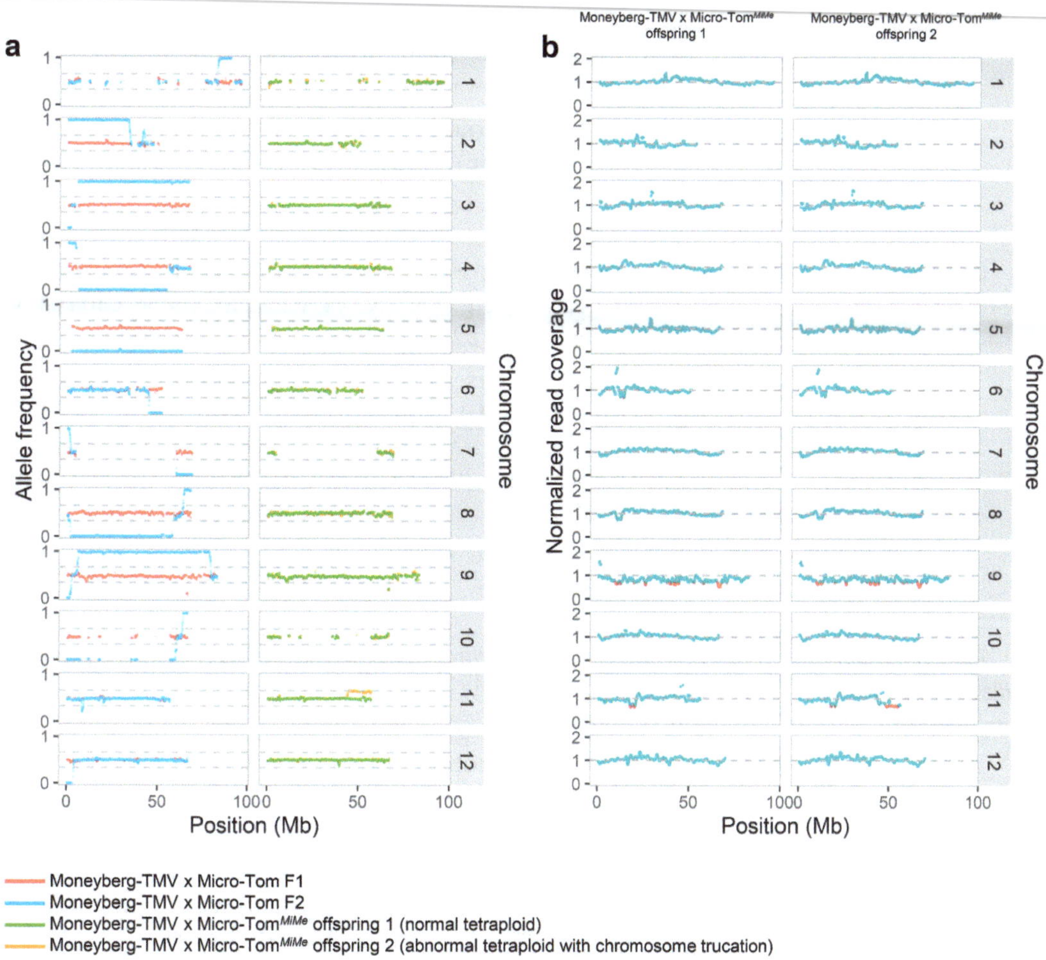

Fig. 5 Whole genome sequencing analysis of allele frequency and normalized read coverage in control hybrid, control F2, and hybrid *MiMe* mutant offspring plants. (**a**) Analysis of allele frequency in Moneyberg-TMV × Micro-Tom F1 (in red), Moneyberg-TMV × Micro-Tom F2 offspring (in blue), Moneyberg-TMV × Micro-Tom^{MiMe} offspring 1 (in green), and Moneyberg-TMV × Micro-Tom^{MiMe} offspring 2 (in orange). On the *y*-axis, 0 means homozygous Moneyberg-TMV (MbTMV), 1 means Micro-Tom (MT) and 0.5 means heterozygous between MB-TMV and MT. The *MiMe* plants maintain the heterozygosity (green and orange are similar to red), while in the F2 plant (blue) genetic segregation of alleles is observed (due to random segregation and meiotic recombination). Moneyberg-TMV × Micro-Tom^{MiMe} offspring 2 (in orange) has a region at the end of chromosome 11 with 0.66 allele frequency indicating a potential loss of a small part of a MbTMV chromosome 11. (**b**) Normalized read coverage of two Moneyberg-TMV × Micro-Tom^{MiMe} offspring among the whole chromosomes. There is chromosome truncation at the end of chromsome 11 (region in red), in Moneyberg-TMV × Micro-Tom^{MiMe} offspring 2

Fig. 4 (continued) pollen. (**b**) Seeds collected after selfing of hybrid Moneyberg-TMV × Micro-Tom (control) and hybrid Moneyberg-TMV × Micro-Tom^{MiMe} mutants. Scale bars = 3 mm. (**c**) Flow cytometry histogram of hybrid Moneyberg-TMV × Micro-Tom^{MiMe} (left), the offspring (middle) and the merged result (right)

3. 2.7% NaClO: dilute the 12% Sodium hypochlorite solution (NaClO) (e.g., Roth #9062.3) to a working solution (2.7% NaClO) with sterile water.
4. Saturated Na_3PO_4 solution: dissolve 28 g $Na_3PO_4 \cdot 12H_2O$ in 100 mL (final volume) water.
5. Glass vessels and leucopore tape.

2.3 Cloning of CRISPR-Cas9 Plasmids

1. A vector expressing *35S::AtCas9* with cloning sites to add multiple sgRNAs, e.g., 35S:Csy4-P2A-AtCas9 + CmYLCV: gRNAs (pDIRECT_22C, Addgene Plasmid #91135).
2. Enzymes and buffers for Golden Gate cloning: BsaI-HFv2 (e.g., NEB #R3733L); SapI (e.g., NEB #R0569L); T4 DNA Ligase (e.g., NEB #M0202L); Adenosine 5′-Triphosphate (ATP) (e.g., NEB #P0756S9) and rCutSmart™ Buffer (e.g., NEB # B6004S).
3. Oligos encoding the sgRNA sequences with overhangs for Golden Gate cloning.

2.4 Agrobacterium-Mediated Tomato Transformation

1. *Agrobacterium tumefaciens*, strain GV3101, carrying the final CRISPR-Cas9 genetic construct.
2. YEB Agrobacterium growth medium and antibiotics specific for the GV3101 Agrobacterium strain and CRISPR construct used (working solutions: 50 μg/mL kanamycin, 20 μg/mL rifampicin and 20 μg/mL gentamicin).
3. Basic MS medium: 4.3 g/L MS salt (incl. vitamins) (e.g., M0222.0050, Duchefa Biochemie), 100 mg/L myo-inositol (e.g., I0609.0100, Duchefa Biochemie), 30 g/L sucrose (e.g., S0809.5000, Duchefa Biochemie), 7 g/L phyto agar (e.g., P1003.1000, Duchefa Biochemie) (pH 5.9 adjusted with 1 M KOH). Autoclaved (*see* **Note 1**).
4. MS-0 medium (seed germination): 4.3 g/L MS salt (incl. vitamins), 100 mg/L myo-inositol, 20 g/L sucrose, 7 g/L phyto agar (pH 5.9 adjusted with 1 M KOH). Autoclaved (*see* **Note 2**).
5. MS-1 medium (transformation and callus induction): Basic MS medium (as above point 3). After autoclaving, add zeatin riboside (e.g., Z0937.0050, Duchefa Biochemie), acetosyringone (e.g., D134406, Sigma), and IAA (I2886, Sigma) to a final concentration of 2 mg/L, 73 mg/L, and 0.1 mg/L, respectively.
6. MS-2 medium (callus differentiation and shoot initiation): Basic MS medium (as above point 3). After autoclaving, add zeatin riboside, carbenicillin (e.g., C0109.0005, Duchefa Biochemie), and kanamycin (e.g., K0126.0001, Duchefa Biochemie) to a final concentration of 2 mg/L, 500 mg/L, and 100 mg/L, respectively (*see* **Note 3**).

7. MS-3 medium (root development): MS-0 medium (as above point 2). After autoclaving, add carbenicillin and kanamycin to a final concentration of 250 mg/L and 50 mg/L, respectively.
8. Vented Petri dishes (9-cm diameter) containing MS-1 and MS-2 medium and tissue culture containers (e.g., jars or boxes) containing MS-0 and MS-3 medium.
9. Stacks of filter paper (10 × 10 cm) wrapped in tin foil and autoclaved.
10. 0.7% phyto agar. Autoclaved.
11. Flame-sterilized scalpels and/or scissors, and a flat-beaked tweezer.
12. Leucopore and sterile glass plates.
13. 50-mL Falcon tubes and 1-mL sterile pipette tips.

2.5 Chromosome Spreading

1. Carnoy's fixation buffer: 3:1 (v/v) absolute ethanol: glacial acetic acid.
2. 60% acetic acid.
3. 10 mM of citric buffer (pH 4.6): mix 5 mM of citric acid (e.g., Sigma, C0759) with 5 mM of sodium citrate dihydrate (e.g., Sigma, W302600) and adjust the pH to 4.6 with hydrochloric acid or sodium hydroxide.
4. 3x digestion enzyme mix: 0.1% (w/v) cellulase Onozuka R-10 (e.g., Sigma, C8883), 0.1% (w/v) cytohelicase from *Helix pomatia* (e.g., Sigma, C8274), and 0.1% (w/v) pectolyase from *Aspergillus japonicus* (e.g., Sigma, P5936) in 10 mM of citric buffer (pH 4.6). This 3x stock solution can be stored at –20 °C. Before use, dilute the 3x stock solution (containing 0.3% (w/v) of each enzyme) into a 1x digestion enzyme mix using 10 mM of citric buffer (pH 4.6).
5. 2 μg/mL DAPI staining solution: 40 μL 100 μg/mL DAPI stock (e.g., DAPI powder, Roche 10236276001, dissolved in water), in an antifade mounting medium (e.g., 1960 μL VECTASHIELD® Antifade Mounting Medium (H-1000-10)).

2.6 Pollen Analysis

1. 100 mL Alexander staining solution: 10 mL 96% ethanol, 1 mL 1% malachite green in 96% ethanol (10 mg Malachite green), 50 mL distilled water, 25 mL glycerol, 5 mL 1% acid fuchsin in distilled water (50 mg acid fuchsin), and 0.5 mL 1% Orange G solution in distilled water (5 mg Orange G), 4 mL glacial acetic acid, and 4.5 mL distilled water (stored in the dark).
2. The alternative commercial Alexander staining solution: Morphisto, Labochemikalien & Histologieservice 13441.00250.
3. Mulitisizer 4e instrument with 100-μm Aperture Tube.
4. ISOTON II Diluent particle characterization for Multisizer 4e.

2.7 Flow Cytometry

1. Galbraith's buffer: 45 mM MgCl2 (4.574 g MgCl2·6H2O, e.g., Sigma, M2670), 30 mM sodium citrate (4.412 g sodium citrate, e.g., Sigma, W302600), 20 mM MOPS (2.093 g MOPS, e.g., Sigma, M1254), 0.1% (v/v) Triton X-100 (0.5 mL Triton X-100, e.g., T9284, Sigma), pH 7.0 (Final volume: 500 mL).
2. QC beads: Three drops of mixed original QC beads (CytoFLEX Daily QC fluorospheres, B53230) into 1 mL distilled water (vortex beads in advance).
3. DAPI, 100 μg/mL stock solution (e.g., 6335.1, Roth).
4. CellTrics disposable filters-green 30 μm (e.g., 04-004-2326, Sysmex).
5. Sharp razors (e.g., Wilkinson Sword) for chopping.
6. CytoFLEX Sheath Fluid, B51503.
7. FlowClean Cleaning Agent, A64669.

2.8 Equipment

1. Tomato transformation: Laminar flow cabinet and growth chamber.
2. Particle size detector (for pollen size): the Mulitisizer 4e instrument with 100 μm Aperture Tube. The Multisizer 4e Coulter Counter (Beckman Coulter, Germany).
3. Anthers separation: Stereo Microscope (e.g., Stemi 508, Zeiss).
4. DIVAC (Diaphram Vacuum Pumps) machine: used for vacuum treatment of anthers.
5. Ploidy checking: CytoFLEX LX Flow Cytometer (Beckman Coulter, Germany).
6. Seeds imaging: digital stereomicroscope (Fig. 4b), e.g., Leica M205 FA (Leica Microsystems).
7. Chromosome imaging: Zeiss Axio Imager Z2 upright microscope.
8. Pollen imaging: Zeiss Axioplan 2 imaging fluorescence microscope.

3 Methods

3.1 Generation of the MiMe Genotype by CRISPR-Cas9 Gene Editing

1. Identify the tomato homologs of the Arabidopsis genes *SPO11-1*, *REC8*, and *TAM* based on the Arabidopsis gene information (https://www.arabidopsis.org/), using the SolGenomics Network database (https://solgenomics.net/) or a comparable database (*see* **Note 4**).
2. Perform gene annotations according to the gene and protein information against NCBI database (https://blast.ncbi.nlm.nih.gov/Blast.cgi).

3. Design specific sgRNAs to knock-out *SlSPO11-1*, *SlREC8*, and *SlTAM* using CRISPR-P (http://crispr.hzau.edu.cn/CRISPR2/) (*see* **Note 5**).
4. Verify the conservation of the sgRNA target sequence among different tomato accessions/species and confirm they can target different hybrid lines (https://solgenomics.net/projects/tomato13/).
5. Clone the CRISPR vector using the Golden Gate assembly approach with sgRNA products (PCR-purified) and also as outlined in Čermák et al., [12] (*see* **Note 6**). Prepare the Golden Gate reaction as follows:

Golden gate assembly reaction (Total volume: 20 μL)	
pDIRECT22C (50 ng)	1 μL
Diluted purified PCR products (4 samples)	0.5 μL
SapI	1 μL
BsaI-HF	1 μL
T4 DNA ligase	1 μL
ATP (10 mM)	2 μL
10x CutSmart	2 μL
ddH2O	10 μL

6. Run reaction in a thermocycler with the following steps:

Temperature (°C)	Time	Cycles
37 16	10 min 10 min	x8
37	10 min	x1
85	5 min	x1
4	Hold	

7. Perform *E. coli* transformation with the reaction product, single colony identification, and subsequent plasmid sequencing (Sanger sequencing or whole-plasmid sequencing).

3.2 Hybrid Tomato Seed Germination

1. Incubate hybrid tomato seeds in 1 mL sterile water in a 2-mL Eppendorf tube for 2 h (*see* **Note 7**). Germinate around 50 or 180 seeds when using true leaves or cotyledons as explants, respectively.
2. Remove sterile water and incubate seeds for 15 min in a saturated Na_3PO_4 solution to degrade spores and viruses.

3. From here onwards, work in a laminar flow cabinet to ensure a sterile working environment.
4. Wash seeds with sterile water three times.
5. Incubate seeds in 2.7% NaClO for 20 min.
6. Wash seeds with sterile water three times.
7. Place the wet seeds on 0.7% agar (seal the glass vessels with leucopore tape) and let them germinate and grow at 25 °C in long-day conditions.
8. Transfer the seedlings into MS-0 solid medium, and cotyledons or true leaves can be used for tomato transformations (*see* **Note 8**).

3.3 Agrobacterium-Mediated Transformation for Hybrid Tomatoes

1. Grow the *Agrobacterium* strain containing the CRISPR construct in 100 mL liquid YEB containing antibiotics (20 μg/mL rifampicin, 20 μg/mL gentamycin, and the selection antibiotic from the CRISPR construct, which is 50 μg/mL kanamycin for the plasmid described in Subheading 2.2) for about 2 days at 28 °C, 220 rpm (*see* **Note 9**).
2. Re-inoculate 100 mL liquid YEB with antibiotics with various amounts of *Agrobacterium* original culture (100 μL, 200 μL, and 400 μL). Be sure to include 100 mL liquid YEB (50 μg/mL kanamycin, 20 μg/mL rifampicin, and 20 μg/mL gentamicin) as baseline control for the OD_{600}. Incubate around 12 h on 28 °C, 220 rpm (roughly from 18:00 to 08:00).
3. From here onwards, work in a laminar flow cabinet to ensure a sterile working environment.
4. On the day before the transformation, cut about 200 explants using flame-sterilized scalpels and/or scissors, and a flat-beaked tweezer. Explants can be obtained from the true leaves of 4–6-week-old plants grown in vitro on MS-0 medium. Alternatively, cotyledons can be used (*see* **Note 10**). Explants can be cut in a sterile environment on sterile glass plates, should have two or three cut sides, and be approximately 6 × 6 mm in size.
5. To avoid drying of the explants, transfer them adaxially onto MS-1 plates directly after cutting, using a flat-beaked tweezer.
6. Incubate the plates, sealed with leucopore, overnight at 25 °C in the dark to adapt the leaves to the medium and avoid stress by photosynthesis.
7. On the day of the transformation, measure the absorbance value (OD_{600}) of *Agrobacterium* culture medium with different concentration gradients. Choose the appropriate overnight culture (OD_{600} = 0.6–0.8) and dilute 1:20 with LB without antibiotics (100 mL total volume). This can be done outside the laminar flow cabinet.

8. Collect the explants from the MS-1 plates using a flat-beaked tweezer in a 50-mL Falcon tube and keep the empty MS-I plates. Add the *Agrobacterium* solution to the tube and incubate the explants for 15 min on a roller-bench.
9. Place the leaves on sterile filter paper to remove the liquid and place the explants adaxially on the MS-1 plates (kept in **step 8**) and seal plates with leucopore tape.
10. Incubate the plates, containing explants infected with *Agrobacterium*, for 48 h in the dark at 25 °C.
11. Transfer the explants adaxially onto MS-2 plates and place them in long-day conditions at 25 °C and seal the plates with leucopore tape. *See* **Note 11** for regeneration control.
12. After 7 days, transfer explants to fresh MS-2 plates.
13. Transfer the explants to fresh MS-2 plates every 2–4 weeks for callus and shoot formation.
14. Cut off emerging shoots at the base and place on MS-3 medium in tissue-culture jars for root development. Take care to keep the shoot apical meristem intact and to not include callus.
15. Transfer well-developed tomato plants with roots into nonsterile conditions in a growth chamber or greenhouse. Let the plants gradually adjust to decreased humidity using a plastic cover or cling film. Continue with further genotyping and phenotyping.

3.4 Chromosome Spreading to Visualize Chromosome Dynamics During Meiosis

1. Harvest developing unopened meiotic buds of various sizes (separated meiotic anthers length about 3–4 mm) into fresh Carnoy's fixation buffer (*see* **Note 12**).
2. Deeply infiltrate the tissue with the fixate by placing them in a diaphragm vacuum pumps (DIVAC) machine and incubate at room temperature under a vacuum for at least 30 min.
3. Refresh the Carnoy's fixation buffer so the fixed materials can be stored for longer term at −20 °C.
4. Manually dissect unopened flower buds using a stereomicroscope and tweezers, and separate the anthers from the other tissues. Keep the tissue moist in a droplet of Milli-Q while dissecting. Store the dissected anthers in Milli-Q.
5. Pipet off Milli-Q from the anthers and add 200 μL of 10 mM citric buffer (pH 4.6) and mix gently by flicking the tube or pipetting. The anthers should stay intact.
6. Remove the citric buffer and add 200 μL of 1x digestion enzyme mix. Mix by pipetting carefully and incubate the anthers for 3 h at 37 °C.

7. Prepare ethanol dilutions (75%, 90%, and 100%) and grab liquid nitrogen.
8. After incubation, remove the 1x digestion enzyme mix and add 75% ethanol.
9. Take three or four anthers of the size of interest and place them on a microscope slide. Gently remove the ethanol from the microscope slide with filter paper and add 5 μL of Milli-Q.
10. Cut the anthers in half using tweezers and press/squash the anthers into a mixed soup.
11. Add 50 μL of 60% acetic acid and crush the anthers by using tweezers until no visible intact pieces are left.
12. Mix and transfer 10 μL to a new slide.
13. Put on a coverslip. Remove air bubbles by gently tapping with the forceps.
14. Put the slide and coverslip between filter paper and press firmly for 5 s.
15. Put the slide and coverslip in liquid nitrogen until the sizzling stops.
16. Remove coverslip (in a single go) with a knife.
17. Dehydrate the slides by putting them in 75% ethanol, 90% ethanol, and 100% ethanol for 5 min each (*see* **Note 13**).
18. Add DAPI and visualize chromosomes using a fluorescent microscope (e.g., Zeiss Axio Imager Z2 upright microscope) (Fig. 3a, b). Besides meiocytes, the slide will contain somatic cells going through mitosis. *See* **Note 14** (and Fig. 3a, b) on how to differentiate meiotic and mitotic cells.

3.5 Pollen Viability and Single Pollen Size Determination

3.5.1 Alexander Staining

1. Select a blooming inflorescence from the tomato plants.
2. Add 20 μL Alexander staining solution into 2.0-mL Eppendorf tube.
3. Use an electric pollinator to release pollen from a single flower into the 2.0-mL Eppendorf tube (*see* **Note 15**).
4. 10 s centrifugal spin down and mix the contents evenly by pipetting up and down using a cut 200-μL tip.
5. Add 10 μL of mixed solution on the slide and cover with a cover slip.
6. Capture images of the pollen staining, for example, using Axioplan 2 Imaging Microscope (Zeiss, Germany) using 10x objective. *See* Fig. 3c for Alexander staining results of wild-type Moneyberg-TMV x Micro-Tom, and Moneyberg-TMV x Micro-Tom^{MiMe}.
7. Use ImageJ software to automatically measure the single pollen diameter and size.

3.5.2 Single-Particle Diameter Determination

1. Collect a fresh blooming inflorescence.
2. Install an aperture tube (100 μm) for pollen measurement according to Beckman Coulter Multisizer 4 User Manual.
3. Edit Standard Operating Method (SOM) settings (12,000 events per sample) and enter the low and high limits (Sizing threshold: 8–60 μm) for the range of particle diameter sizes. *See* Fig. 4a for high-throughput single pollen diameter distribution (*see* **Note 16**).
4. Add 10 mL Isoton II diluent fluid into the container and shake the pollen from one flower into the solution using an electric pollinator.
5. Mix the sample well and then measure the particle size.
6. Analyze the data using the Multisizer 4e software according to the manufacturer's manual (https://www.manualslib.com/manual/1221961/Beckman-Coulter-Multisizer-4.html?page=3#manual).

3.6 Ploidy Analysis: Flow Cytometry

1. Collect a small piece of leaf (approximately 1 cm × 1 cm).
2. Chop the collected leaf tissue in 520 μL of Galbraith's buffer using a sharp razor blade (*see* **Note 17**).
3. Filter the slurry through a 30-μm CellTrics green filter (e.g., 04-0042, Sysmex).
4. Add 20 μL DAPI (100 μg/mL) into 500 μL of filtered sample and incubate for 15 min.
5. Use flow cytometer (e.g., cytoFLEX V5-B5-R3) to check the ploidy level of plants according to the manufacturer's instructions.
6. Perform daily clean and quality control by using CytoFLEX Daily QC fluorospheres (e.g., Beckman coulter, B53230).
7. For final data collection, 12,000 events per sample were acquired in fast mode for each independent measurement.
8. Analyze the final data (Fig. 4c). In our case, this was done using CytExpert (Beckman Coulter) and FCS Express 7 software (*see* **Note 18**).

3.7 Whole-Genome Sequencing for Crossover Detection

3.7.1 SNP Marker Identification Using Short- and Long-Read Sequencing

1. To identify the markers between Moneyberg-TMV and Micro-Tom, align the short reads (BGI) and long reads (HiFi) of Micro-Tom [10] against MbTMV reference genome [13] using bwa-mem [14] and minimap2 [15], respectively.
2. Detect SNPs using GATK HaplotypeCaller and keep SNPs that are supported by both long and short reads.

3.7.2 High-Throughput Sequencing and Genomic Profiling of Parent Plants and Offspring Populations

1. Collect small (1 cm × 1 cm) leaf samples from wild-type F1 plants, wild-type F2 plants, and *MiMe* offspring plants.
2. Use the NucleoMag Plant Kit (Macherey Nagel, 744400.4) in 96-well plates to extract genomic DNA, employing a robotic KingFisher device (Thermo) for automated processing.
3. Prepare TPase-based barcoded DNA sequencing libraries from the extracted DNA to enable multiplexed sequencing.
4. Sequence the pooled libraries using an Illumina device (Paired end 150-bp mode) and subsequently separate raw reads of individual libraries by demultiplexing. Map the reads to the Moneyberg-TMV genome using bwa-mem [14].

3.7.3 Analysis of Sequencing Data of Parent Plants and Offspring Populations

1. Illumina reads of wild-type F1 plants, wild-type F2 plants, and *MiMe* offspring plants are mapped against the MbTMV reference genome [13] using bwa-mem [14] and SNPs are called using GATK HaplotypeCaller.
2. Using the marker set developed in Subheading 3.7.1, compute the average allele frequency in a sliding 1-Mb window with 50 kb step size per sample and visually inspect deviations from the expected frequency to differentiate F2 offspring from F1 hybrid and *MiMe* samples and to infer presence or absence of recombination.
3. Using the short-read mapping, compute the genome-wide mean read coverage per sample and inspect regions with significant fold change. Deviation in both read coverage and allele frequency, which span from one genomic location to the end of the chromosome, may indicate chromosome truncation or aneuploidy. *See* Fig. 5a, b for illustration of WGS analysis of allele frequency and normalized read coverage in control hybrid, control F2, and hybrid *MiMe* mutant offspring plants.

4 Notes

1. It is important to work sterile in tissue culture. Therefore, antibiotics and hormones should be filter sterilized before use. All media should be autoclaved.
2. This protocol has been optimized for Moneyberg-TMV x Micro-Tom hybrid plant transformations. Sucrose and hormone concentrations deviate from other protocols specifically for Micro-Tom transformation.
3. The carbenicillin is added to kill the *Agrobacterium* and prevent contamination of the tissue culture; kanamycin is a construct-specific antibiotic selection agent and may be replaced by another agent depending on the construct used.

4. The Sol Genomics Network website (https://solgenomics.net/) provides tools and resources for tomato genomics research. The alternative websites are as follows: Phytozome (https://phytozome-next.jgi.doe.gov/), Ensembl Plants (https://plants.ensembl.org/index.html), and TomAP (https://bioinformatics.cau.edu.cn/TomAP/index.html) [16].
5. Apart from this CRISPR-P website, some other websites can be used to design sgRNAs: CHOPCHOP (https://chopchop.cbu.uib.no/) and CRISPR-PLANT (http://omap.org/crispr/CRISPRsearch.html).
6. For the Golden Gate assembly, it is recommended to use purified PCR products diluted after column purification, rather than gel-purified or unpurified PCR products, as this significantly improves assembly efficiency.
7. For some tomato seeds germination, incubation of seeds with 700 ppm GA3 overnight can increase the germination rate.
8. True leaves of 4–6-week-old plants can be used for the transformation. In our experience, using older leaf materials is not advised, as it increases the chance of generating tetraploid plants.
9. Liquid cultures can be stored in the fridge (4 °C) for 6 weeks. Glycerol stocks (25% glycerol as final concentration) from the liquid culture can be stored at −80 °C for several years.
10. Cotyledons may be used as alternative to true leaves as explants. Per cotyledon, one or two explants can be cut.
11. A few explants can be placed on a regeneration control medium without kanamycin (or other plant selection antibiotic). These explants should regenerate rapidly and confirm that the conditions are good for regeneration.
12. The anther sizes linked with meiosis stages vary between different tomato varieties, and also the growth temperature influences the anther development. For the exact meiotic stage determination, the tip of one anther can be cut and then make the squash to check stage via DAPI staining or acetocarmine staining. To ensure that the microscope slide will contain a complete range of meiotic stages, pool anthers of different sizes (from different sizes of flower buds, thus different developmental stages). Anthers from up to six flower buds can be pooled. Do not pool anthers from different flower buds when interested in a specific meiotic stage.
13. The slides can be stored at −80 °C and used for fluorescence in situ hybridization (FISH) or immunostaining in the future.

14. Meiotic cells can be differentiated from mitotic cells based on size and chromosome behavior such as chromosome pairing and synapsis. Meiocytes are bigger than mitotic cells.
15. A home-made electric pollinator can be made from an electric toothbrush. To do this, the toothbrush head needs to be removed, and a metal rod (length: 10 cm, diameter: 2 mm) needs to be attached to the vibrating tip/shaft. Alternatively, a commercial electric pollinator can be bought.
16. To accurately determine the exact pollen number per flower, the Control Mode (Volumetric-2 mL) can be selected for SOM Wizard. So, the final number of individual pollen will be five times the measured number.
17. To detect the diploid and tetraploid samples, co-chopping is a good approach (select the same size of both leaf materials and put them together for co-chopping, so it will exhibit overlapping tetraploid peaks).
18. The online website https://floreada.io/analysis for analyzing flow cytometry data (.fcs file) is free to use.

Acknowledgments

This work was supported by the Max Planck Society (core funding), the Radboud University executive board, the German Research Foundation (DFG grant UN 442 to C.J.U.), and the European Research Council (ERC starting grant 101076355, "AsexualEmbryo" to C.J.U.).

Conflict of Interest Statement C.J.U. and Y.W. are listed as inventors on European Patent Application no. 23179909.9 entitled "UNREDUCED CLONAL GAMETE FORMATION AND POLYPLOID GENOME DESIGN IN THE SOLANACEAE", owned by the Max-Planck-Gesellschaft zur Förderung der Wissenschaften e.V.

References

1. Wang Y, Sun C, Ye Z et al (2024) The genomic route to tomato breeding: past, present, and future. Plant Physiol 195:2500–2514
2. Krieger U, Lippman ZB, Zamir D (2010) The flowering gene SINGLE FLOWER TRUSS drives heterosis for yield in tomato. Nat Genet 42:459–463
3. Wang Y, Van Rengs WMJ, Zaidan MWAM, Underwood CJ (2021) Meiosis in crops: from genes to genomes. J Exp Bot 72:6091–6109
4. Wang Y, Copenhaver GP (2018) Meiotic recombination: mixing it up in plants. Annu Rev Plant Biol 69:577–609
5. Mercier R, Mézard C, Jenczewski E, Macaisne N, Grelon M (2015) The molecular biology of meiosis in plants. Annu Rev Plant Biol 66:297–327
6. Underwood CJ, Mercier R (2022) Engineering apomixis: clonal seeds approaching the fields. Annu Rev Plant Biol 73:201–225

7. Wang Y, Underwood CJ (2023) Apomixis. Curr Biol 33:R293–R295
8. d'Erfurth I, Jolivet S, Froger N et al (2009) Turning meiosis into mitosis. PLoS Biol 7: e1000124
9. Mieulet D, Jolivet S, Rivard M et al (2016) Turning rice meiosis into mitosis. Cell Res 26: 1242–1254
10. Wang Y, Fuente RR, Van Rengs WMJ et al (2024) Harnessing clonal gametes in hybrid crops to engineer polyploid genomes. Nat Genet 56:1075–1079
11. Li S-Z, Wang J, Jia S-G et al (2023) Synthetic apomixis from genetic basis to agricultural application. Seed Biol 110:2
12. Čermák T, Curtin SJ, Gil-Humanes J et al (2017) A multipurpose toolkit to enable advanced genome engineering in plants. Plant Cell 29:1196–1217
13. van Rengs WMJ, Schmidt MH-W, Effgen S et al (2022) A chromosome scale tomato genome built from complementary PacBio and Nanopore sequences alone reveals extensive linkage drag during breeding. Plant J 110: 572–588
14. Li H (2013) Aligning sequence reads, clone sequences and assembly contigs with BWA-MEM. Preprint at https://doi.org/10.48550/arXiv.1303.3997
15. Cheng H, Concepcion GT, Feng X et al (2021) Haplotype-resolved de novo assembly using phased assembly graphs with hifiasm. Nat Methods 18:170–175
16. Cao Y, She J, Li Z et al (2024) TomAP: a multi-omics data analysis platform for advancing functional genomics research in tomatoes. New Crops 1:100002

Chapter 18

Engineering Synthetic Apomixis in Hybrid Rice

Donaldo Meynard, Aurore Vernet, Imtiyaz Khanday, Laurine Gilles, Raphaël Mercier, Venkatesan Sundaresan, and Emmanuel Guiderdoni

Abstract

The transfer of apomixis, a clonal mode of reproduction by seeds, to crops has the potential to revolutionize agriculture by enabling the generation of one-line F1 hybrids that propagate clonally by seeds from one generation to the next. However, despite nearly four decades of intensive research, all the attempts to transfer the identified genetic determinants of apomixis from naturally apomictic wild plants to their crop relatives have failed to produce apomictic crops. Engineering of apomixis, mimicking the key features of a natural form of apomixis known as gametophytic diplosporous apomixis, has recently been achieved in rice and further improvements by the introduction of a single "all-in-one" T-DNA construct into calli derived from F1 hybrid seed embryos resulted in high frequency of clonal seeds. The T-DNA encodes a constitutively expressed Cas9 protein guided by sgRNAs designed to knock-out the functions of three genes essential for regulating crucial steps involved in meiosis, thereby eliminating meiosis and creating the *Mitosis instead of Meiosis* (*MiMe*) triple mutant. Additionally, the T-DNA contains another gene expression cassette consisting of a parthenogenetic trigger, the BABY BOOM1 (BBM1) transcription factor driven by an egg cell-specific (ECS) promoter. Inactivation of the three *MiMe* genes converts meiosis into mitosis—apomeiosis—yielding unrecombined and unreduced male and female spores developing into gametophytes. The BBM1 expression triggers parthenogenetic development of an embryo from the diploid egg cell of the female gametophyte. The endosperm develops sexually by the fusion of a diploid sperm cell and the central cell of the female gametophyte, which contains two diploid polar nuclei, resulting in an initially hexaploid endosperm. To date, the "all-in-one" T-DNA method has proved to be the most efficient for achieving high frequency (95–100% clonal seeds) synthetic apomixis in rice. Since the original publication, we have successfully generated synthetic apomictic events in three additional F1 hybrids. Here, we describe the methods for designing T-DNA constructs, analyzing mutations in first-generation (T0) *MiMe* mutant plants of F1 hybrids, and ascertaining the apomictic nature of the progenies from confirmed *MiMe* F1 hybrids, which relies on the egg cell-specific accumulation of BBM1.

Key words BABY BOOM, CRISPR/Cas9, *MiMe*, Hybrid rice, Synthetic apomixis

---

First Co-authors = Donaldo Meynard, Aurore Vernet and Imtiyaz Khanday.

Tatyana Radoeva and Peter J. van Dijk (eds.), *Apomixis: Methods and Protocols*, Methods in Molecular Biology, vol. 2987, https://doi.org/10.1007/978-1-0716-5001-1_18,

1 Introduction

The phenomenon of heterosis, also called hybrid vigor, has been harnessed in modern plant breeding through the development of F1 hybrid seeds, well known for their higher yield and yield stability across environments [1]. Cultivated rice of Asian origin, *Oryza sativa* L., exhibits a yield heterosis of 20–30%, with hybrid rice accounting for 20% of world-wide rice acreage, and more than 60% of rice cultivation in China [2]. However, unlike maize, rice is an autogamous species and has a low outcrossing rate. Therefore, the production of commercial F1 hybrid seeds relies on either a geno-cytoplasmic or a genic (photoperiod-sensitive or temperature-sensitive) male (pollen) sterility systems, known as 3-line (male sterile, maintainer, and restorer) and 2-line hybrid systems, respectively [3]. Preparation of female parents and production of pure hybrid seeds are both time- and labor-consuming tasks, making hybrid seeds costly and often unaffordable for small-scale rice farmers. Furthermore, due to the heterozygosity of F1 hybrids, the F2 progeny segregate for parental traits, creating the need to purchase new F1 seeds for cultivation each crop season for farmers.

An alternative to these systems would be the introduction of apomixis, a natural mode of clonal reproduction through seeds, in crops such as rice, to enable the production of self-propagating one-line F1 hybrids. Several forms of apomixis exist in more than 400 angiosperm species, classified as sporophytic or gametophytic, depending on whether the clonal embryo originates from somatic or germinal tissues [4]. A well-studied type of apomixis is gametophytic diplosporous apomixis, in which the specialized meiotic division program is replaced by a simple mitotic-like division in the megaspore mother cell. This process, known as apomeiosis, produces two unreduced and unrecombined cells instead of four reduced and recombined spores, produced during normal meiosis. Following gametophyte development, the egg cell of the apomeiotic embryo sac undergoes parthenogenetic development. In some apomictic species, the central cell can develop autonomously into a viable endosperm, while in others, it undergoes fertilization by the pollen sperm cell (pseudogamy) to produce a regular sexual endosperm.

Despite extensive searches in the 1980s and 1990s, no natural source of apomixis has been found in cultivated rice or its wild relatives [5], although apomict wild relatives exist in other cereal species (e.g., *Pennisetum*, *Tripsacum*) [4]. While the loci responsible for apomeiosis and parthenogenesis have been identified and mapped in some apomictic species, their transfer to crops has not succeeded in producing apomictic crops [6]. An alternative is therefore to use biotechnology to introduce synthetic apomixis by engineering gametophytic diplosporous apomixis [7, 8]. This requires three steps: (1) Turning meiosis into mitosis (apomeiosis) to produce an unreduced and unrecombined embryo sac. (2) Triggering embryo development from the egg cell without fertilization by the sperm cell. (3) Achieving endosperm development with either the fusion of the central cell with the sperm cell or triggering autonomous endosperm development from the unfertilized central cell.

While loss-of-function mutation in *DYAD/SWI1*, a gene required for sister chromatid cohesion and centromere organization, has been shown to produce unrecombined and unreduced embryo sacs [9], the most robust and penetrant form of apomeiosis has been achieved through the simultaneous inactivation of three genes, each essential for key features that differentiate meiosis from mitosis [10]. First, mutation of a member of the recombination initiation complex (e.g., *SPO11-1* or *PAIR1/PRD3*) abolishes parental homolog chromosome pairing and recombination [11, 12]. Second, mutation of the cohesin *REC8* allows separation of the sister chromatids at the first meiotic division [13]. Third, mutation of the cell cycle regulator *OSD1* promotes meiosis exit after the first division, omitting the second division, and generating diploid dyads instead of haploid tetrads. The triple mutant, called *MiMe* for *Mitosis instead of Meiosis*, exhibits a full penetrance in male sporogenesis and near-complete or complete penetrance in female sporogenesis in Arabidopsis and rice, respectively [10, 14].

Arabidopsis and rice *MiMe* mutants were first generated by crossing three individual insertion mutants. Since then, *MiMe* has been efficiently reproduced by knocking out the three genes using CRISPR/Cas technology, establishing a platform to investigate induction of a gynogenetic or a parthenogenetic development of the diploid unrecombined egg cell [15, 16]. Pollination of *MiMe* with GEM [17], a *cenh3* mutant expressing a CENH3 variant protein that promotes loss of paternal genome in the zygote and gynogenetic embryo development, resulted in up to 35% apomictic seeds in Arabidopsis. However, this system is nonautonomous, as it requires crossing [18]. Alternatively, a quadruple mutant of the three *MiMe* genes and that of a pollen tube phospholipase, *MATRILINEAL* (*MATL*), that induces paternal genome elimination and haploid embryo formation of maternal origin [19], produces apomictic seeds, albeit at low frequency (6%). This approach also exhibits reduced fertility (8%) associated with *MATL* inactivation [15]. A more promising alternative involves the egg cell-specific expression of an embryogenic inducer, the AP2 transcription factor *BABY BOOM1* (*BBM1*) that triggers the parthenogenetic embryo development [16]. The egg cell expression can be driven by the Arabidopsis *EC1.2* (*DD45*) promoter [20, 21] or other specific promoters (*see* details below). In wild-type plants, the maternal allele of *BBM1* remains repressed in the egg cell. However, the male allele is expressed in the sperm cells and triggers zygotic embryogenesis after fertilization. The male allele expression in the zygote after fertilization also derepresses the female allele [16]. Four *BBM* genes redundantly regulate zygotic embryogenesis in rice [16, 22]. In rice (cv. Kitaake), transgenic lines expressing *OsBBM1* in the egg cells produced haploid progeny at a frequency of ~30% [16]. When this parthenogenesis triggering capability of *OsBBM1* was combined with the inactivation of the three *MiMe* genes in the same genetic background, up to 29% clonal progeny were obtained, creating the first autonomous synthetic apomixis in any crop [16]. More recently, the egg cell-specific expression of a WOX family transcription factor, *WUSCHEL-LIKE HOMEODOMAIN 9A* (*OsWOX9A*) in the *OsBBM1* background synergistically increased the formation of haploid, parthenogenetic progeny to 86–91% [23].

In a significant advancement, Vernet and co-workers introduced an "all in one" T-DNA construct in hybrid rice, which enables, in a single step, both the inactivation of the three *MiMe* genes and the induction of parthenogenesis in F1 hybrids [24]. Two "all-in-one" constructs were utilized that differed in the origin of the egg cell-specific promoter used to drive *OsBBM1* expression: *EC1.2* gene from Arabidopsis and the *ECA1* gene from rice [25]. This approach achieved more than 95% clonal seed production, stable over several greenhouse generations. Using a similar "all-in-one" strategy *OsBBM4* was substituted for *OsBBM1*, but this resulted in a low (<3%) frequency of parthenogenesis [26]. More recently, replacing *OsBBM1* with the *PARTHENOGENESIS* (*PAR*) coding sequence isolated from the naturally apomictic plant, dandelion [27], achieved a 43–68% parthenogenesis frequency in *MiMe* mutants [28]. Similarly, the use of *OsWUS/MONOCULM3* [29] instead of *OsBBM1* in *MiMe* mutants yielded up to 21.7% parthenogenetic progeny [30].

Here, we describe the methods for implementing the most efficient strategy of inducing synthetic apomixis to date: the *MiMe* + *OsBBM1* "all in one" T-DNA approach. This system was successfully engineered in four different hybrid backgrounds of rice in our laboratory. Specifically, we provide detailed protocols for designing gRNAs, analyzing mutations in the first-generation (T0) *MiMe* CRISPR/Cas9 mutants in F1 hybrids, and ascertaining the apomictic/clonal nature of the progenies obtained from *MiMe* F1 hybrids, expressing *OsBBM1* in the egg cells.

1.1 Design of Apomeiosis and Apomixis "All-in-One" T-DNA Vectors

All-in-One T-DNA Design

The "all-in-one" apomixis inducer T-DNA construct (Fig. 1) consists of the following components: (1) three single-guide (sg) RNAs, each targeting *PAIR1* (or *SPO11-1*), *OsREC8,* and *OsOSD1*, driven by rice U6 Pol III promoters, (2) a codon-optimized Cas9 nuclease coding sequence (e.g., [31]) driven by

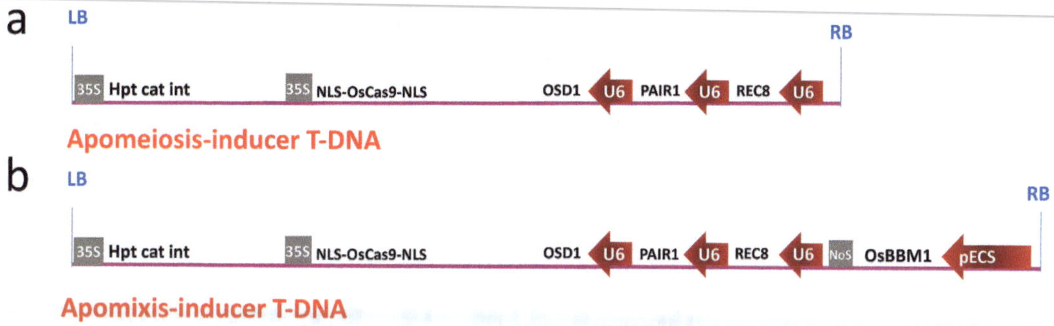

Fig. 1 Schematic representation of "all-in-one" T-DNAs introduced in rice hybrids for inducing apomeiosis and apomixis: (**a**) Apomeiosis T-DNA. (**b**) Apomixis T-DNA. LB and RB: left and right borders of the T-DNA; p35S: promoter of the Cauliflower Mosaic Virus (CaMV); 35S: polyadenylation signal of CaMV; hpt cat int: hygromycin phosphotransferase II with castor bean catalase intron; ZmUbi: promoter, first intron and first exon of the maize Ubiquitin 1 gene; "Os"Cas9: rice-optimized Cas9 coding sequence; NLS: nucleus localization signal fused to Cas9; *OsOSD1*, *PAIR1*, and *OsREC8*: Three cassettes including sgRNAs (20 bp crRNA specific to the target gene + 82 bp tracR RNA) driven by the Pol III U6 promoter targeting *OsOSD1*, *PAIR1*, and *OsREC8*; EC1.2: egg cell-specific promoter from Arabidopsis

the maize *UBIQUITIN1* constitutive promoter [32], and (3) the rice AP2 transcription factor *BABY BOOM1* (*OsBBM1*) controlled by an egg cell-specific promoter (either the *EC1.2* promoter from Arabidopsis or its homolog in rice, ECA1) [16, 24]. Additionally, a *HYGROMYCIN PHOSPHOTRANSFERASE* (*HPT*) selectable marker cassette is included for efficient selection of transformed plant cells. The castor bean *CATALASE1* intron was also included in *HPT* coding sequences, which enhances its expression in rice cells and allows efficient decontamination of *Agrobacterium* following co-cultivation [33].

1.2 Choice of MiMe Genes

The original rice *MiMe* mutant was generated by combining *pair1*, *Osrec8*, and *Ososd1* mutations through crossing of insertion lines carrying individual mutations [14]. At the time of this study, *PAIR1*, another gene of the recombination initiation complex, was used in place of *OsSPO11-1*, which was originally used in Arabidopsis *MiMe*, as no confirmed *Osspo11-1* insertion mutant lines were available in rice. However, *Osspo11-1* can be used as a substitute for *pair1*. Conversely, mutations in the *PAIR1* ortholog in Arabidopsis, *PRD3*, has recapitulated the *MiMe* phenotype in the triple *prd3 rec8 osd1* mutant [14]. In principle, any of the genes essential for meiotic recombination initiation can be used instead of *pair1*. To date, there is no known alternative to *REC8*. The cyclin *TAM* can be targeted instead of *OSD1*, as shown in Arabidopsis—but with lower penetrance than *osd1*—and tomato [34, 35].

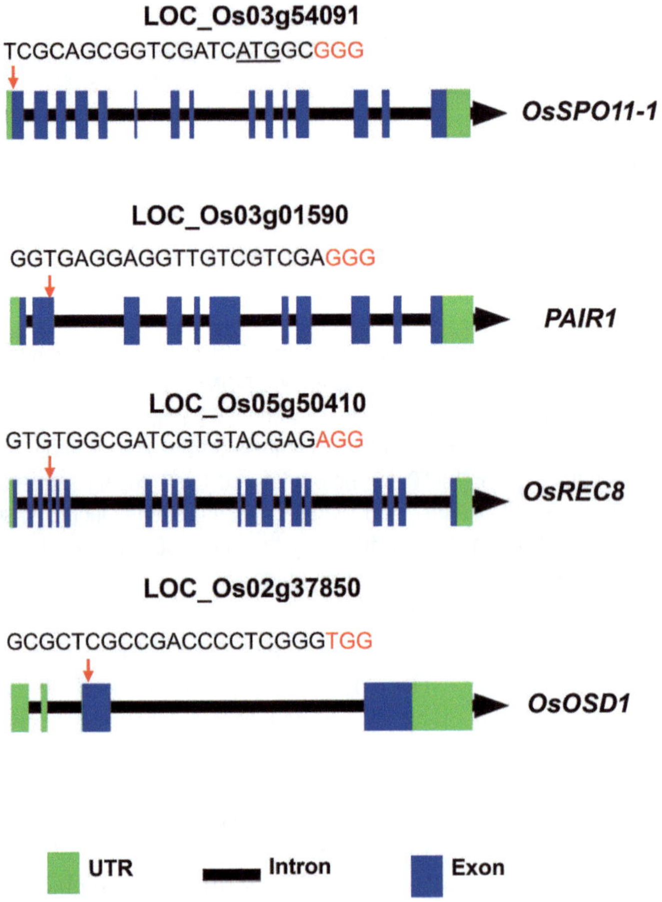

Fig. 2 Single-guide RNAs (sgRNAs) for *MiMe* genes. Schematic diagrams illustrate the gene structures of the *MiMe* loci. Blue boxes represent exons, green boxes indicate untranslated regions (UTRs), and thick black lines denote introns. Arrows indicate the 5′ to 3′ orientation of each locus. The sequences above each schematic represent the gRNA protospacer sequence in black, with the protospacer adjacent motif (PAM) highlighted in red. A red arrow above each schematic marks the position of the gRNA and the targeted exon for editing in each gene. The genes shown (from top to bottom) are OsSpo11-1, PAIR1, OsREC8, and OsOSD1

1.3 Choice of gRNA Targets

Efficient gRNA targeting *OsPAIR1*, *OsREC8,* and *OsOSD1* have been generated using http://omap.org/crispr/webtool (Fig. 2; Table 1) [36]. They target sequences close to the start codon of the gene (Fig. 2), enabling the editing machinery to induce frameshift mutations, leading to gene inactivation through the downstream formation of a premature STOP codon, resulting in a truncated protein. These gRNAs have been shown to generate high-frequency biallelic or homozygous edits of the three *MiMe* genes in the genomes of the four rice hybrids tested so far, without requiring modifications of the crRNA sequences. This is likely due

Table 1
Sequences of gRNA targets in *OsSPO11-1*, *PAIR1*, *OsREC8*, and *OsOSD1*. *OsSPO11-1* can be used as alternative to *PAIR1* to prevent chromosome pairing and recombination

crRNA	Target sequence (PAM)
OsSPO11-1	TCGCAGCGGTCGATCATGGC **GGG**
PAIR1	GGTGAGGAGGTTGTCGTCGA **GGG**
OsREC8	GTGTGGCGATCGTGTACGAG **AGG**
OsOSD1	GCGCTCGCCGACCCCTCGGG **TGG**

to the conservation of gRNA target sequences in the *MiMe* genes among different rice cultivars. However, it is advisable to check that there are no polymorphisms in the target sequences of the parents (both the 20 nt crRNA target and the PAM sequences) before undertaking a transformation experiment. gRNA targeting *OsREC8* tends to generate a higher frequency of non-frameshift mutations (-3 nt or multiples thereof) compared to the gRNAs targeting *OsSPO11-1*, *PAIR1,* and *OsOSD1*. Fortunately, these mutations still result in loss-of-function of alleles, as T0 plants carrying such an *OsREC8* allele in a *pair1 ososd1* genetic background still exhibit high fertility. Here, we also provide an efficient guide RNA, which targets *OsSPO11-1* close to ATG sequences and generates high-frequency biallelic/homozygous frameshift mutations in rice, resulting in the desired asynaptic phenotype (Fig. 2, Table 1) [37].

1.4 Choice of the Promoter Driving Egg Cell-Specific (ECS) Expression

Two different ECS promoters have been used for driving *OsBBM1* expression in egg cells of apomictic rice hybrids: the promoter of the Arabidopsis *EC1.2* gene (also called *DD45*) [20, 21] and the promoter of the homologous gene from rice, ECA1 rice [25]. Both promoters were found to be effective resulting in 95–100% parthenogenesis frequency in *MiMe,* which was stable across generations [24]. The choice of the ECS promoter is critical, as its activity must ensure stringent cell specificity and precise timing to allow sufficient accumulation of the parthenogenesis trigger proteins. This ensures timely development of the parthenogenetic embryo without interfering with the fertilization of polar nuclei by the sperm cell. Furthermore, its activity must remain stable across generations and under challenging environmental conditions. Insufficient or absence of expression of the parthenogenesis trigger would lead to an increased frequency of unwanted tetraploid progeny plants in the field.

1.5 Choice of the Parthenogenesis Inducer Gene

Three rice genes, *OsBBM1*, *OsBBM4*, and *OsWOX/MOC1*, have been independently used to trigger parthenogenetic development in egg cells of rice *MiMe* hybrids with maximum frequencies of 97%, 2.3%, and 21.7%, respectively [24, 26, 30]. More recently, egg cell-specific expression of *OsWOX9A* in an *OsBBM1* expressing background demonstrated a high parthenogenesis frequency of 86–91% [23]. Alternatively, egg cell-specific expression of the parthenogenesis inducing PAR gene from a natural apomict, dandelion (*Taraxacum*), has enabled parthenogenetic embryogenesis in a rice *MiMe* context, with frequencies ranging from 32% to 54% [28]. Thus, the preferred inducer gene for rice is *OsBBM1*, which has the highest reported frequencies of parthenogenesis so far.

1.6 Generating and Analyzing MiMe Mutations in T0 Events and Progeny Plants

Transformation of Apomeiosis and Apomixis T-DNA Constructs

Commercial F1 seeds of the *indica/indica* hybrid BRS-CIRAD 302 (1F x D24) and F1 seeds resulting from manual crossing of the *japonica/japonica* hybrid DongJin/Nipponbare and of the *indica/japonica* hybrids KalingaIII/Kitaake and IR64/Kitaake have been used for generating both apomeiotic and apomictic events (Table 2). Seeds of the standard rice cv. Kitaake have also been used, to a lesser extent, to test the constructs.

Secondary calli derived from scutellum of the embryo of mature rice seeds were co-cultured with a suspension of *Agrobacterium tumefaciens* strain EHA105 cells carrying either the Apomeiosis or the Apomixis T-DNA binary plasmid. Transgenic plants were raised on hygromycin-containing selection media according to Sallaud et al. [38] with a few modifications like extended subculture durations due to slower development of resistant calli particularly in *indica/japonica* hybrids.

After the introduction into rice cells by *Agrobacterium* infection, both Arabidopsis and rice egg cell promoter-containing constructs were able to induce biallelic/homozygous triple *MiMe* mutations and achieve high parthenogenetic egg cell development due to *BBM1* expression [24]. Only a low frequency of editing chimaeras (more than two mutations in a gene) in regenerated T0 events is observed, suggesting that allele editing occurs early after delivery of the CRISPR/Cas9 machinery in rice cells.

1.7 Analysis of MiMe Mutations in T0 Plants

PCR amplification and sequencing of the genomic regions containing the crRNA targets enable the determination of mutations in the three genes responsible for the *MiMe* phenotype (Tables 3 and 4) (Figs. 3 and 4). It is recommended to test the primers for amplification of target regions on WT DNA of the two parent inbred lines and the F1 hybrid to make sure that both the parental alleles can be amplified simultaneously. Leaves can be harvested for genomic DNA extraction during the plant transfer from tissue culture vessels

Table 2
Efficiencies of transformation using the all-in-one apomeiosis and apomixis T-DNA constructs in 4 hybrids of rice

T-DNA construct	Genotype	Co-cultured, F1 seed embryo-derived calli (*A*)	Resistant cell lines (*B*)	Regeneration frequency (%) (*C*)	Transformation efficiency (Plant per co-cultured callus) = (((*B*) × (*C*))/(*A*))
Apomeiosis	BRS-CIRAD 302[a]	22	85	58.0	2.2
	KalingaIII/Kitaake	100	335	70.4	2.4
	IR64/Kitaake	80	127	82.7	1.3
	Kitaake	50	160	63.7	2.0
Apomixis	BRS-CIRAD 302[a]	48	194	71.0	1.5
	KalingaIII/Kitaake	100	261	83.9	2.2
	IR64/Kitaake	100	85	90.6	0.8
	DongJin/Nipponbare	60	83	97.6	1.3
	Kitaake	57	84	66.7	1.0

[a]Data from [1]

Table 3
Primers used to amplify the gRNA target regions in *OsSPO11-1*, *PAIR1*, *OsREC8*, and *OsOSD1*

Primer	Sequence
OsSPO11-1 Seq F	5′-CCAAAATTCTTGTGGGTGCT-3′
OsSPO11-1 Seq R	5′-CGGAGGAGCAGTAGTTCTGG-3′
PAIR1 Seq F	5′-GTGGTGTGGTGTGTTCAGGAG-3′
PAIR1 Seq R	5′-TGGAATCCCCAATCAGTAAGGCAC-3′
OsREC8 Seq F	5′-GCACTAAGGCTCTCCGGAATTCTC-3′
OsREC8 Seq R	5′-AATGGATCAAGGAGGAGGCACC-3′
OsOSD1 Seq F	5′-TTACTTGGAAGAGGCAGGAGCC-3′
OsOSD1 Seq R	5′-ACCTTGACGACTGACGTGATGTC-3′

Table 4
PCR conditions for amplifying gRNA target regions in *OsSPO11-1*, *PAIR1*, *OsREC8*, and *OsOSD1*

Primers	*OsSPO11-1*	*PAIR1*	*OsREC8*	*OsOSD1*
Number of cycles	35	35	35	35
Annealing T°	61 °C	58 °C	58 °C	58 °C
Expected amplicon size (bp)	800	326	324	364

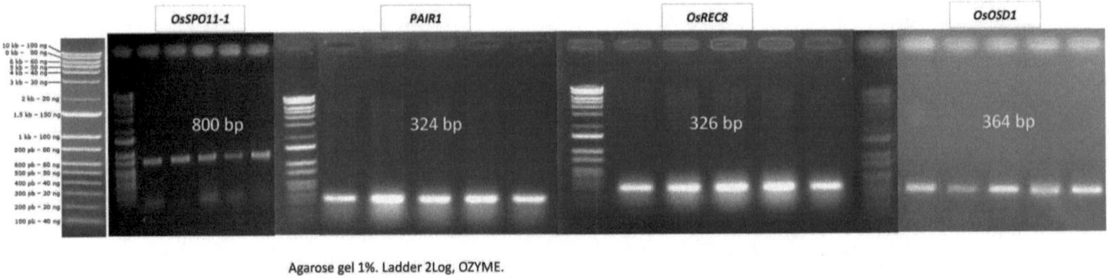

Fig. 3 Analyzing mutations in *MiMe* genes: PCR amplification of target regions in *OsSPO11-1*, *PAIR1*, *OsREC8*, and *OsOSD1* genes

to the containment greenhouse to allow early identification of mutations. CRISPR/Cas9 gene editing is highly efficient in rice and mutant alleles of target genes can be identified at this stage. Although rare, it is possible to detect persisting unedited WT alleles that may be edited later during the T0 plant development.

Fertile T0 plants are observed with a 55–67% frequency range among the primary T0 transformants generated with apomeiotic and apomixis T-DNAs constructs (Table 5). Genotyping these plants for mutation at the three *MiMe* genes could reveal two outcomes: they are either WT at the three loci or exhibit triple mutations that are biallelic or homozygous. An additional possibility is the recovery of *osd1*-only single fertile regenerants, which are capable of producing recombined diploid spores and gametes, ultimately giving rise to tetraploid T1 progeny. The frequency of confirmed *MiMe* mutants among the total primary transformants transferred to the greenhouse is high, ranging from 50% to 90% in F1 hybrid materials (Table 5).

Populations of T0 plants generated with either the apomeiosis or apomixis constructs also contain a fraction of fully sterile plants. These sterility events correspond to cases in which at least one of the alleles in *OsSPO11-1/PAIR1*, *OsREC8*, and/or *OsOSD1* has no mutation (i.e., is WT) or has an in-frame (non-frameshift) mutation resulting in at least one functional allele (Table 2b). Notably, unlike mutations in *PAIR1* and *OsOSD1*, the position of the gRNA in

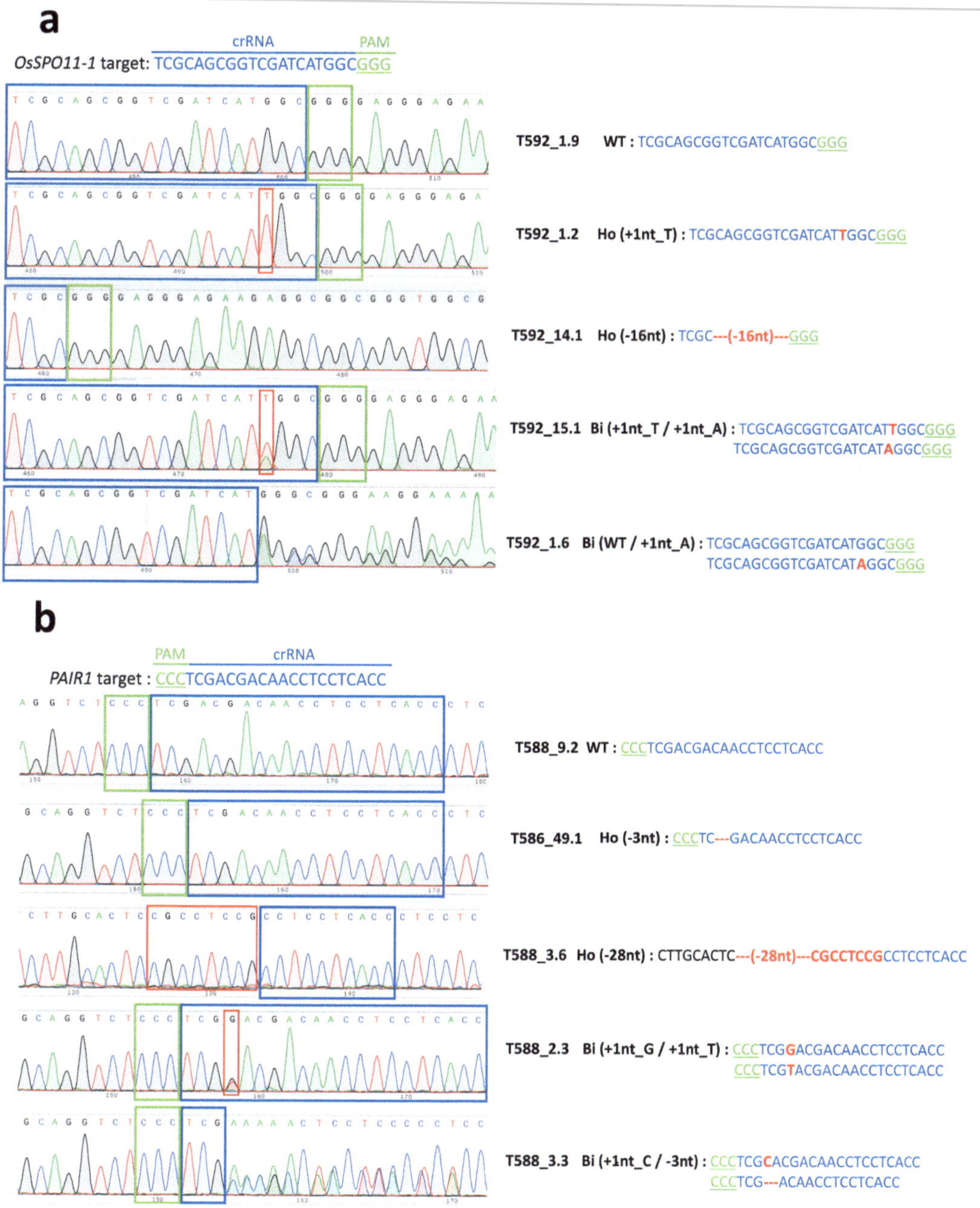

Fig. 4 (**a**–**d**) Examples of chromatogram interpretations for mutations in *OsSPO11-1*, *PAIR1*, *OsREC8*, and *OsOSD1*. (Ho = homozygous mutations; Bi = biallelic mutations)

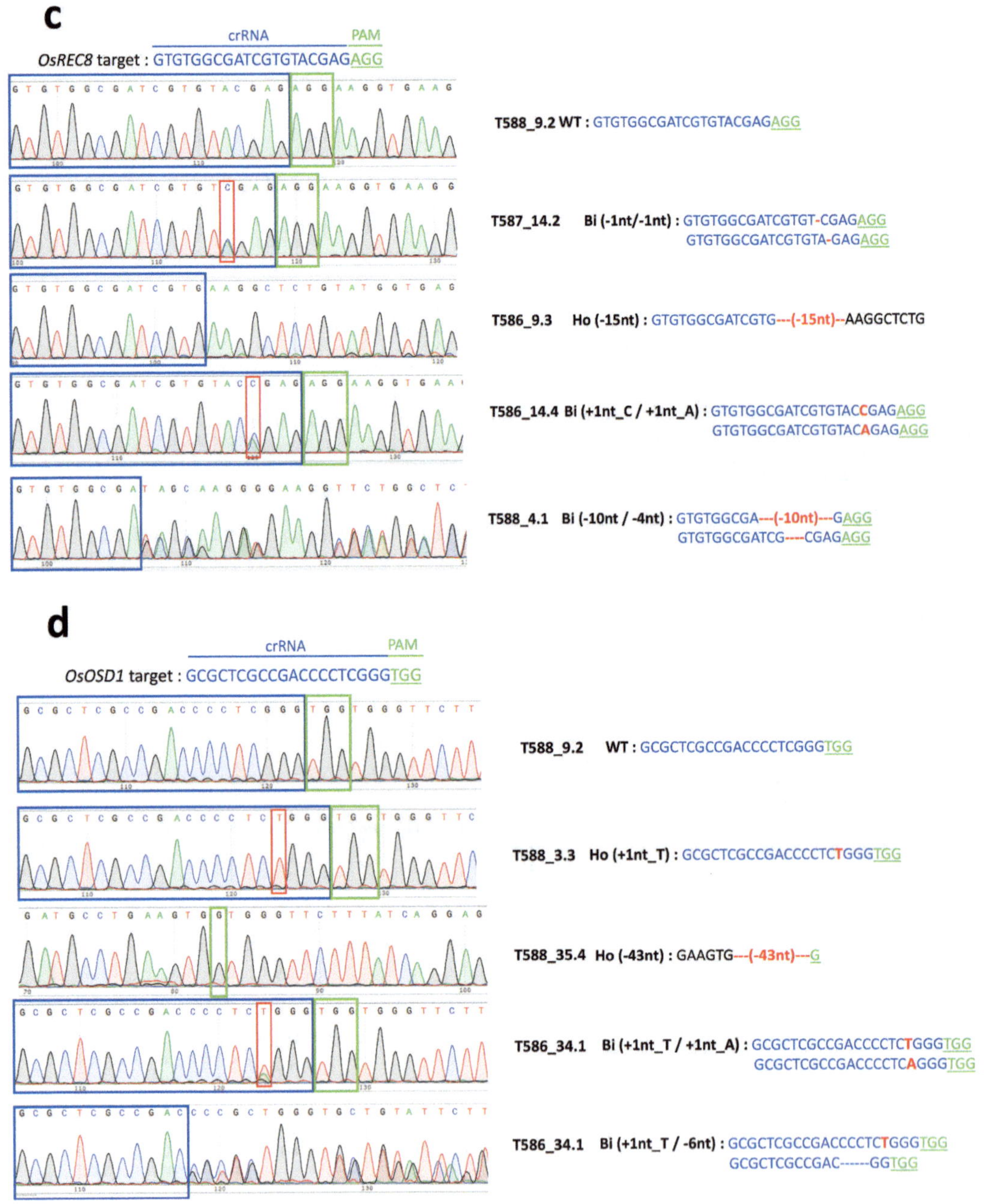

Fig. 4 (continued)

OsREC8 appears to be located in a critical domain for protein function as even a non-frameshift deletion as small as 3 nt leads to allele inactivation.

When both the alleles of all three *MiMe* genes are mutated, no recombination or segregation occurs and the mutant genotype is

Table 5

Frequencies of fertile primary transformants (T0) harboring the apomeiosis or apomixis T-DNA in 4 different rice F1 hybrids and confirmation of homozygous/biallelic mutations in the 3 *MiMe* genes in fertile primary transformants

T-DNA construct	Genotype	Number of T0 plants transferred to the greenhouse	Number (%) of fertile T0	Fertile T0 plants			
				Analyzed	Confirmed *MiMe* (%)	Confirmed wild-type[a]	Confirmed *ososd1*-only
Apomeiosis	BRS-CIRAD 302[b]	41	20 (48.8)	20	18 (90.0)	2	0
	KalingaIII/Kitaake	221	143 (54.5)	80	60 (75.0)	20	0
	IR64/Kitaake	101	56 (55.4)	56	51 (91.1)	2	0
	Kitaake	52	33 (63.5)	33	29 (87.9)	2	2
Apomixis	BRS-CIRAD 302[b]	49	33 (67.3)	49	26 (53.1)	7	0
	KalingaIII/Kitaake	205	137 (66.8)	76	61 (80.3)	14	1
	IR64/Kitaake	75	42 (56.0)	38	30 (78.9)	8	0
	DongJin/Nipponbare	90	55 (61.1)	55	43 (78.2)	12	0
	Kitaake	36	13 (36.1)	36	9 (19.5)	4	1

[a]Includes some monoallelic mutations at *PAIR1*, *OsREC8*, and/or *OsOSD1*
[b]Data from [24]

Table 6
Examples of *MiMe* mutations in near-fully and fully penetrant apomictic events

Genotype	Event	Frequency of 2n seedlings in T1 or T3	Panicle fertility (% filled grains)	*PAIR1*	*OsREC8*	*OsOSD1*
BRS-CIRAD 302	15.1	99	33.8	+A/+A	-5 nt + C/-5 nt + C	+AG/+T
	37.7	94	27.0	+T/+T	-3 nt/+G	+A/-5 nt
KalingaIII/Kitaake	34.4	100	40.2	+A/+G	+C/-5 nt	-5 nt/+T
	40.2	98	34.3	-1 nt/-49 nt	-2 nt/rearrangement	+A/-11 nt
DongJin/Nipponbare	4.1	95	30.5	-1 nt/+T	+C/-24 nt	+T/+T
	13.1	95	24.4	-73 nt/+C	-24 nt/-26 nt	+2 nt/+T
	14.1	100	25.3	+A/+C	-12 nt/-17 nt	+2 nt/+T
IR64/Kitaake	1.3	96	24.5	+A/+T	-4 nt/-3 nt	+T/-4 nt
	17.5	91	30.9	+A/-2 nt	-15 nt/-10 nt	-2 nt/-53 nt

transmitted to the progeny without segregation of alleles, due to the apomeiotic/apomictic mode of reproduction. So far, we have tested four different rice hybrids with the apomixis T-DNA construct and have successfully identified events with high-frequency synthetic apomixis in each hybrid (Table 6).

2 Materials

2.1 Plant Materials and DNA Isolation

Regenerated T0 plants in tissue culture container before their transfer to the greenhouse or T1 progeny seedlings at the 4–5 leaf stage grown under containment greenhouse conditions in trays or pots.

Genomic DNA is isolated from leaf blades of young transgenic plants (T0 regenerant or progeny seedlings of next generation). One hundred mg of a young leaf, i.e., a 3-cm segment of leaf blade, is immediately frozen in liquid nitrogen in a rack containing 1.2-mL Eppendorf tubes (column of 8, plate 96, Qiagen) containing a steel ball (2 mm). They are then ground in a ball mill (Genogrinder) (frequency = 30, 30 s, twice). The powder is resuspended in 900 μL MATAB (100 mM Tris–HCl, 1.5 mM NaCl, 20 mM EDTA, 2% MATAB, 0.5% Na2SO3, 1% PEG6000, pH = 8) at

72 °C for 1 h. Tubes are mixed by inversion every 15–20 min. A volume of 900 μL CIAA (chloroform-isoamyl alcohol, proportions: 24/1) is added, and then vortexed. The mixture is centrifuged for 30 min at 6000 rpm at room temperature. The 250 μL supernatant is removed and transferred to a new tube, and the DNA is precipitated by adding 200 μL isopropanol. After centrifugation for 30 min at 6000 g, the supernatant is removed and the pellet washed 2 times with 300 μL of 70% ethanol, and then centrifuged at 6000 g for 30 min. The DNA pellet is dried briefly, resuspended in 100 μL milliQ H2O, and preserved at 4°.

2.2 Reagents

1. Taq'OZYME containing Taq DNA polymerase and MgCl2 in 10X Buffer (OZYME OZYA001-5000 for 5 × 1000 units) or equivalent.
2. dNTP bundle (EUROMEDEX NU-1005L) 100 mM for preparing the mix of 2 mM dNTP each.
3. Synthesized forward (F) and reverse (R) oligonucleotides from a DNA synthesis company (*see* Table 1b).
4. ExactLadder DNA PreMix 2 Log (OZYME OZYC002-100) for electrophoresis gel.

2.3 Equipment

1. Eppendorf™ Mastercycler Nexus Gradient Eco (Eppendorf™ 6334000018).
2. Sub-Cell Model 192 Cell Electrophoresis device, BIORAD.

2.4 Software and Algorithms

1. SYNTHEGO ICE Analysis.

3 Methods

3.1 Genotyping of (Transgenic) Plants for MiMe Mutations

1. Adjust genomic DNA (gDNA) concentration at approximately 20 ng/μL with distilled water.
2. For 50 μL of final PCR reaction volume, mix together 2 μL of diluted gDNA, 5 μL of 10X Ozyme-Taq buffer (1X final), 3 μL of MgCl2 25 mM (1.5 mM final), 5 μL of 2 mM solution of each dNTP (200 μM of each final), 1 μL of 10 μM solution of each primer (reverse and forward), and 0.5 μL solution of 5 unit/μL Taq'Ozyme solution (2.5 units final). Bring up the volume to 50 μL with distilled water.
3. Set the thermocycler conditions for 2 mn at 95 °C followed by 35 cycles of amplification.

   Each cycle encompasses 30 s at 95 °C (denaturation step), 1 mn at either 58 °C (annealing step) (for PAIR1, OsREC8, and OsOSD1 primers) or 61 °C (if OsSPO11-1 primers are used instead of PAIR1 primers), 1 mn per kbp at 72 °C (elongation step) and is completed by a final elongation step of 5 mn at 72 °C (Table 1c).

4. Prepare 400 mL of 1% agarose gel in TAE Buffer and run electrophoresis for 1 h for two 96 well PCR plates, at 130 V.
5. The PCR products (Fig. 3) are sent to a sequencing facility and sequenced using the forward primer.
6. The sequences obtained from the sequencing company as *.ab1* files are analyzed using ICE-SYNTHEGO software. The sequences with mutations are initially identified through alignment with the wild-type sequence. Subsequently, chromatograms are individually analyzed to determine the type of mutation (deletion and/or insertion; homozygous/biallelic/monallelic) within the gRNA target region (Fig. 4).

3.2 Characterization of Apomeiotic and Apomictic Progenies for Ploidy Level and Morphological Traits

3.2.1 Ploidy Analysis of Progenies of MiMe Plants

Progenies obtained from apomictic parents are generally a mixture of diploid and tetraploid plants, depending on if the progeny develops by apomixis or sexual reproduction. The frequency of tetraploid progeny depends on the penetrance of parthenogenesis in the transformation event. This frequency ranges from 100% when *OsBBM1* does not express (e.g., due to truncated T-DNA transgene and silencing of the ECS promoter) to 0% in fully penetrant events. T0 *MiMe* apomeiotic and *osd1*-only mother plants produce exclusively tetraploid T1 progeny.

In progenies of apomictic rice plants, diploid and tetraploid phenotype can be easily distinguished at flowering. Tetraploid plants typically exhibit darker green leaves with larger and thicker blades and reduced tillering compared to diploid plants. However, the most reliable distinguishing traits are panicle characteristics: tetraploid plants have larger, awned spikelets and fewer spikelets per panicle compared to diploids. Additionally, grain filling in tetraploids is generally very low or negligible.

For short-cycle hybrids, it is possible to grow progeny seedlings in 30 × 45 cm trays at high density (8 rows × 10 columns = 80 progeny plants) until flowering to estimate the frequency of parthenogenesis based on the panicle phenotype without requiring flow cytometry. However, for long-cycle hybrids with significant biomass, this approach may not be feasible.

Progeny ploidy analysis can be carried out by chromosome counting using various chromatin staining agents. However, an easier alternative to chromosome counting is flow cytometry analysis of somatic cell nuclei released from young leaf blades of Tx progeny seedlings.

4 Materials

4.1 Plant Materials

Fresh 1-cm leaf segment fresh tissues or tissues preserved in 2-mL Eppendorf tubes containing 0.5 mL of Murashige and Skoog salts diluted to 1/4th in distilled water. The samples can be stored at 4 ° C for up to 1 week without affecting the outcome of flow cytometry analysis (*see* **Notes 1**, **2** and **3**).

4.2 Reagents and Consumables

1. Sysmex Partec GmbH Germany UV ploidy (DAPI) staining solution REF 05-5001.
2. Sysmex Partec GmbH Germany sheath fluid REF 04-4007_R.
3. Partec CellTrics Germany disposable filters 30 μM REF 04-0042_2316.
4. SARSTEDT AG Germany sample analysis tube REF 55 484.

4.3 Equipment

1. Sysmex Cyflow cell analyzer or equivalent.

4.4 Software

1. Partec FloMax software v2.3 (Partec, Germany) or equivalent.

5 Methods

1. Finely chop a 1-cm segment of the antepenultimate leaf from a seedling using a sharp razor blade in a 60-mm Petri dish containing 500 μL of commercial 05-5001 buffer containing DAPI, to release nuclei from leaf cells (Fig. 5). Note: Use young leaf tissue from seedlings at the 5–6 leaf stage, as it is preferred to older tissues. Replace the razor blade if debris accumulation becomes significant.
2. Add 1 mL of flow cytometer sheath fluid to dilute the preparation.
3. Filter the nuclei suspension through a 30-μm filter mesh into a sample tube to remove debris.
4. Run the stained nuclei through a flow cytometer (e.g., PARTEC).
5. Set the relative fluorescent units (RFUs) of control nuclei from a diploid WT control to either 50 or 100 grade.
6. Run samples on the flow cytometer: diploid nuclei will form a sharp and high peak on the fluorescence histogram, while tetraploid nuclei form a broader and comparatively lower peak (Fig. 6) (*see* **Notes 4** and **5**).

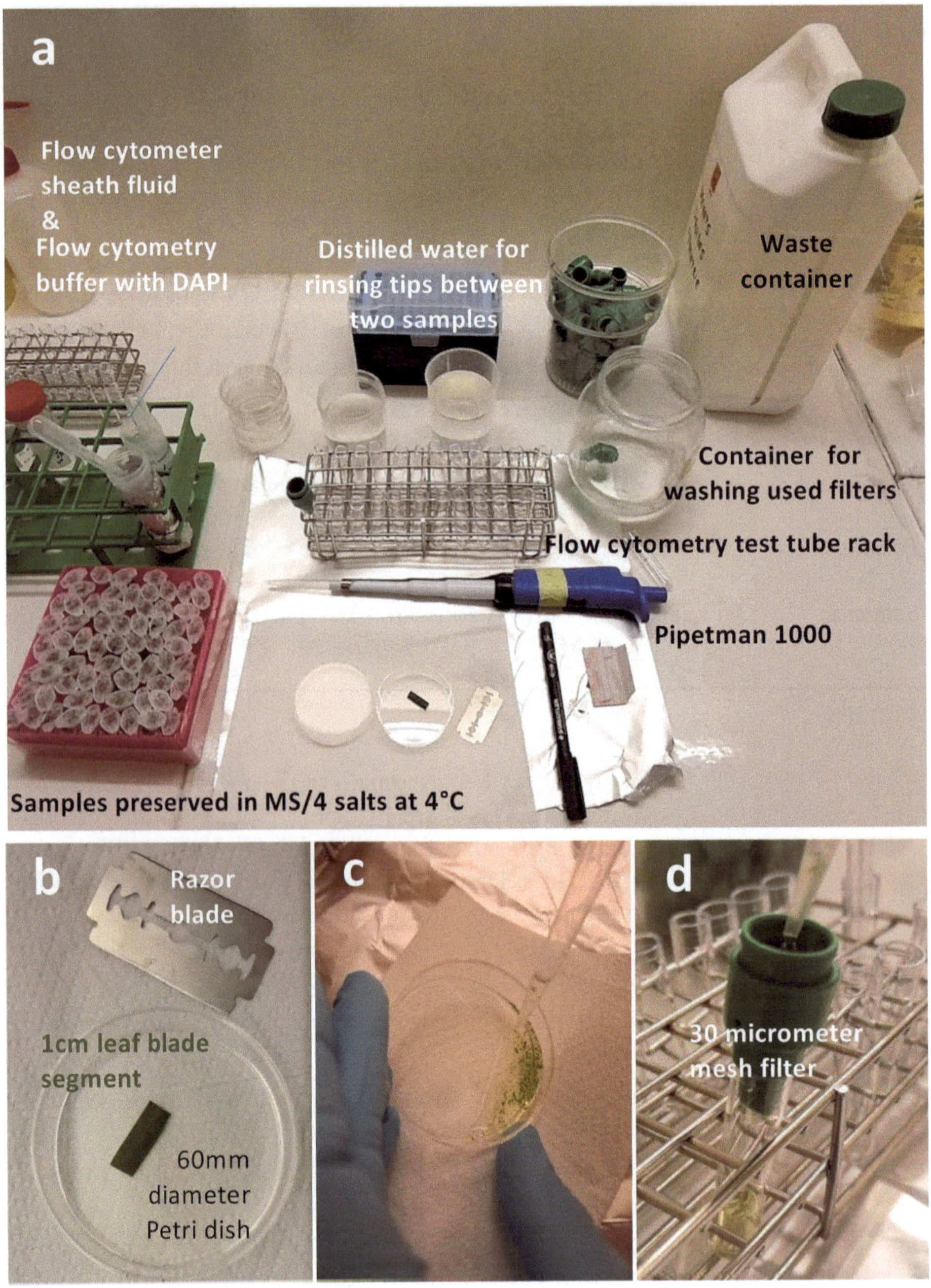

Fig. 5 Preparation of cell nuclei suspension released from young leaf blade segments for flow cytometry determination of ploidy level. (**a**) Getting ready for the analysis of samples. (**b**) Segment of leaf blade to be chopped with a razor blade in buffer. (**c**) Pipetting of released nucleus + tissue debris suspension. (**d**) Filtering a purified nucleus suspension into a flow cytometry sample tube through a 30-μm CellTrics filter retaining debris

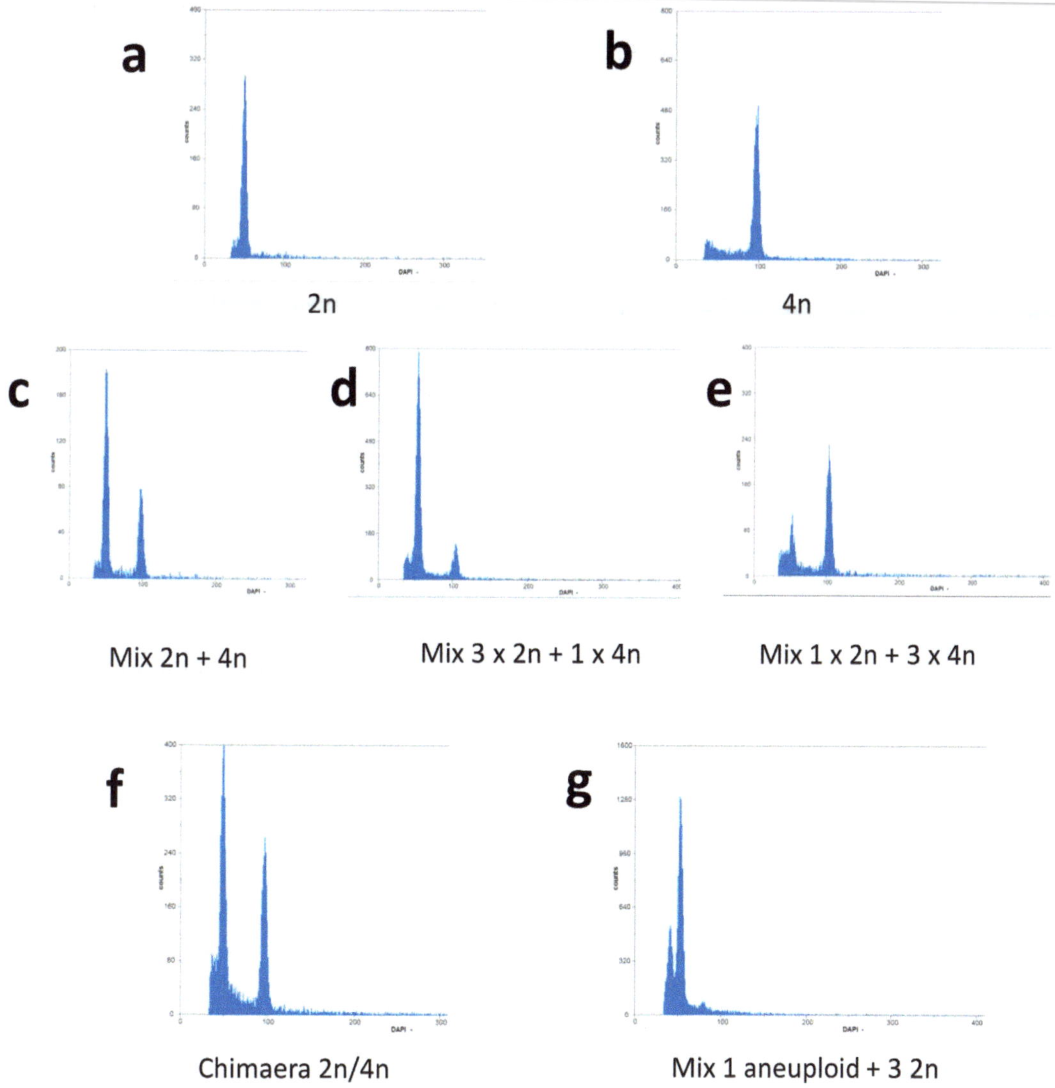

Fig. 6 Representative histograms of flow cytometry counts of DAPI-marked nuclei released from leaf tissues of apomictic progeny plants. Diploid control peak is set at 50 arbitrary fluorescence units (afu), Tetraploid peak forms at 100 afu. (**a**) Diploid progeny plant nuclei form a sharp peak at 50 afu. (**b**) Tetraploid progeny plant forms a spreader and lower peak at 100 afu. (**c**) Equal pool of leaf tissues of a diploid and a tetraploid progeny plant. (**d**) Equal pool of leaf tissue of 3 diploid and 1 tetraploid plants (**e**) Equal pool of leaf tissue of 1 diploid and 3 tetraploid plants. (**f**) Leaf tissues of a chimeric 2n/4n plant (**g**) Equal pool of leaf tissue of one aneuploid and 3 diploid plants

6 Notes

1. A total of 30–50 progeny seeds are sufficient to provide a precise estimate of the penetrance of parthenogenesis. However, increasing the sample size to 100 or more provides confirmatory frequencies.

If the plants are not intended for further growth or harvest, and confidence in peak analysis has been established, it is possible to pool leaf segments from 2, 3, and even 4 progeny plants. This allows to determine the ratio of diploid versus tetraploid in a mix of samples by simply observing the relative heights of the 2C and 4C peaks.

2. Progeny seeds occasionally form two shoots, referred to as twins, with frequencies ranging from a few per cent to as high as 50%. In partially penetrant events (e.g., 30–60% clonal seeds), the two shoots can be either both diploids or one diploid and one tetraploid, whereas in near fully or fully penetrant events (e.g., 95–100% clonal seeds), the two shoots are generally diploid. The possible origins of the twin embryos have been recently reviewed [39].
3. When the leaf tissue is immature, a smaller S/G2 peak at 4C position is also observed in diploid nuclei preparations. It is important to regularly run diploid WT controls between test samples, especially if a drift from the diploid peak position becomes noticeable.
4. Chimeric plants with tillers or leaf sectors exhibiting different ploidy levels have also been observed with a very low frequency. The origin of these chimera (endoreduplication of cells in an embryo of single origin or fusion of two embryos of independent origins) remains to be elucidated.
5. Flow cytometry can also be used to determine the ploidy of endosperm nuclei in immature spikelets of apomictic plants by simply pipetting the milky endosperm through piercing the ovary. In contrast to triploid endosperm in control wild types, the endosperm of apomictic plants is initially hexaploid. Later in the development, it undergoes endoreduplication, resulting in doubled (12C) DNA content [24].

6.1 Genotype of Apomictic Progenies

F1 hybrids exhibit inter-parental polymorphisms (SNP, indels, SSRs) that can be utilized to develop markers for confirming the maintenance of the heterozygous status in apomeiotic and apomictic progenies derived from hybrid plants. For example, analyzing four loci spread over four rice chromosomes in 30 progeny is sufficient to verify the transmission of heterozygosity. The probability of obtaining only heterozygous plants by chance in a segregating F2 of 30 is 1/240, and therefore very unlikely. Whole-genome sequencing of progenies from apomictic parents has further verified the faithful transmission of F1 hybrid genotype (*see*, e.g., [24]).

6.2 Phenotype of Apomictic Progenies

Apomictic progenies typically exhibit a faithful F1 hybrid phenotype (Fig. 7). Traits such as plant height, tiller number, panicle length, leaf width, leaf length, and days to flower can be used to confirm the transmission of hybrid phenotype through seeds in apomictic progenies grown alongside control seed-derived F1 hybrid plants. On the other hand, plants regenerated through tissue culture are known to be prone to phenotypic alterations due to somaclonal variation. When such variations have a genetic basis, they become fixed by the apomictic mode of reproduction in the next generations.

6.3 Fertility of Apomeiotic and Apomictic Plants

Confirmed apomeiotic and apomictic *MiMe* plants exhibit a range of spikelet filling/panicle fertility. However, while apomeiotic *MiMe* T0 plants and lowly penetrant apomictic *MiMe* T0 plants and diploid progenies can reach a fertility close to that of their control seed-derived F1 hybrid counterparts, near-fully to fully penetrant apomictic events exhibit a consistently reduced panicle fertility (Table 6, our data, and Table 7). The underlying causes of fertility variation in *MiMe* apomeiotic mutants remain to be fully elucidated. However, insufficient penetrance of *Ososd1* during female meiosis has been suggested as a likely cause for this variation [24]. In apomictic plants, a relationship has been observed between the penetrance of parthenogenesis (percentage of clonal diploid progeny seedlings) and the frequency of grain filling (unpublished results). Apomictic plants that achieve a 98–100% frequency of clonal diploid seeds, levels suitable for agricultural application, exhibit a reduced seed set. The cause(s) of this major bottleneck remain(s) to be fully understood and, if possible, addressed. However, the use of an "all-in-one" T-DNA with *OsBBM1* expression directed by the rice ECA1 promoter region the activity of which is enhanced through a Suntag technology approach recently enabled the generation—albeit at very low frequency—of both highly penetrant and fully fertile apomictic lines in hybrid rice [41]. This encouraging result makes even closer the perspective of deployment of apomictic hybrid rice in the field.

Conflict of Interest Statement

This study was funded by the France-Berkeley exchange fund #55-2018, Berkeley, USA to VS and EG and the Program of Excellence i-Site (PEI) PE24PR01-RA11-APOMIZING project from the University of Montpellier, France to EG, the Innovative Genomics Institute (IGI) grant to VS, core funding from the Max Planck Society to RM, and funding support from Department of Plant Sciences, UC Davis, and USDA AES CA-D-PLS-2736-H and CA-D-PLS-2812-RR projects to IK.

Fig. 7 Phenotypes of plants grown under controlled greenhouse conditions. (**a**) BRS-CIRAD 302 hybrid plant (middle) flanked by its maternal (1F, left) and paternal (D24, right) parents. (**b**) Five F2 progeny plants derived from the self-fertilization of BRS-CIRAD 302 compared to a BRS-CIRAD 302 F1 plant. (**c**) Four T1 progenies from an apomictic event (15.1) compared to a BRS-CIRAD 302 F1 plant

Table 7
Respective levels of penetrance of synthetic apomixis and panicle fertility using egg cell-specific expression of different parthenogenesis triggers in *MiMe*

Parthenogenesis trigger	Maximum penetrance of parthenogenesis (% clonal, diploid seedlings)	Panicle fertility of the most penetrant event	References
OsBBM1	98.7	33.8	[24]
OsBBM1	98.2	15.7	[40]
BBM4	2.3	86.4	[26]
OsWUS/MOC3	21.7	80.8	[30]
PAR	67.7	82.8	[28]

EG, IK, RM, and VS are co-inventors of patent WO2024074888 (PCT/IB2023/000562)
CIRCUMVENTING BARRIERS TO HYBRID CROPS FROM GENETICALLY DISTANT CROSSES.

References

1. Schnable PS, Springer NM (2013) Progress toward understanding heterosis in crop plants. Annu Rev Plant Biol 64:71–88. https://doi.org/10.1146/annurev-arplant-042110-103827
2. Liao C, Yan W, Chen Z et al (2021) Innovation and development of the third-generation hybrid rice technology. Crop J 9:693–701. https://doi.org/10.1016/j.cj.2021.02.003
3. Chen Z, Wu J, Deng XW, Tang X (2023) Establishment and advances of third-generation hybrid rice technology: a review. Rice 16:56. https://doi.org/10.1186/s12284-023-00670-z
4. Hand ML, Koltunow AMG (2014) The genetic control of apomixis: asexual seed formation. Genetics 197:441–450. https://doi.org/10.1534/genetics.114.163105
5. Khush GS, Brar DS, Bennett J (1998) Apomixis in rice and prospects for its use in heterosis breeding. In: Advances in hybrid rice technology. IRRI, Manila, pp 297–310
6. Ozias-Akins P, van Dijk PJ (2007) Mendelian genetics of apomixis in plants. Annu Rev Genet 41:509–537. https://doi.org/10.1146/annurev.genet.40.110405.090511
7. Khanday I, Sundaresan V (2021) Plant zygote development: recent insights and applications to clonal seeds. Curr Opin Plant Biol 59:101993. https://doi.org/10.1016/j.pbi.2020.101993
8. Underwood CJ, Mercier R (2022) Engineering apomixis: clonal seeds approaching the fields. Annu Rev Plant Biol 73:201–225
9. Ravi M, Marimuthu MPA, Siddiqi I (2008) Gamete formation without meiosis in Arabidopsis. Nature 451:1121–1124. https://doi.org/10.1038/nature06557
10. d'Erfurth I, Jolivet S, Froger N et al (2009) Turning meiosis into mitosis. PLoS Biol 7: e1000124. https://doi.org/10.1371/journal.pbio.1000124
11. Grelon M, Vezon D, Gendrot G, Pelletier G (2001) AtSPO11-1 is necessary for efficient meiotic recombination in plants. EMBO J 20: 589–600. https://doi.org/10.1093/emboj/20.3.589
12. Nonomura K-I, Nakano M, Fukuda T et al (2004) The novel gene HOMOLOGOUS PAIRING ABERRATION IN RICE MEIOSIS1 of rice encodes a putative coiled-coil protein required for homologous chromosome pairing in MEIOSIS. Plant Cell 16:1008–1020. https://doi.org/10.1105/tpc.020701
13. Chelysheva L, Diallo S, Vezon D et al (2005) AtREC8 and AtSCC3 are essential to the monopolar orientation of the kinetochores during meiosis. J Cell Sci 118:4621–4632. https://doi.org/10.1242/jcs.02583
14. Mieulet D, Jolivet S, Rivard M et al (2016) Turning rice meiosis into mitosis. Cell Res 26:

1242–1254. https://doi.org/10.1038/cr.2016.117

15. Wang C, Liu Q, Shen Y et al (2019) Clonal seeds from hybrid rice by simultaneous genome engineering of meiosis and fertilization genes. Nat Biotechnol 1. https://doi.org/10.1038/s41587-018-0003-0
16. Khanday I, Skinner D, Yang B et al (2019) A male-expressed rice embryogenic trigger redirected for asexual propagation through seeds. Nature 565:91–95. https://doi.org/10.1038/s41586-018-0785-8
17. Ravi M, Chan SWL (2010) Haploid plants produced by centromere-mediated genome elimination. Nature 464:615–618. https://doi.org/10.1038/nature08842
18. Marimuthu MPA, Jolivet S, Ravi M et al (2011) Synthetic clonal reproduction through seeds. Science 331:876. https://doi.org/10.1126/science.1199682
19. Kelliher T, Starr D, Richbourg L et al (2017) MATRILINEAL, a sperm-specific phospholipase, triggers maize haploid induction. Nature 542:105–109. https://doi.org/10.1038/nature20827
20. Steffen JG, Kang I-H, Macfarlane J, Drews GN (2007) Identification of genes expressed in the Arabidopsis female gametophyte. Plant J 51: 281–292. https://doi.org/10.1111/j.1365-313X.2007.03137.x
21. Sprunck S, Rademacher S, Volger F et al (2012) Egg cell–secreted EC1 triggers sperm cell activation during double fertilization. Science 338:1093–1097. https://doi.org/10.1126/science.1223944
22. Khanday I, Santos-Medellín C, Sundaresan V (2023) Somatic embryo initiation by rice BABY BOOM1 involves activation of zygote-expressed auxin biosynthesis genes. New Phytol 238:673–687. https://doi.org/10.1111/nph.18774
23. Ren H, Shankle K, Cho M-J et al (2024) Synergistic induction of fertilization-independent embryogenesis in rice egg cells by paternal-genome-expressed transcription factors. Nat Plants. https://doi.org/10.1038/s41477-024-01848-z
24. Vernet A, Meynard D, Lian Q et al (2022) High-frequency synthetic apomixis in hybrid rice. Nat Commun 13:7963. https://doi.org/10.1038/s41467-022-35679-3
25. Ohnishi T, Takanashi H, Mogi M et al (2011) Distinct gene expression profiles in egg and synergid cells of rice as revealed by cell type-specific microarrays. Plant Physiol 155:881–891. https://doi.org/10.1104/pp.110.167502
26. Wei X, Liu C, Chen X et al (2023) Synthetic apomixis with normal hybrid rice seed production. Mol Plant 16:489–492. https://doi.org/10.1016/j.molp.2023.01.005
27. Underwood CJ, Vijverberg K, Rigola D et al (2022) A PARTHENOGENESIS allele from apomictic dandelion can induce egg cell division without fertilization in lettuce. Nat Genet 54:84–93. https://doi.org/10.1038/s41588-021-00984-y
28. Song M, Wang W, Ji C et al (2024) Simultaneous production of high-frequency synthetic apomixis with high fertility and improved agronomic traits in hybrid rice. Mol Plant 17:4–7. https://doi.org/10.1016/j.molp.2023.11.007
29. Lu Z, Shao G, Xiong J et al (2015) MONOCULM 3, an ortholog of WUSCHEL in rice, is required for tiller bud formation. J Genet Genomics 42:71–78. https://doi.org/10.1016/j.jgg.2014.12.005
30. Huang Y, Meng X, Rao Y et al (2024) OsWUS-driven synthetic apomixis in hybrid rice. Plant Commun:101136. https://doi.org/10.1016/j.xplc.2024.101136
31. Miao J, Guo D, Zhang J et al (2013) Targeted mutagenesis in rice using CRISPR-Cas system. Cell Res 23:1233–1236. https://doi.org/10.1038/cr.2013.123
32. Christensen AH, Quail PH (1996) Ubiquitin promoter-based vectors for high-level expression of selectable and/or screenable marker genes in monocotyledonous plants. Transgenic Res 5:213–218
33. Wang M-B, Upadhyaya NM, Brettell RIS, Waterhouse PM (1997) Intron-mediated improvement of a selectable marker gene for plant transformation using Agrobacterium tumefaciens. J Genet Breed 51:325–334
34. d'Erfurth I, Cromer L, Jolivet S et al (2010) The CYCLIN-A CYCA1;2/TAM is required for the meiosis I to meiosis II transition and cooperates with OSD1 for the prophase to first meiotic division transition. PLoS Genet 6: e1000989. https://doi.org/10.1371/journal.pgen.1000989
35. Wang Y, Fuentes RR, van Rengs WMJ et al (2024) Harnessing clonal gametes in hybrid crops to engineer polyploid genomes. Nat Genet 56:1075–1079. https://doi.org/10.1038/s41588-024-01750-6
36. Xie K, Zhang J, Yang Y (2014) Genome-wide prediction of highly specific guide RNA spacers for CRISPR–Cas9-mediated genome editing in model plants and major crops. Mol Plant 7: 923–926. https://doi.org/10.1093/mp/ssu009

37. Fayos I, Meunier AC, Vernet A et al (2020) Assessment of the roles of SPO11-2 and SPO11-4 in meiosis in rice using CRISPR/Cas9 mutagenesis. J Exp Bot 71:7046–7058. https://doi.org/10.1093/jxb/eraa391
38. Sallaud C, Meynard D, van Boxtel J et al (2003) Highly efficient production and characterization of T-DNA plants for rice (Oryza sativa L.) functional genomics. Theor Appl Genet 106:1396–1408. https://doi.org/10.1007/s00122-002-1184-x
39. Peha H, Ren H, Skinner D, Sundaresan V (2024) Twin embryo formation by induced parthenogenesis. Plant Reprod 38:3. https://doi.org/10.1007/s00497-024-00512-y
40. Dan J, Xia Y, Wang Y et al (2024) One-line hybrid rice with high-efficiency synthetic apomixis and near-normal fertility. Plant Cell Rep 43:79. https://doi.org/10.1007/s00299-024-03154-6
41. Song M, Li F, Chen Z et al (2024) Engineering high-frequency apomixis with normal seed production in hybrid rice. iScience 27:111479. https://doi.org/10.1016/j.isci.2024.111479

Index

A

B

C

D

E

F

G

H

L

Tatyana Radoeva and Peter J. van Dijk (eds.), *Apomixis: Methods and Protocols*, Methods in Molecular Biology, vol. 2987, https://doi.org/10.1007/978-1-0716-5001-1,

M

O

P

Q

R

S

T

MIX
Papier aus verantwortungsvollen Quellen
Paper from responsible sources
FSC® C105338

If you have any concerns about our products,
you can contact us on
ProductSafety@springernature.com

In case Publisher is established outside the EU,
the EU authorized representative is:
Springer Nature Customer Service Center GmbH
Europaplatz 3, 69115 Heidelberg, Germany

Printed by Libri Plureos GmbH
in Hamburg, Germany